MW00736397

Optical Network Design and Planning

Optical Networks

Series Editor: Biswanath Mukherjee
University of California, Davis
Davis, CA

Optical Network Design and Planning
Jane M. Simmons
ISBN 978-0-387-76475-7

Quality of Service in Optical Burst Switched Networks
Kee Chaing Chua, Mohan Gurusamy, Yong Liu, and Minh Hoang Phung
ISBN 978-0-387-34160-6

Optical WDM Networks
Biswanath Mukherjee
ISBN 978-0-387-29055-3

Traffic Grooming in Optical WDM Mesh Networks
Keyao Zhu, Hongyue Zhu, and Biswanath Mukherjee
ISBN 978-0-387-25432-6

Survivable Optical WDM Networks
Canhui (Sam) Ou and Biswanath Mukherjee
ISBN 978-0-387-24498-3

Optical Burst Switched Networks
Jason P. Jue and Vinod M. Vokkarane
ISBN 978-0-387-23756-5

Jane M. Simmons

Optical Network Design and Planning

 Springer

Jane M. Simmons
Monarch Network Architects
Holmdel, NJ
USA

Series Editor
Biswanath Mukherjee
University of California, Davis
Davis, CA
USA

ISBN: 978-0-387-76475-7   e-ISBN: 978-0-387-76476-4
DOI: 10.1007/978-0-387-76476-4

Library of Congress Control Number: 2008920263

To my beautiful mother, Marie

Foreword

The huge bandwidth demands predicted at the start of the millennium have finally been realized. This has been sparked by the steady growth of a variety of new broadband services such as high-speed Internet applications, residential video-on-demand services, and business virtual private networks with remote access to huge databases. In response, carriers are undergoing widespread upgrades to their metro and backbone networks to greatly enhance their capacity. Carriers are demanding WDM optical networking technologies that provide both low capital expenses and low operational expenses. This need has been satisfied by automatically reconfigurable optical networks that support optical bypass. Automatic reconfigurability enables the carriers, or their customers, to bring up new connections and take down existing ones to meet fluctuating bandwidth requirements in near real-time. It also enables rapid automatic restoration from network failures. The optical-bypass property of the network, coupled with long-reach WDM optics, greatly reduces the need for optical-electrical-optical conversion, thus resulting in huge savings in capital and operational expenses.

This book provides a timely and thorough coverage of the various aspects of the design and planning of automatically reconfigurable optical networks in general, with special emphasis on optical-bypass-enabled networks. While the reality of such networks today is somewhat different from the earlier research visions of a purely all-optical network that is transparent to signal format and protocol, the goals of greatly improved economics, flexibility, and scalability have been realized. The optical-bypass networking paradigm has been adopted by many of the major carriers around the world, in both metro and backbone networks. Moreover, efficient optical networking algorithms have emerged as one of the critical components that have enabled this technology to work in practice.

This book provides broad coverage of the architecture, algorithms, and economics of optical networks. It differs from other books on this general subject in that it focuses on real-word networks and it provides good perspective on the practical aspects of the design and planning process. The book serves as a valuable guide to carriers, vendors, and customers to help them better understand the intricacies of the design, planning, deployment, and economics of optical networks. The book also provides practitioners, researchers, and academicians with a wealth of knowledge and ideas on efficient and scalable optical networking algorithms that are suitable for a broad range of optical networking architectures and technologies.

Jane Simmons has been actively working in this area since the mid 1990s. In this time-frame, there was much activity covering all aspect of optical networking - technology, architecture, algorithms, control, and applications. A particularly influential research effort that started in the United States around this time, in which Jane participated, was the government-supported Multiwavelength Optical Networking (MONET) consortium among the telecommunications giants AT&T, Lucent, Verizon and SBC. Just a short time later, much of the vision generated by this research was turned into reality. In the 2000 time-frame, Corvis Corporation became the first company to commercialize the 'all-optical' backbone-network vision when it introduced a product with 3,200-km optical reach and the associated optical switching equipment. Jane played a key role at Corvis, where she developed efficient and scalable networking algorithms to support and exploit this technology. This culminated in the first commercial deployment of the 'all-optical' vision with Broadwing's backbone network, in 2001. Jane performed the network design for the Broadwing network, from link engineering to network architecture. Jane also performed network designs for a broad array of North American and European carriers. She successfully showed in these diverse and real environments that 'all-optical', or more accurately, optical-bypass-enabled networks are architecturally viable in terms of achieving high network efficiency.

She has continued to work on optical network architecture and algorithms, as a founding partner of Monarch Network Architects, which provides architectural services and networking tools to carriers and system vendors. With this vast experience, and being in the right place at the right time, Jane has developed a unique perspective in the field of optical networking, which she brings forward in this book. I thoroughly enjoyed reading it and I learned a lot from it. I am sure that the reader is in for a real treat!

Dr. Adel A. M. Saleh
Program Manager
DARPA Strategic Technology Office

Preface

I have been involved with the research and development of optical networks for the past 15 years. More specifically, I have worked on the architecture and algorithms of networks with optical bypass, where much of the electronic regeneration is removed from the network. These networks are referred to in this book as optical-bypass-enabled networks.

Optical bypass has progressed from a research topic to a commercial offering in a relatively short period of time. I was fortunate to be in the midst of the activity, as a member of Bell Labs/AT&T Labs Research and Corvis Corporation. There are a few key lessons learned along the way, which I hope have been successfully captured in this book.

First, algorithms are a key component of optical networks. It is not hard to produce studies where poor algorithms lead to inefficient network utilization. Conversely, armed with a good set of algorithms, one can generate efficient designs across a range of network topologies, network tiers, and traffic distributions. It is also important to stress that while replacing electronics with optics in the network poses unique challenges that require algorithms, which is often cited as a concern by the opponents of such networking technology, the design of electronic-based networks requires algorithms as well. Processes such as shared protection or subrate-traffic-grooming are complex enough that algorithms are needed regardless of the nature of the underlying technology.

Second, there should be a tight development relationship between the system engineers, hardware designers, and the network architects of any system vendor developing optical networking equipment. The mantra of many a hardware developer when dealing with the potentially messy consequences of a design decision is often 'the algorithms will take care of it'. While their confidence in the algorithms may be flattering, this is not always the wisest course of action. It is the responsibility of the network architects to push back when appropriate to ensure that the overall system complexity does not grow unwieldy. Based on experience, when challenged, much more elegant solutions were forthcoming. Of course, there are times when the physics of the problem, as opposed to expediency, dictates a solution; it is important to recognize the difference.

This leads to the last point in that the algorithms in a well-designed system do not need to be overly complex. Much effort has been put into algorithm development, which has been successful in producing efficient and scalable algorithms.

Furthermore, it is not necessary that the algorithms take many hours or days to run. With well-honed heuristics, a design that is very close to optimal can often be produced in seconds to minutes.

The primary goal of this book is to cover the aspects of optical network design and planning that are relevant in a practical environment. The emphasis is on planning techniques that have proved to be successful in actual networks, as well as on potential trouble areas that must be considered. While the algorithms and architecture are the core of the content, the various enabling optical network elements and the economics of optical networking are covered as well. The book is intended for both practitioners and researchers in the field of optical networking.

The first two chapters should be read in order. Chapter 1 puts the book in perspective and reviews the terminology that is used throughout the book. Chapter 2 covers the various optical network elements; it is important to understand the functionality of the elements as it motivates much of the remainder of the book. If desired, Section 2.7 and Sections 2.10 through 2.12 can be skipped without affecting the readability of the subsequent chapters.

Chapters 3, 4, and 5 cover routing, regeneration, and wavelength assignment algorithms, respectively. Chapter 3 is equally applicable to O-E-O networks and optical-bypass-enabled networks; Chapters 4 and 5 are relevant only to the latter. The first three sections of Chapter 4 are more focused on physical-layer issues and can be skipped if desired.

Chapters 6 and 7 are standalone chapters on grooming and protection, respectively. Much of these chapters apply to both O-E-O networks and optical-bypass-enabled networks, with an emphasis on the latter. Finally, Chapter 8 presents numerous economic studies.

Acknowledgements

The nucleus of this book began as a Short Course taught at the Optical Fiber Communication (OFC) conference. I would like to thank the students for their suggestions and comments over the past five years that the course has been taught.

I am indebted to Dr. Adel Saleh, with respect to both my career and this book. As a leader of MONET, AT&T optical networking research, and Corvis, he is recognized as one of the foremost pioneers of optical networking. I appreciate the time he put into reading this book and his numerous helpful suggestions and encouragement.

I thank the editor of the Springer Optical Networks Series, Prof. Biswanath Mukherjee, for providing guidance and enabling a very smooth publication process. He provided many useful comments that improved the readability and utility of the book.

The team from Springer, Alex Greene and Katie Stanne, has been very professional and a pleasure to work with. Their promptness in responding to all my questions expedited the book.

Table of Contents

Chapter 1. Introduction to Optical Networks... 1
  1.1 Brief Evolution of Optical Networks... 1
  1.2 Geographic Hierarchy of Optical Networks 3
  1.3 Layered Architectural Model 5
  1.4 Interface to the Optical Layer 7
  1.5 Configurable Optical Networks 8
  1.6 Terminology 10
  1.7 Network Design and Network Planning 14
  1.8 Focus on Practical Optical Networks.. 15

Chapter 2. Optical Network Elements... 17
  2.1 Basic Optical Components... 18
  2.2 Optical Terminal... 19
    2.2.1 Slot Flexibility 20
  2.3 Optical-Electrical-Optical (O-E-O) Architecture 22
    2.3.1 O-E-O Architecture at Nodes of Degree-Two 22
    2.3.2 O-E-O Architecture at Nodes of Degree-Three or Higher 23
    2.3.3 Advantages of the O-E-O Architecture............................ 25
    2.3.4 Disadvantages of the O-E-O Architecture 26
  2.4 OADMs (ROADMs)... 27
    2.4.1 Configurability.. 29
    2.4.2 Wavelength vs. Waveband Granularity 30
    2.4.3 Wavelength Reuse 30
    2.4.4 Automatic Power Equalization... 32
    2.4.5 Edge Configurability 32
    2.4.6 Multicast 33
    2.4.7 Slot Flexibility 33
    2.4.8 East/West Separability.. 34
    2.4.9 Broadcast-and-Select and Wavelength-Selective Architectures 34
  2.5 Multi-Degree OADMs 36
    2.5.1 Optical Terminal to OADM to OADM-MD Upgrade Path............... 40
    2.5.2 First-Generation OADM-MD Technology 41
    2.5.3 Second-Generation OADM-MD Technology...................... 42
  2.6 Optical Switches.. 44

2.6.1 O-E-O Optical Switch.. 44
2.6.2 Photonic Switch 46
2.6.3 All-Optical Switch 46
2.7 Hierarchical Switches.. 49
2.8 Adding Edge Configurability to a Node 50
2.8.1 Adjunct Edge Switch 51
2.8.2 Flexible Transponders... 52
2.9 Optical Reach... 53
2.10 Integrating WDM Transceivers in the Client Layer 55
2.11 Photonic Integrated Circuits... 56
2.12 Multi-Fiber-Pair Systems 57

Chapter 3. Routing Algorithms 61
3.1 Shortest Path Algorithms.. 63
3.2 Routing Metrics 65
3.2.1 Fewest-Hops Path vs. Shortest-Distance Path 65
3.2.2 Shortest-Distance Path vs. Minimum-Regeneration Path.......... 67
3.3 Generating a Set of Candidate Paths... 68
3.3.1 K-Shortest Paths Strategy 68
3.3.2 Bottleneck-Avoidance Strategy... 69
3.4 Routing Strategies 71
3.4.1 Fixed-Path Routing.. 71
3.4.2 Alternative-Path Routing 71
3.4.3 Dynamic-Path Routing 73
3.5 Avoiding Infeasible Paths 75
3.5.1 Capturing the Available Equipment in the Network Model 75
3.5.2 Predeployment of Equipment 78
3.6 Diverse Routing for Protection 79
3.6.1 Shortest Pair of Disjoint Paths.. 82
3.6.2 Shortest Pair of Disjoint Paths: Dual-Sources/Dual-Destinations...... 84
3.6.3 Shared Risk Link Groups (SRLGs)... 87
3.6.4 Routing Strategies With Protected Demands 91
3.7 Routing Order 92
3.8 Multicast Routing 93
3.8.1 Multicast Protection... 97
3.9 Routing with Inaccurate Information... 97

Chapter 4. Regeneration.. 101
4.1 Factors That Affect Regeneration 102
4.1.1 Optical Impairments 102
4.1.2 Mitigation of Optical Impairments.. 103
4.1.3 Network Element Effects.. 104
4.1.4 Transmission System Design.. 105
4.1.5 Fiber Plant Specifications 107
4.1.6 System Regeneration Rules 107

4.2 Routing with Noise Figure as the Link Metric 108
    4.2.1 Network Element Noise Figure 110
    4.2.2 Impact of the OADM without Wavelength Reuse 111
4.3 Link Engineering 113
    4.3.1 Cohesive System Design 113
4.4 Regeneration Strategies 114
    4.4.1 Islands of Transparency 114
    4.4.2 Designated Regeneration Sites 116
    4.4.3 Selective Regeneration 118
4.5 Regeneration Architectures 119
    4.5.1 Back-to-Back WDM Transponders 120
    4.5.2 Regenerator Cards 122
    4.5.3 All-Optical Regenerators 125

Chapter 5. Wavelength Assignment **127**
5.1 Role of Regeneration in Wavelength Assignment 128
5.2 Multi-Step RWA 130
    5.2.1 Alleviating Wavelength Contention 131
5.3 One-Step RWA 132
5.4 Wavelength Assignment Strategies 136
    5.4.1 First-Fit 137
    5.4.2 Most-Used 139
    5.4.3 Relative Capacity Loss 139
    5.4.4 Qualitative Comparison 141
5.5 Subconnection Ordering 141
5.6 Bi-directional Wavelength Assignment 143
5.7 Wavelengths of Different Optical Reach 144
5.8 Wavelength Contention and Network Efficiency 146
    5.8.1 Backbone Network Study 147
    5.8.2 Metro Network Study 149
    5.8.3 Study Conclusions 151

Chapter 6. Grooming **153**
6.1 End-to-End Multiplexing 155
6.2 Grooming 157
6.3 Grooming Node Architecture 159
    6.3.1 Grooming Switch at the Nodal Core 159
    6.3.2 Grooming Switch at the Nodal Edge 160
6.4 Selection of Grooming Sites 163
6.5 Backhaul Strategies 164
6.6 Grooming Tradeoffs 166
    6.6.1 Cost vs. Path Distance 166
    6.6.2 Cost vs. Capacity 168
    6.6.3 Grooming Design Guidelines 169
6.7 Grooming Strategies 169

6.7.1 Initial Bundling and Routing 169
6.7.2 Grooming Operations 170
6.8 Grooming Network Study 176
6.8.1 Grooming Switches at All Nodes 176
6.8.2 Grooming Switches at a Subset of the Nodes 178
6.9 Evolving Techniques for Grooming Bursty Traffic 179
6.9.1 Selective Randomized Load Balancing 180
6.9.2 Optical Flow Switching (OFS) 180
6.9.3 Optical Burst Switching (OBS) 181
6.9.4 TWIN 181
6.9.5 Lighttrail 182
6.9.6 Optical Packet Switching (OPS) 182

Chapter 7. Optical Protection **183**
7.1 Dedicated vs. Shared Protection 185
7.1.1 Dedicated Protection 186
7.1.2 Shared Protection 186
7.1.3 Comparison of Dedicated and Shared Protection 187
7.2 Client-Side vs. Network-Side Protection 190
7.3 Ring Protection vs. Mesh Protection 194
7.3.1 Ring Protection 194
7.3.2 Mesh Protection 198
7.4 Fault-Dependent vs. Fault-Independent Protection 200
7.5 Protection Against Multiple Concurrent Failures 205
7.6 Effect of Optical Amplifier Transients on Protection 207
7.7 Shared Protection Based on Predeployed Subconnections 209
7.7.1 Cost vs. Spare Capacity Tradeoff 211
7.8 Shared Protection Based on Pre-cross-connected Bandwidth 213
7.9 Protection Planning Algorithms 217
7.9.1 Algorithms for Dedicated Protection 217
7.9.2 Algorithms for Shared Protection 218
7.10 Protection of Subrate Demands 224
7.10.1 Wavelength-Level Protection 224
7.10.2 Subrate-Level Protection 226
7.10.3 Wavelength-Level vs. Subrate-Level Protection 228
7.10.4 Multilayer Protection 229
7.11 Fault Isolation 231

Chapter 8. Economic Studies **233**
8.1 Assumptions 234
8.1.1 Reference Network Topology 234
8.1.2 Reference Traffic Set 235
8.1.3 Cost Assumptions 236
8.2 Prove-In Point for Optical-Bypass Technology 238
8.2.1 Comments on Comparing Costs 240

8.2.2 O-E-O Technology with Extended Optical Reach............................ 240
8.3 IP Transport Architectures 242
8.4 Optimal Optical Reach 246
    8.4.1 Add/Drop Percentage as a Function of Optical Reach 249
8.5 Architecture of Higher-Degree Nodes 251
8.6 Reduced-Reach Transponders... 253
8.7 Optimal Topology from a Cost Perspective 256
8.8 Optimal Line-Rate... 259
    8.8.1 Study Assumptions 260
    8.8.2 Study Results 262
    8.8.3 Less Aggressive Cost Assumptions 264
8.9 Optical Grooming in the Edge Networks 265
8.10 General Conclusions 268

Appendix A. Suggestions for RFI/RFP Network Design Exercises............... 269

Appendix B. C-Code For Routing Routines.................................... 271

Abbreviations.. 293

Bibliography... 297

Index 309

Chapter 1

Introduction to Optical Networks

1.1 Brief Evolution of Optical Networks

While the basic function of a network is quite simple – enabling communications between the desired endpoints – the underlying properties of a network can greatly affect its value. Network capacity, reliability, cost, scalability, and operational simplicity are some of the key benchmarks on which a network is evaluated. Network designers are often faced with tradeoffs among these factors, and are continually looking for technological advances that have the power to improve networking on a multitude of fronts.

One such watershed development came in the 1980s as the telecommunications industry began migrating much of the physical layer of their inter-city networks to fiber-optic cable. Optical fiber is a lightweight cable that provides low-loss transmission; but clearly its most significant benefit is its tremendous potential networking capacity. Not only did fiber optics open the possibilities of a huge vista for transmission, it also gave rise to optical networks and the field of optical networking.

An optical network is composed of the fiber-optic cables that carry channels of light, combined with the equipment deployed along the fiber to process the light. The capabilities of an optical network are necessarily tied to the physics of light and the technologies for manipulating lightstreams. As such, the evolution of optical networks has been marked with major paradigm shifts as exciting break-through technologies are developed.

One of the earliest technological advances was the ability to carry multiple channels of light on a single fiber-optic cable. Each lightstream, or wavelength[1],

---

[1] The term 'wavelength' is commonly used in two different contexts: first, it refers to a channel of light; second, it refers to the specific point in the spectrum of light where the channel is centered (e.g., 1550 nanometers). The context should be clear from its usage; however, when necessary, clarifying text is provided.

J.M. Simmons, *Optical Network Design and Planning*,
DOI: 10.1007/978-0-387-76476-4_1, © Springer Science+Business Media, LLC 2008

is carried at a different optical frequency and multiplexed (i.e., combined) onto a single fiber, giving rise to Wavelength Division Multiplexing (WDM). The earliest WDM systems supported fewer than ten wavelengths on a single fiber. Since 2000, this number has rapidly grown to over one hundred wavelengths per fiber, providing a tremendous growth in network capacity.

A key enabler of cost-effective WDM systems was the development of the Erbium Doped Fiber Amplifier (EDFA). Prior to the deployment of EDFAs, each wavelength on the fiber had to be individually regenerated at roughly 40-km intervals, using costly electronic equipment. The EDFA optically amplifies all of the wavelengths on a fiber at once, allowing optical signals to be transmitted on the order of 500 km before needing to be regenerated.

A more subtle innovation was the migration from an architecture where the optical network served simply as a collection of static pipes to one where it was viewed as another networking layer. In this optical networking paradigm, network functions such as routing and protection are supported at the granularity of a wavelength, which can be operationally very advantageous. A single wavelength may carry hundreds of circuits. If a failure occurs in a fiber cable, restoring service by processing individual wavelengths is operationally simpler than rerouting each circuit individually.

The benefits of scale provided by optical networking have been further accelerated by the increasing capacity of a single wavelength. In the mid 1990s, the maximum capacity of a wavelength was roughly 2.5 Gb/s (Gb/s is 10^9 bits/sec). This has ramped up to 10 Gb/s and 40 Gb/s, with much discussion regarding evolution to 100 Gb/s per wavelength, or higher.

Increased wavelength rate combined with a greater number of wavelengths per fiber has expanded the capacity of optical networks by several orders of magnitude over a period of 25 years. However, transmission capacity is only one important factor. Historically, the contents of each wavelength have undergone electronic processing at numerous points in the network. As networks exploded in size, this necessitated the use of a tremendous amount of electronic terminating and switching equipment, which presented challenges in cost, power consumption, heat dissipation, physical space, and maintenance.

This bottleneck was greatly reduced by the development of *optical-bypass* technology. This technology eliminates much of the required electronic processing and allows a signal to remain in the optical domain for all, or much, of its path from source to destination. Because optical technology can operate on a spectrum of wavelengths at once, and can operate on wavelengths largely independently of their data-rate, maintaining signals in the optical domain allows a significant amount of equipment to be removed from the network and provides a scalable trajectory for network growth.

Achieving optical bypass required advancements in areas such as optical amplification, optical switching, transmission formats, and techniques to counteract optical impairments. Commercialization of optical-bypass technology began in the mid-1990s, leading to its deployment in the networks of several major telecommunications carriers. While reducing the amount of electronic processing addressed many of the impediments to continued network growth, it also brought new challenges. Most notably, it required the development of new algorithms to assist in operating the network so that the full benefits of the technology could be attained. Overall, the advent of optical-bypass technology has transformed the architecture, operation, and economics of optical networks, all of which is covered in this book.

1.2 Geographic Hierarchy of Optical Networks

When considering the introduction of new networking technology, it can be useful to segment the network into multiple geographic tiers, with key differentiators among the tiers being the number of customers served, the required capacity, and the geographic extent. One such partitioning is shown in Fig. 1.1. (In this section, the standalone term 'network' refers to the network as a whole; when 'network' is used in combination with one of the tiers, e.g., 'backbone network', it refers to the portion of the overall network in that particular tier.)

At the edge of the network, closest to the end-users, is the *access* tier, which distributes/collects traffic to/from the customers of the network. Access networks generally serve tens to hundreds of customers and span a few kilometers. (One can further subdivide the access tier into business-access and residential-access, or into metro-access and rural-access.) The *metro-core* tier is responsible for aggregating the traffic from the access networks, and generally interconnects a number of telecommunications central offices or cable distribution head-end offices. A metro-core network aggregates the traffic of thousands of customers and spans tens to hundreds of kilometers.

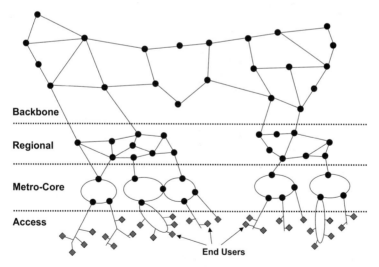

Fig. 1.1 Networking hierarchy based on geography.

Moving up the hierarchy, multiple metro-core networks are interconnected via *regional* networks. A regional network carries the portion of the traffic that spans multiple metro-core areas, and is shared among hundreds-of-thousands of customers, with a geographic extent of several hundred to a thousand kilometers. Inter-regional traffic is carried by the *backbone* network[2]. Backbone networks may be shared among millions of customers and typically span thousands of kilometers.

While other taxonomies may be used, the main point to be made is that the characteristics of a tier are important in selecting an appropriate technology. For example, whereas the backbone network requires optical transport systems with very large capacity over long distances, that same technology would not be appropriate for, nor would it be cost effective in, an access network.

As one moves closer to the network edge, the cost of a network in a particular tier is amortized over fewer end users, and is thus a more critical concern. Because of this difference in price sensitivity among the tiers, there is often a trend to deploy new technologies in the backbone network first. As the technology matures and achieves a lower price point, it gradually extends closer towards the edge. A good example of this trend is the deployment of WDM technology, as represented by the timeline in Fig. 1.2 (the costs and dates in this figure are only approximate).

Even as a technology permeates a network, the particular implementation may differ across tiers. For example, with respect to WDM technology, backbone networks generally have 80 to 160 wavelengths per fiber, regional networks have roughly 40 to 80 wavelengths per fiber, metro-core WDM networks have anywhere from 8 to 40 wavelengths per fiber, and access networks typically have no more than 8 wavelengths.

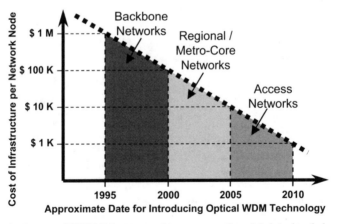

Fig. 1.2 As the cost of WDM infrastructure decreases over time, it is introduced closer to the network edge. (Adapted from [Sale98b].)

---

[2] Other common names for this tier are the long-haul network or the core network. These terms are used interchangeably throughout the book.

A similar pattern is emerging with the introduction of optical-bypass technology. Commercial deployment began in backbone networks in the 2000 timeframe, and has gradually spread closer to the network edge. The capabilities of optical-bypass-based systems are tailored to the particular network tier. For example, the distance a signal can be transmitted before it suffers severe degradation is a fundamental attribute of such systems. In backbone networks, technology is deployed where this distance is a few thousand kilometers; in metro-core networks, it is several hundred kilometers.

While optical networking is supported to varying degrees in the different tiers of the network, the architecture of access networks (especially residential access) is very distinct from that of the other portions of the network. For example, one type of access network is based on passive devices (i.e., the devices in the field do not require power); these systems, aptly named Passive Optical Networks (PONs), would not be appropriate for larger-scale networks. Because the topological characteristics, cost targets, and architectures of access networks are so different from the rest of the network, they are worthy of a book on their own; hence, access networks are not covered here. Detailed treatment of access technologies can be found in [Lin06]. Suffice it to say that as optics enters the access network, enabling the proliferation of high-bandwidth end-user applications, there will be increased pressure on the remainder of the network to scale accordingly.

It should be noted that there is a recent trend in the telecommunications industry to 'blur the boundaries' between the tiers. Carriers are looking for technology platforms that are flexible enough to be deployed in multiple tiers of the network, with unified network management and provisioning systems to simplify operations [ChSc07].

1.3 Layered Architectural Model

Another useful network stratification is illustrated by the three-layered architectural model of Fig. 1.3. At the top of this model is the applications layer, which includes all types of services, such as voice, video, and data. The intermediate layer encompasses multiplexing, transport, and switching based on electronic technology. For example, this layer includes Internet Protocol (IP) routers, Ethernet switches, Asynchronous Transfer Mode (ATM) switches, and Synchronous Optical Network/Synchronous Digital Hierarchy (SONET/SDH) switches. Each of these protocols has a particular method for partitioning data and moving the data from source to destination.

The payloads of the electronic layer are passed to the optical layer, where they are packed into wavelengths. In the model of interest, the optical layer is based on WDM technology and utilizes optical switches that are capable of dynamically routing wavelengths. Thus, the bottom tier of this particular model can also be referred to as the 'configurable WDM layer'.

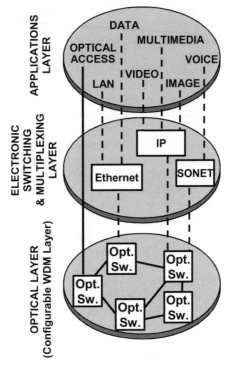

Fig. 1.3 Three-layered architectural model. In the systems of interest, the optical layer is based on WDM technology with configurable optical switches. (Adapted from [WASG96]. © 1996 IEEE)

From the viewpoint of the electronic layer, the wavelengths form a *virtual topology*. This concept is illustrated in Fig. 1.4 by a small network interconnecting five points. In Fig. 1.4(a), the solid lines represent fiber optic cables, or the physical topology, and the dotted lines represent the paths followed by two of the wavelengths. This arrangement of wavelengths produces the virtual topology shown in Fig. 1.4(b); i.e., this is the network topology as seen by the electrical layer. In contrast to the fixed physical topology, the virtual topology can be readily modified by reconfiguring the paths of the wavelengths.

Note that it is possible for the application layer to directly access the optical layer, as represented in Fig. 1.3 by the optical access services. This capability could be desirable, for example, to transfer very large streams of protocol-and-format-independent data. Because the electronic layers are bypassed, no particular protocol is imposed on the data. By transporting the service completely in the optical domain, the optical layer potentially provides what is known as *protocol and format transparency*. While such transparency has often been touted as another benefit of optical networking, thus far these services have not materialized in a major way in practical networks.

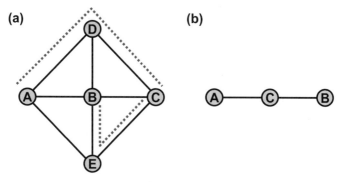

Fig. 1.4 In (a), the solid lines represent the physical fiber optic links and the dotted lines represent the paths of two routed wavelengths. The two wavelength paths create the virtual topology shown in (b), where the solid lines represent virtual links. The virtual topology can be modified by setting up different wavelength paths.

1.4 Interface to the Optical Layer

One difficulty with carrying services directly in wavelengths is that the network can be difficult to manage. Network operations can be simplified by using standard framing that adds overhead for management. For example, the SONET and SDH specifications define a standard framing format for optical transmission, where the frame includes overhead bytes for functionality such as performance monitoring, path trace, and Operations, Administration, and Maintenance (OAM) communication. SONET/SDH is commonly used as the interface to the optical layer; standards exist to map services such as ATM and IP into SONET/SDH frames. (In addition to using SONET/SDH for framing, it is often used for switching and multiplexing in the electronic domain, as was shown in Fig. 1.3.)

The SONET and SDH specifications are very closely related: SONET is the American National Standards Institute (ANSI) standard, whereas SDH is the International Telecommunication Union (ITU) standard. SONET defines a base signal with a rate of 51.84 Mb/s, called Synchronous Transport Signal level-1 or STS-1 (Mb/s is 10^6 bits/sec). Multiple STS-1 signals are multiplexed together to form higher rate signals, giving rise to the SONET rate hierarchy. For example, three STS-1 signals are multiplexed to form an STS-3 signal. The optical instantiation of a general STS-N signal is called Optical Carrier level-N, or OC-N. SDH is similar to SONET, although the framing format is somewhat different. The SDH base signal is defined as Synchronous Transport Module level-1, or STM-1, which has a rate equivalent to an STS-3. Some of the most commonly used SONET and SDH rates are shown in Table 1.1. The bit-rates shown in parentheses for some of the signals are the nominal rates commonly used in reference to these signals. For more details on SONET/SDH technology, see [Tekt01, Gora02, Telc05].

Table 1.1 Commonly Used SONET/SDH Signal Rates

SONET Signal	SDH Signal	Bit-Rate
STS-1, OC-1	-	51.84 Mb/s
STS-3, OC-3	STM-1	155.52 Mb/s
STS-12, OC-12	STM-4	622.08 Mb/s
STS-48, OC-48	STM-16	2.49 Gb/s (2.5 Gb/s)
STS-192, OC-192	STM-64	9.95 Gb/s (10 Gb/s)
STS-768, OC-768	STM-256	39.81 Gb/s (40 Gb/s)

The SONET/SDH standards were initially developed in the 1980s with a focus on voice traffic, although features have been added to make them more suitable for data traffic. More recently, the ITU has developed a new architectural paradigm to better address the needs of optical networking, called the Optical Transport Network (OTN). The associated transport hierarchy and formats are defined in ITU standard G.709, with the basic frame called an Optical channel Transport Unit (OTU). The bit-rate of the OTU hierarchy is slightly higher than the SONET/SDH rates, with an OTU1 having a rate of 2.67 Gb/s, OTU2 at 10.71 Gb/s, and OTU3 at 43.02 Gb/s. Some of the relevant standards documents are [ITU01, ITU03].

Compared with SONET/SDH, OTN provides benefits such as more efficient multiplexing and switching of high-bandwidth services, enhanced monitoring capabilities, and stronger forward error correction (FEC). FEC allows bit errors picked up during signal transmission to be corrected when the signal is decoded. Enhanced FEC can be used to compensate for more severe transmission conditions. For example, it potentially allows more wavelengths to be multiplexed onto a single fiber, or allows a signal to remain in the optical domain for longer distances, which is important for optical-bypass systems.

OTN provides a uniform method of multiplexing a range of protocol types, essentially by providing a generic 'digital wrapper' for the payload. It is envisioned as a step towards network convergence, where carriers can support multiple services with a single network rather than deploying parallel networks. One of its main drivers for acceptance is it potentially offers carriers a managed, cost-effective means of adapting their networks to the growing demand for Ethernet services. OTN has gradually entered commercial applications; however, there is still a great deal of deployed legacy SONET/SDH-based equipment.

1.5 Configurable Optical Networks

Networks undergo continuous cycles of evolution where the requirements of the applications drive the development of innovative technology, and where improved technology encourages the development of more advanced applications. One such example is the automated configurability of an optical network.

The initial driver for automated configurability was the normal forecast uncertainty and churn that occur in a network (churn is the process of connections being established and then later taken down as the demand patterns change). It is difficult to forecast the precise endpoints and bandwidth of the traffic that will be carried in a network. Furthermore, while most traffic has historically been fairly static, with connection holding times on the order of months or longer, there is a subset of the traffic that has a much shorter lifetime, leading to network churn. In addition, as carriers move away from protection schemes where the backup paths are pre-established, reconfigurability[3] is needed to dynamically create a new path for failure recovery.

Thus, it is necessary that the network be able to adapt to inaccurate forecasts, changing demand patterns, and network failures; moreover, it is desirable that the process be automated to eliminate the labor cost and potential errors involved with manual configuration. The infrastructure and distributed intelligence to enable automated reconfigurability are collectively known as the *control plane*. (This is in contrast to the typically centralized *management plane* that has historically been responsible for network operations such as fault management and security.)

Various organizations have developed standards in support of the control plane. For example, the ITU has developed the Automatically Switched Optical Networks (ASON) architecture, and the Internet Engineering Task Force (IETF) has developed the Generalized Multi-Protocol Label Switching (GMPLS) paradigm. These specifications include signaling protocols to automate control of the optical network and enable features such as discovery of the network topology and network resources, and connection establishment. Some of the relevant standards and specifications can be found in [SDIR04, Mann04, ITU06]. A more detailed discussion of the topic is provided in [BeRS03].

GMPLS includes three models for interacting with the optical layer: peer, overlay, and augmented. For concreteness, the discussion will focus on the interaction of the IP and optical layers, but the principles apply to other electronic layers. In the peer (or integrated) model, the IP and optical layers are treated as a single administrative domain, with IP routers having full knowledge of the optical topology. The IP routers can determine the entire end-to-end path of a connection including how it should be routed through the optical layer. In the overlay model, the IP and optical layers are treated as distinct domains, with no exchange of routing and topology information between them. The IP layer is essentially a *client* of the optical layer and requests bandwidth from the optical layer as needed. The augmented model is a hybrid approach where a limited amount of information is exchanged between layers.

Given the amount of information that needs to be shared in the peer model, and the potential trust issues between the layers (e.g., the IP and optical layers may be operated by different organizations), the overlay and augmented models are generally more favored by carriers. In the overlay model, which is more established, the boundary between the client and the optical layers is called the

---

[3] The terms reconfigurability and configurability are used interchangeably in this book.

User-Network Interface (UNI). Signaling specifications for the UNI have been developed by the IETF as well as the Optical Internetworking Forum (OIF) [SDIR04, OIF04].

As these protocols for automated configurability have begun to make their way into carrier networks, the need to support more advanced dynamic services has emerged. In one flavor of dynamic service, the application requests a connection and requires that it be established very rapidly (e.g., in less than a second). For example, in large-scale distributed computing, there may be hundreds of computers that continually need to change their interconnection pattern as the computation evolves. In a second type of dynamic application, very-high-bandwidth transmission is periodically required but only for a short time. The need for the bandwidth is often known in advance, providing the opportunity to schedule the network resources as needed. One example of this is grid computing, which is a means of sharing distributed processing and data resources in order to achieve very high performance. This may require that huge datasets be disseminated to multiple locations in a very short period of time.

The stringent requirements of these applications will require the development of more advanced cross-layer bandwidth optimization, where the bandwidth allocation is dynamically optimized across multiple layers [ElMW06]. For example, the IP layer may automatically initiate a request for more bandwidth from the optical layer via the control plane. Additionally, more sophisticated provisioning protocols that can establish connections across multiple domains are also needed. (A domain is defined as an area of the network under the control of a single entity. The interface between domains is known as the External Network-Network Interface (E-NNI), whereas the interface between networks within a domain is the Internal NNI (I-NNI) [ITU06].)

This book focuses on the optical layer and does not consider topics such as cross-layer bandwidth management. While this approach is more in-line with the overlay model, the general network design principles discussed would need to be incorporated in any of the models.

1.6 Terminology

This section introduces some of the terminology that is used throughout the book. Refer to the small network shown in Fig. 1.5. The circles represent the network *nodes*. These are the points in the network that source/terminate and switch traffic. The lines interconnecting the nodes are referred to as *links*. While the links are depicted with just a single line, they typically are populated by one or more fiber-pairs, where each fiber in a pair carries traffic in just one direction. (It is possible to carry bi-directional traffic on a single fiber, but not common.) Optical amplifiers may be periodically located along each fiber, especially in regional and backbone networks. Sites that solely perform amplification are not considered nodes. The portion of a link that runs between two amplifier sites, or between a node and an amplifier site, is called a *span*.

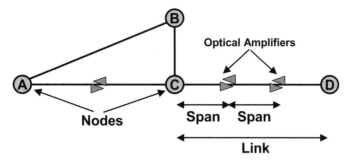

Fig. 1.5 Nodes are represented by circles and links are represented by solid lines. Nodes A and B have a degree of two, Node C has a degree of three, and Node D has a degree of one.

A very important concept is that of *nodal degree*. The degree of a node is the number of links incident on that node. Thus, in the figure, Nodes A and B have a degree of two, Node C has a degree of three, and Node D has a degree of one. Nodal degree is very important in determining the type of equipment appropriate for a node.

The specific arrangement of nodes and links constitutes the network topology. Early networks were almost always based on ring topologies due to the simple restoration properties of rings. More recently, networks, especially those in the backbone, have migrated to more flexible *mesh* topologies. In mesh networks, the nodes are arbitrarily interconnected, with no specific routing pattern imposed on the traffic. In Fig. 1.1, the topologies in the metro-core tier are shown as rings, whereas the regional and backbone topologies are mesh. While it is possible to develop network design techniques that are specifically optimized for rings, the approach of this book is to present algorithms and design methodologies that are general enough to be used in any topology.

The *traffic* in the network is the collection of services that must be carried. The term *demand* is used to represent an individual traffic request. For the most part, demands are between two nodes and are bi-directionally symmetric. That is, if there is a traffic request from Node A to Node B, then there is equivalent traffic from Node B to Node A. In any one direction, the originating node is called the source and the terminating node the destination. In multicast applications, the demands have one source and multiple destinations; such demands are typically one-way only. (It is also possible to have demands with multiple sources and one or more destinations, but not common.)

The term *connection* is used to represent the path allocated through the network for carrying a demand. The process of deploying and configuring the equipment to support a demand is called *provisioning,* or *turning up,* the connection. The rate of a demand or a connection will be referred to in either absolute terms (e.g., 10 Gb/s) or using SONET terminology (e.g., OC-192), depending on the context.

The optical networks of interest in this book are based on WDM technology. Figure 1.6 shows the portion of the light spectrum where WDM systems are generally based, so chosen because of the relatively low fiber attenuation in this region (as shown in the figure, the loss is typically between 0.20 and 0.25 dB/km). This spectrum is broken into three regions: the conventional band or C-band; the long wavelength band or L-band; and the short wavelength band or S-band. Most WDM systems make use of the C-band, however, there has been expansion into the L and S bands to increase system capacity.

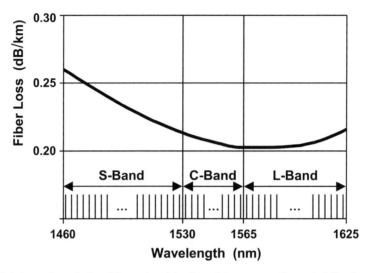

Fig. 1.6 Approximate S, C, and L wavelength bands, and the corresponding typical fiber loss.

An optical channel can be referred to as operating at a particular wavelength, in units of nanometers (nm), or equivalently at a particular optical frequency, in units of Terahertz (THz). The term *lambda* is frequently used to refer to the particular wavelength on which an optical channel is carried; *lambda i*, or λ_i, is used to represent the ith wavelength in the WDM system. The distance between adjacent channels in the spectrum is generally noted in frequency terms, in units of Gigahertz (GHz). For example, a 40-channel C-band system is achieved with 100-GHz spacing between channels, whereas an 80-channel C-band system is obtained using 50-GHz spacing.

An important piece of equipment is the WDM *transponder*, which is illustrated in Fig. 1.7(a). One side of the transponder is termed the *client side*, which takes a signal from the client of the optical network, e.g., an IP router. The client optical signal is generally carried on a 1310-nm wavelength. (1310-nm is outside the WDM region; WDM is usually not used for intra-office[4] communication.) Various interfaces can be used on the client side of the transponder, depending on how much optical loss is encountered by the client signal. For example, *short-reach*

---

[4] Office refers to a building that houses major pieces of telecommunications equipment, such as switches and client equipment.

interfaces tolerate up to 4 dB or 7 dB of loss depending on the signal rate, whereas *intermediate-reach interfaces* tolerate up to 11 dB or 12 dB of loss[5]. The interface converts the client optical signal to the electronic domain; the electronic signal modulates (i.e., drives) a WDM-compatible laser such that the client signal is converted to a particular wavelength (i.e., optical frequency) in the WDM region. The WDM side of the transponder is also called the *network side*. In the reverse direction, the WDM-compatible signal enters from the network side and is converted to a 1310-nm signal on the client side.

A single WDM transponder is shown in more detail in Fig. 1.7(b), to emphasize that there is a client-side receiver and a network-side transmitter in one direction and a network-side receiver and a client-side transmitter in the other direction. For simplicity, the transponder representation in Fig. 1.7(a) is used in the remainder of the book; however, it is important to keep in mind that a transponder encompasses separate devices in the two signal directions.

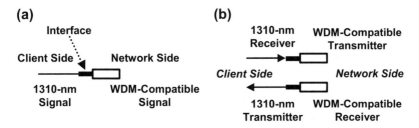

Fig. 1.7 (a) A simplified depiction of a WDM transponder that converts between a 1310-nm signal and a WDM-compatible signal. (b) A more detailed depiction of the WDM transponder that emphasizes its bi-directional composition. There is both a 1310-nm transmitter/receiver and a WDM-compatible transmitter/receiver.

In fixed-tuned transponders, the client signal can be converted to just one particular optical frequency. In transponders equipped with tunable lasers, the client signal can be converted to any one of a range of optical frequencies. Some architectures require that the transponder have an optical filter on the network side to receive a particular frequency. Tunable filters allow any one of a range of optical frequencies to be received. Since the early 2000s, most networks have been equipped with transponders with tunable lasers; transponders with both tunable lasers and filters were commercially available some time later. While they cost somewhat more, tunable transponders greatly improve the flexibility of the network as well as simplify the process of maintaining spare equipment for failure events.

The signal rate carried by a wavelength is called the *line-rate*. It is often the case that the clients of the optical network generate traffic that has a lower rate than the wavelength line-rate. This is referred to as *subrate* traffic. For example,

---

[5] The loss increases with fiber distance and the number of fiber connectors; thus, these various types of interfaces determine the allowable interconnection arrangements within an office.

an IP router may generate 10 Gb/s signals but the line-rate may be 40 Gb/s. This mismatch gives rise to the need to *multiplex* or *groom* traffic, where multiple client signals are carried on a wavelength in order to improve the network efficiency. (End-to-end multiplexing bundles together subrate traffic with the same endpoints; grooming uses more complex aggregation than multiplexing and is thus more efficient, though more costly.) It is also possible, though less common, for the client signal rate to be higher than the wavelength line-rate. In this scenario, *inverse multiplexing* is used, where the client signal is carried over multiple wavelengths.

1.7 Network Design and Network Planning

As indicated by the title of the book, both network design and network planning are covered. Network design encompasses much of the up-front work such as selecting which nodes to include in the network, laying out the topology to interconnect the nodes, selecting what type of transmission and switching systems to deploy (e.g., selecting the line-rate and whether to use optical bypass), and what equipment to deploy at a particular node. Network planning is more focused on the details of how to accommodate the traffic that will be carried by the network. For example, network planning includes selecting how a particular demand should be routed, protected, and groomed, and what wavelength(s) in the system spectrum should be assigned to carry it.

Network planning is carried out on two time scales, both of which are covered in this book. In *long-term network planning*, there is sufficient time between the planning and provisioning processes such that any additional equipment required by the plan can be deployed. In the long-term planning that typically occurs before a network is deployed, there is generally a large set of demands to be processed at one time. In this context, the planning emphasis is on determining the optimal strategy for accommodating the whole traffic set. After the network is operational, long-term planning is performed for the traffic that does not need to be provisioned immediately; however, typically it is performed for a smaller number of demands at one time. Again, the focus is on determining optimal strategies, as there is enough time to deploy equipment to accommodate the design.

In *real-time network planning*, there is little time between planning and provisioning, and demands are generally processed one at a time. It is assumed that the traffic must be accommodated using whatever equipment is already deployed in the network. Thus, the planning process must take into account any constraints posed by the current state of deployed equipment, which, for example, may force a demand to be routed over a sub-optimal path. (A related topic is *traffic engineering*, which in this context is a process where traffic is controlled to meet specific performance objectives; e.g., a demand may be routed over a specific path to meet a particular availability metric, or a demand may be routed such that it avoids a heavily utilized region of the network. Traffic-engineering support for real-time routing has been incorporated in several protocols; e.g., see [ABGL01, KaKY03].)

1.8 Focus on Practical Optical Networks

This book examines the design and planning of state-of-the-art optical networks, with an emphasis on the ramifications of optical-bypass technology. It expands on the aspects of optical network design and planning that are relevant in a practical environment, as opposed to taking a more theoretical approach. Much research has focused on idealized optical-bypass systems where all intermediate electronic processing is removed; such networks are often referred to as 'all-optical'. However, in reality, a small amount of intermediate electronic processing may still be required, for example, to improve the quality of the signal or to more efficiently pack data onto a wavelength. This small deviation from the idealized 'all-optical' network can have a significant impact on the network design, as is covered in later chapters. Thus, rather than use the term 'all-optical network', this book uses the term '*optical-bypass-enabled network*'.

Many of the principles covered in the book are equally applicable in metro-core, regional, and backbone networks. However, it will be noted when there are significant differences in the application of the technology to a particular tier.

The foundation of today's optical networks is the network elements; i.e., the major pieces of equipment deployed at a node. Chapter 2 discusses the various network elements in detail, with a focus on functionality and architectural implications. The underlying technology will be touched on only to the level that it affects the network architecture. Traditional network elements are covered as well as the new elements that enable optical bypass. From the discussion of the network elements, it will be apparent why algorithms play an important role in optical-bypass-enabled networks. Chapters 3 through 5 focus on the algorithms that are an integral part of operating an efficient and cost-effective optical network. The goal is not to cover all possible optical networking algorithms, but to focus on techniques that have proved useful in practice. Chapter 3, on routing algorithms, is equally applicable to optical-bypass-enabled networks as well as more traditional networks. Chapters 4 and 5, on regeneration and wavelength assignment, respectively, are relevant just to optical-bypass-enabled networks.

As mentioned earlier, treating the optical network as another networking layer can be very advantageous. However, networking at the wavelength level can potentially be at odds with operating an efficient network if the wavelengths are not well packed. Chapter 6 looks at efficient grooming of subrate demands, with an emphasis on various grooming architectures and methodologies that are compatible with optical bypass. It also considers strategies for grooming in the optical domain. Chapter 7 discusses protection in the optical layer. Rather than covering the myriad variations of optical protection, the discussion is centered on how protection in the optical layer is best implemented in a network with optical-bypass technology.

From the viewpoint of the network operator, perhaps the single most important characteristic of a network is its cost, both capital cost (i.e., equipment cost) and operating cost. Chapter 8 includes a range of economic studies that probe how and when optical networking can improve the economics of a network. These studies can serve as a guideline for network architects planning a network evolution strategy, as well as equipment vendors analyzing the potential benefits of a new technology. The emphasis of the studies in this chapter, as well as the book as a whole, is on real-world networks. For general books on optical networking, see reference texts such as [StBa99, RaSi01, Mukh06].

Chapter 2

Optical Network Elements

The dramatic shift in the architecture of optical networks is chiefly due to the development of advanced optical network elements. These elements are based on the premise that the majority of the traffic that enters a node is being routed *through* the node en-route to its final destination as opposed to being *destined for* the node. This transiting traffic can potentially remain in the optical domain as it traverses the node rather than be electronically processed. By deploying technology that enables this so-called *optical bypass,* a significant reduction in the amount of required nodal electronic equipment can be realized.

After briefly discussing some basic optical components and switch terminology, Sections 2.2 and 2.3 review the traditional network architecture where all traffic entering a node is electronically processed. This architecture is based on a simple optical network element, the *Optical Terminal.* The economic and operational challenges of these legacy networks motivated the development of optical-bypass technology. The three major network elements that are capable of optical bypass are the *Optical Add/Drop Multiplexer* (OADM), the *Multi-Degree OADM* (OADM-MD), and the *All-Optical Switch*, all of which are discussed in Sections 2.4 through 2.7. These elements come in many flavors; the chief attributes that affect their efficiency, cost, and flexibility are covered in these sections.

The configurability of the network elements in response to changing traffic is one of the most important attributes. In some implementations, the network element does not provide sufficient configurability at the edge of the optical network (i.e., the interface between the client network and the optical network). Alternative methods of achieving the desired edge flexibility are presented in Section 2.8. Note that the terms 'configurability' and 'reconfigurability' are used interchangeably throughout the book.

While the network elements are a necessary ingredient for enabling optical bypass, they are only half of the story. Another key requirement is extended *optical reach*, which is the distance an optical signal can travel before it degrades to a level that necessitates it be 'cleaned up', or regenerated. The interplay of optical reach and optical-bypass-enabled elements is presented in Section 2.9.

J.M. Simmons, *Optical Network Design and Planning,*
DOI: 10.1007/978-0-387-76476-4_2, © Springer Science+Business Media, LLC 2008

Integration of elements or components within a node is a more recent development, motivated by the desire to eliminate individual components and reduce cost. There can be a range of levels of integration as illustrated by the discussions of Sections 2.10 and 2.11.

Throughout this chapter, it is implicitly assumed that there is one fiber-pair per link; e.g., a degree-two node has two incoming and two outgoing fibers. Due to the large capacity of current transmission systems, single-fiber-pair deployments are common. However, the last section of the chapter briefly addresses multi-fiber-pair scenarios.

Throughout this chapter, the focus is on the functionality of the network elements. The underlying technology of the elements is discussed only when it has a major impact on how a particular element is used.

2.1 Basic Optical Components

Some of the optical components that come into play throughout this chapter are discussed here. One very simple component is the *wavelength-independent optical splitter*, which is typically referred to as a *passive splitter*. A splitter has one input port and N output ports, where the input optical signal is sent to all of the output ports. Note that if the input is a WDM signal, then the output signals are WDM as well. In many splitter implementations, the input power level is split equally across the N output ports, such that each port receives $1/N$ of the original signal power level. This corresponds to a nominal optical loss of $10 \cdot \log_{10} N$, in units of decibels (dB). Roughly speaking, for every doubling of N, the optical loss increases by another 3 dB. It is also possible to design optical splitters where the power is split non-uniformly across the output ports so that some ports suffer lower loss than others.

The inverse device is called a *passive optical coupler* or *combiner*. This has N input ports and one output port, such that all of the inputs are combined into a single output signal. The input signals are usually at different optical frequencies to avoid interference when they are combined. The coupler losses are the same as for the splitter.

Another important component is the 1xN demultiplexer, which may be built using an *arrayed waveguide grating* (AWG) [Okam98, RaSi01, DoOk06]. (AWGs are also called *wavelength grating routers*.) For large N, the loss through an AWG is on the order of 4 dB to 6 dB. This device can be viewed as having one input port and N output ports. With this configuration, an input WDM signal is internally demultiplexed into its constituent wavelengths, and specific wavelengths are sent to each output port. In a common implementation, the number of output ports and the number of wavelengths in the WDM signal are the same, such that exactly one wavelength is sent to each output port. The inverse device is an Nx1 multiplexer, with N input ports and one output port. Only specific wavelengths from each of the input ports are multiplexed together on the output port. Again, in a common implementation where the number of input ports equals the number of wavelengths, just one wavelength from each input port is multiplexed in the outgoing WDM signal.

Throughout this chapter, various types of switches are mentioned. For the most part, the relevant details of the switch are presented when the application is discussed; however, it is advantageous to introduce some terminology here. First, there is a broad class of switches known as *optical switches*. Contrary to what the name implies, these switches do not necessarily have a switch fabric that operates on optical signals. (The switch fabric is the 'guts' of the switch, where the interconnection between the input and output ports is established.) Rather, the term 'optical switch' is used to indicate a switch where the ports operate on the granularity of a wavelength or a group of wavelengths.

Wavelength-selective is a term used to classify devices that are capable of treating each wavelength differently. A wavelength-selective optical switch of size 1xN is known as a *wavelength-selective switch* (WSS) [MMMT03, Maro05]. A WSS can direct any wavelength on the input port to any of the N output ports. A wavelength-selective optical switch of size MxM is known as a *wavelength-selective cross-connect* (WSXC) [KYJH06]. This device can direct any wavelength from any of the M input ports to any of the M output ports.

Micro-electro-mechanical-system (MEMS) technology [WuSF06] is often used to fabricate switches with an optical switch fabric. This technology essentially uses tiny movable mirrors to direct light from input ports to output ports. Note that an individual MEMS switching element is not wavelength selective; it simply switches whatever light is on the input port without picking out a particular wavelength. However, when combined with multiplexers and demultiplexers that couple the individual wavelengths of a WDM signal to the ports of the MEMS switch, the combination is wavelength-selective, capable of directing any wavelength to any output port.

Various optical switch configurations are discussed in the chapter.

2.2 Optical Terminal

In traditional optical network architectures, optical terminals are deployed at the endpoints of each link. Figure 2.1 illustrates a single optical terminal equipped with several WDM transponders. An optical terminal is typically depicted in figures as a trapezoid to capture its multiplexing/demultiplexing functionality; i.e., there are individual wavelengths on the client side of the terminal and a WDM signal on the network side. Unfortunately, a trapezoid is often used to specifically represent a 1xN AWG. While an optical terminal can be based on AWG technology, there are other options as well, some of which are discussed in Section 2.2.1. *Throughout this book, the trapezoid is used to represent a general optical terminal, not necessarily one based on a specific technology.*

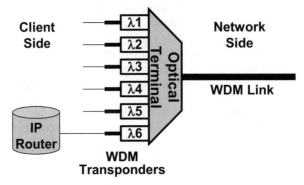

Fig. 2.1 An optical terminal equipped with WDM transponders.

Tracing the flow from left to right in the figure, a client signal enters a transponder, which converts the signal to a WDM-compatible optical frequency. (Only one client is shown in the figure, an IP router.) The optical terminal multiplexes the signals from all of the transponders onto a single network fiber. In general, the transponders plugged into an optical terminal generate different optical frequencies; otherwise, the signals would interfere with each other after being multiplexed together by the terminal.

In the reverse direction, a WDM signal is carried by the network fiber into the optical terminal, where it is demultiplexed into its constituent frequencies. Each transponder receives its signal on a particular optical frequency and converts it to a client-compatible signal.

Recall from Fig. 1.7(b) in Section 1.6 that the WDM transponder encompasses both a client-side receiver/network-side transmitter in one direction and a network-side receiver/client-side transmitter in the other direction. Similarly, the optical terminal is composed of both a multiplexer and a demultiplexer. Note that it is possible for the network-side signal transmitted by a transponder to be at a different optical frequency than the network-side signal received by the transponder; however, in most scenarios the frequencies are the same.

2.2.1 Slot Flexibility

An optical terminal is deployed with equipment shelves in which the transponders are inserted. One figure of merit of an optical terminal is the density of the transponders on a shelf, where higher density is preferred as it requires less space. For example, if a shelf holds up to sixteen 10 Gb/s transponders, the shelf density is 160 Gb/s.

The flexibility of the individual slots in the transponder shelves is another important attribute. In the most flexible optical terminal architecture, any slot can accommodate a transponder of any frequency. This 'colorless' slot architecture allows for the most efficient deployment of optical terminals; i.e., the number of slots deployed needs to be only as large as the maximum number of transponders required at the node (subject to the shelf granularity). This architecture also maximizes the benefits of tunable transponders for dynamic networking, as it allows a transponder to tune to a different frequency without needing to be manually moved to a different slot.

One implementation of a colorless optical terminal is based on a passive splitter and coupler. The received WDM signal is passively split and sent to each of the slots. The transponder receiver (on the network-side) is equipped with an optical filter to select the desired optical frequency; for maximum transponder flexibility, this filter should be tunable. In the reverse direction, the signals from the transponders are passively coupled together into a WDM signal; again, for maximum flexibility, the transponders should be equipped with tunable lasers. Because passive splitters and couplers can result in significant optical loss, this architecture often requires optical amplifiers to boost the signal level, especially if the number of supported slots is large.

Another flexible optical-terminal architecture is based on a 1xN wavelength-selective-switch (WSS), where the switch can be configured to direct any wavelength from the input WDM signal to any of the optical terminal slots. A separate Nx1 WSS is used in the reverse direction to multiplex the signals. In this architecture, the transponder receiver does not need to have an optical filter because the wavelength selection is carried out in the WSS (i.e., the transponder is capable of receiving whatever optical frequency is directed to its slot). One drawback of this approach is the limited size of the WSS, which limits the number of supported slots on the terminal. Commercially available WSSs originally had a maximum size on the order of 1x8, although they have been continually increasing in size. Another disadvantage is the cost of the WSS relative to passive splitters and couplers.

In contrast to these colorless optical terminals, there are also fixed optical terminals where each slot can accommodate a transponder of only one particular frequency. This type of terminal is often implemented using AWG technology. Again, the transponder receiver does not need an optical filter. Though relatively cost effective and low loss, this architecture can lead to inefficient shelf packing and ultimately higher cost in networks where the choice of optical frequencies is very important. A new shelf may need to be added to accommodate a desired frequency even though the shelves that are already deployed have available slots. This architecture also negates the automated configurability afforded by tunable transponders.

In an intermediary optical-terminal architecture, the WDM spectrum is partitioned into groups, and a particular slot can accommodate transponders only from one group [ChLH06]. This type of terminal can be architected with lower loss or cost than a completely colorless design, but it has only limited configurability.

2.3 Optical-Electrical-Optical (O-E-O) Architecture

2.3.1 O-E-O Architecture at Nodes of Degree-Two

The traditional (non-configurable) optical-terminal-based architecture for a node of degree-two is shown in Fig. 2.2. There are two network links incident on the node, where it is common to refer to the links as the 'East' and 'West' links (there is not necessarily a correspondence to the actual geography of the node). As shown in the figure, the node is equipped with two optical terminals arranged in a 'back-to-back' configuration. The architecture shown does not support automated reconfigurability. Connectivity is provided via a manual patch-panel; i.e., a panel where equipment within an office is connected via fiber cables to one side (typically in the back), and where short patch cables are used on the other side (typically in the front) to manually interconnect the equipment as desired. Providing automated reconfigurability is discussed in the next section in the context of higher-degree nodes.

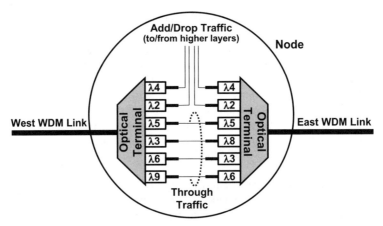

Fig. 2.2 O-E-O architecture at a degree-two node (without automated reconfigurability). Nodal traffic is characterized as either add/drop traffic or through traffic. All traffic entering and exiting the node is processed by a transponder. Note that the through traffic can undergo wavelength conversion, as indicated by interconnected transponders of different wavelengths (e.g., the bottom pair of interconnected transponders converts the signal from $\lambda6$ to $\lambda9$ in the East-to-West direction).

Tracing the path from right to left, the WDM signal enters the East optical terminal from the East link. This WDM signal is demultiplexed into its constituent wavelengths, each of which is sent to a WDM transponder that converts it to a 1310-nm optical signal. (Recall that 1310-nm is the typical wavelength of the client-side optical signal.) At this point, it is important to distinguish two types of traffic with respect to the node. For one type of traffic, the node serves as the exit point from the optical layer. This traffic 'drops'[6] from the optical layer and is sent to a higher layer (the higher layers, e.g., IP, are the clients of the optical layer). The other type of traffic is transiting the node en-route to its final destination. After this transiting traffic has been converted to a 1310-nm optical signal by its associated transponder, it is sent to a second transponder located on the West optical terminal. This transponder converts it back into a WDM-compatible signal, which is then multiplexed by the West optical terminal and sent out on the West link. There are also transponders on the West terminal for traffic that is being 'added' to the optical layer, from higher layers, that needs to be routed on the West link.

In the left to right direction of the figure, the operation is similar. Some of the traffic from the West link drops from the optical layer and some is sent out on the East link. Additionally, there are transponders on the East terminal for traffic that is added to the optical layer at this node that needs to be routed on the East link.

The traffic that is being added to or dropped from the optical layer at this node is termed *add/drop* traffic; the traffic that is transiting the node is called *through* traffic. Regardless of the traffic type, note that all of the traffic entering and exiting the node is processed by a WDM transponder. In the course of converting between a WDM-compatible optical signal and a client optical signal, the transponder processes the signal in the electronic domain. Thus, all traffic enters the node in the optical domain, is converted to the electronic domain, and is returned to the optical domain. This architecture, where all traffic undergoes optical-electrical-optical conversion, is referred to as the *O-E-O architecture*.

2.3.2 O-E-O Architecture at Nodes of Degree-Three or Higher

The O-E-O architecture readily extends to a node of degree greater than two. In general, a degree-N node will have N optical terminals. Figure 2.3 depicts a degree-three node equipped with three optical terminals, with the third link referred to as the 'South' link. The particular architecture shown does not support automated reconfigurability.

As with the degree-two node, all of the traffic entering a node, whether add/drop or through traffic, is processed by a transponder. The additional wrinkle with higher-degree nodes is that the through traffic has multiple possible path directions. For example, in the figure, traffic entering from the East could be directed to the West or to the South; the path is set by interconnecting a transponder on the East optical terminal to a transponder on the West or the South optical terminal, respectively. In many real-world implementations, the transponders are interconnected using a manual patch-panel. Modifying the through path of a connection requires that a technician manually rearrange the patch-panel, a process that is not conducive to rapid reconfiguration and is subject to operator error.

---

[6] While 'drop' often has a negative connotation in telecommunication networks (e.g., dropped packets, dropped calls), its usage here simply means a signal is exiting from the optical layer.

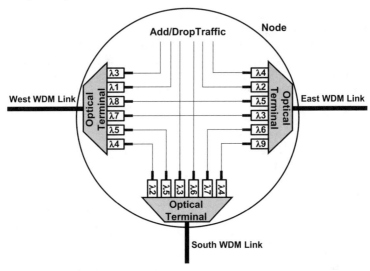

Fig. 2.3 O-E-O architecture at a degree-three node (without automated reconfigurability). There are three possible directions through the node. The path of a transiting connection is set by inter-connecting a pair of transponders on the associated optical terminals.

The reconfiguration process can be automated through the addition of an optical switch, as shown in Fig. 2.4. Each transponder at the node feeds into a switch port, and the switch is configured as needed to interconnect two transponders to create a through path. Additionally, the add/drop signals are fed into ports on the switch, so that they can be directed to transponders on any of the optical terminals. Furthermore, the switch allows any transponder to be flexibly used for either add/drop or through traffic, depending on how the switch is configured. Note that the degree-two architecture of Fig. 2.2 could benefit from a switch as well with respect to these latter two applications; i.e., redirecting add/drop traffic to any optical terminal and flexibly using a transponder for either add/drop or through traffic.

While deploying a switch enhances the network flexibility and reduces operational costs due to less required manual intervention, the downside is the additional equipment cost. There are generally two types of optical switches that are used in such applications. Most commonly, a switch with an electronic switching fabric is used, where each port is equipped with a short-reach interface to convert the 1310-nm optical signal from the transponder to an electrical signal (this is the option shown in the figure). A second option is to use a switch with an optical switch fabric such as a MEMS-based switch. This technology can directly switch an optical signal, thereby obviating the need for short-reach interfaces on the switch ports (although it may require that the transponders be equipped with a special interface that is tolerable to the optical loss through the switch). As MEMS technology comes down in price, this type of switch may be the more cost-effective option. (Electronic switches and MEMS switches are revisited in Sections 2.6.1 and 2.6.2, respectively.)

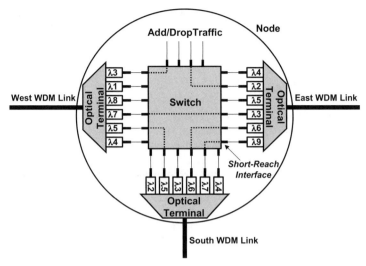

Fig. 2.4 A switch is used to automate node reconfigurability. The particular switch shown has an electronic switch fabric and is equipped with short-reach interfaces on all of its ports.

2.3.3 Advantages of the O-E-O Architecture

The fact that the O-E-O architecture processes all traffic entering the node in the electronic domain does offer some advantages. First, converting the signal to the electronic domain and back to the optical domain 'cleans up' the signal. Optical signals undergo degradation as they are transmitted along a fiber. The O-E-O process re-amplifies, reshapes, and retimes the signal, a process that is known as *3R-regeneration.*

Second, it readily allows for performance monitoring of the signal. For example, assuming the SONET framing format is being used, a SONET-compatible transponder can examine the overhead bytes using electronic processing to determine if there are errors in the signal. Because this process can be done at every node, it is fairly straightforward to determine the location of a failure.

Third, the O-E-O architecture is very amenable to a multi-vendor architecture because all communication within a node is via a standard 1310-nm optical signal. Whereas the WDM transmission characteristics on a link may be proprietary to a vendor, the intra-nodal communication adheres to a well-defined standard. This allows the transmission systems on the links entering a node to be supplied by different vendors. Furthermore, the vendors of the switch (if present) and the transmission system can be different as well.

The O-E-O architecture also affords wavelength assignment independence to the traffic passing through a node. Here, the term wavelength is used to indicate a particular frequency in the WDM spectrum. As noted above, the through traffic enters the node on one transponder and exits the node using a second transponder. These two transponders communicate via the 1310-nm optical signal; there is no requirement that the WDM-compatible wavelengths of these two transponders be the same. This is illustrated in Fig. 2.2 through Fig. 2.4, where the wavelengths of two interconnected transponders are not necessarily the same. This process accomplishes what is known as *wavelength conversion*, where the signal enters and exits the node on two different wavelengths. There is complete freedom in selecting the wavelengths of the two transponders, subject to the constraint that the same wavelength cannot be used more than once on any given fiber. The most important implication is that the choice of wavelength is local to a particular link; the wavelength assignment on one link does not affect that on any other link.

2.3.4 Disadvantages of the O-E-O Architecture

While the O-E-O architecture does have advantages, terminating every wavelength entering every node on a transponder poses many challenges in scaling this architecture to larger networks. For example, as capacities grow to over 100 wavelengths per fiber, this architecture could potentially require several hundred transponders at a node. The first barrier is the cost of all this equipment, although transponder costs do continue to decrease (see Section 2.11). Second, there are concerns regarding the physical space at the sites that house the equipment. More transponders translate to more shelves of equipment, and space is already at a premium in many carrier offices. Furthermore, providing the power for all of this electronics and dissipating the heat created by them is another operational challenge that will only worsen as networks continue to grow.

Another barrier to network evolution is the fact that electronics are often tied to a specific technology. For example, a short-reach interface that supports a 10-Gb/s signal typically does not also support a 40-Gb/s signal. Thus, if a carrier upgrades its network from a 10-Gb/s line-rate to a 40-Gb/s line-rate, a great deal of equipment needs to be replaced.

Provisioning a connection can be cumbersome in the O-E-O architecture. A technician may need to visit every node along the path of a connection to install the required transponders. Even if transponders are predeployed in a node, a visit to the node may be required, for example, to manually interconnect two transponders via a patch-panel. As noted above, manual intervention can be avoided through the use of a switch; however, as the node size increases, the switch size must grow accordingly. A switch, especially one based on electronics, will have its own scalability issues related to cost, size, power, and heat dissipation.

Finally, all of the equipment that must be deployed along a connection is potentially a reliability issue. The connection can fail, for example, if any of the transponders along its path fails.

2.4 OADMs (ROADMs)

The scalability challenge of the O-E-O architecture was a major impetus to develop alternative technology where much of electronics could be eliminated. Given that through traffic is converted from a WDM-compatible signal to a 1310-nm signal only to be immediately converted back again to a WDM-compatible signal, removing the need for transponders for the through traffic was a natural avenue to pursue.

This gave rise to the Optical Add/Drop Multiplexer (OADM) network element for nodes of degree two, as shown in Fig. 2.5. With an OADM, the through traffic remains in the optical domain as it transits the node; transponders are needed only for the add/drop traffic. The through traffic is said to *optically bypass* the node. Studies have shown that in typical carrier networks, on average, over 50% of the traffic entering a node is through traffic; thus, the amount of transponders that can be eliminated with optical bypass is significant. While the OADM itself costs more than two optical terminals, the reduction in transponders results in an overall lower nodal cost, assuming the level of traffic is high enough. The economics of optical bypass are explored further in Chapter 8.

OADMs have been commercially available since the mid 1990s, although significant deployment did not start until after 2000. The name of the element derives from a SONET/SDH Add/Drop Multiplexer (ADM), which is capable of adding/extracting lower-rate SONET/SDH signals to/from a higher-rate signal without terminating the entire higher-rate signal. Similarly, the OADM adds/extracts wavelengths to/from a fiber without having to electronically terminate all of the wavelengths comprising the WDM signal.

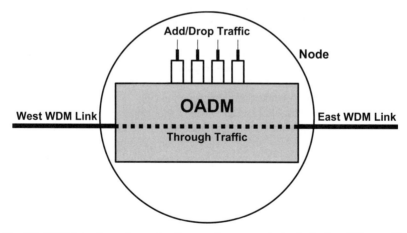

Fig. 2.5 OADM at a degree-two node. Transponders are required only for the add/drop traffic. The through traffic remains in the optical domain as it transits the node. (Adapted from [Simm05]. © 2005 IEEE)

Networks equipped with elements that support optical bypass are referred to here as *optical-bypass-enabled* networks. Other terms used in the literature to describe this type of network are 'all-optical' or 'transparent'. However, given that O-E-O-based regenerators are typically not entirely eliminated from all end-to-end paths (see Section 2.9) and that networks supporting protocol-and-format transparency have not materialized in a major way (see Section 1.3), these terms are not completely accurate.

The advantages and disadvantages of the optical-bypass-enabled architecture are diametrically opposite to those that were discussed for the O-E-O architecture. First, as the network traffic level increases, optical-bypass technology is potentially more scalable in cost, space, power, and heat dissipation because much of the electronics is eliminated. Second, optics is more agnostic to the system bit-rate as compared to electronics. For example, an OADM should typically function whether the wavelengths are carrying 10-Gb/s signals or 40-Gb/s signals (as long as the wavelength spacing and the signal spectrum are compatible with the OADM). Third, provisioning a connection is operationally simpler; in many scenarios, a new connection requires that a technician visit just the source and destination nodes to install the add/drop transponders. Fourth, the elimination of much of the electronics also improves the overall reliability; though the optical-bypass equipment itself may have higher failure rates than optical terminals, the removal of much of the equipment in the signal path typically leads to an overall lower failure rate for the connection [MaLe03].

Conversely, removing the transponders from the through path eliminates the functions that they provided. First, the optical signal of the through traffic is not regenerated, thereby requiring extended optical reach, as described in Section 2.9. Second, removing the transponders from some or all of the intermediate nodes of a path also eliminates the node-by-node error-checking functionality they provided. In the absence of electronic performance monitoring, optical monitoring techniques are needed, as discussed in Chapter 7. Third, it may be challenging to support a multi-vendor environment in an optical-bypass-enabled network because not all intra-nodal traffic is converted to a standard optical signal as it is in an O-E-O network. Standards for extended-reach WDM transmission have not been defined yet, such that interoperability between multiple vendors is not guaranteed. For example, the transmission system of one vendor may not be compatible with the OADM of another vendor. Furthermore, without standard performance guidelines, it may be difficult to isolate which vendor's equipment is malfunctioning under a failure condition. Thus, in optical-bypass-enabled networks, it is common for a single vendor to provide both the transmission system and the optical networking elements. (The notion of 'islands of transparency', where designated vendors operate within non-overlapping subsets of the network, is discussed in Chapter 4.)

Finally, and most important, a transiting connection enters and exits the node on the same wavelength; there is not the same opportunity for wavelength conversion as with the O-E-O architecture. Given that two signals on the same fiber cannot be assigned the same wavelength, this implies that the assignment of wavelengths on one link potentially affects the assignment on other links in the network. This *wavelength continuity constraint* is the major reason why advanced algorithms are required to efficiently operate a network based on optical-bypass technology. Such algorithms are covered in Chapters 3, 4 and 5. (Note that in the future, technology may be commercially available that all-optically performs wavelength conversion. With this technology, a path that remains in the optical domain may not necessarily be carried on the same wavelength. However, all-optical wavelength converters represent an additional cost and are not likely to be deployed at all nodes on all wavelengths. Thus, the wavelength continuity constraint is likely to remain a relevant factor in optical-bypass-enabled networks.)

There are numerous important properties that characterize the various types of OADMs. These properties, which affect how the element is used in a network, are discussed in the following sections.

2.4.1 Configurability

One of the most important properties of an OADM is its degree of reconfigurability. The earliest commercial OADMs were not configurable. Carriers needed to specify up front which particular wavelengths would be added/dropped at a particular node, with all remaining wavelengths transiting the node. Once installed, the OADM was fixed in that configuration. Clearly, this rigidity limits the ability of the network to adapt to changing traffic patterns.

Today, however, most OADMs are configurable. This implies that any wavelength can be added/dropped at any node, and that the choice of add/drop wavelengths can be readily changed without impacting any of the other connections terminating at or transiting the node. Furthermore, it is highly desirable that the OADM be remotely configurable through software as opposed to requiring manual intervention. Such fully configurable OADMs are often called Reconfigurable OADMs, or ROADMs; however, there were fully configurable OADMs deployed in carrier networks prior to this term being coined.

One limitation of some OADMs is that they are fully configurable provided that the amount of add/drop does not exceed a given threshold. A typical threshold in such OADMs is a maximum of 50% of the wavelengths supportable on a fiber can be added/dropped (e.g., a maximum of 40 add/drops from a fiber that can support 80 wavelengths). For many nodes in a network, this is sufficient flexibility. However, there is typically a small subset of nodes that would ideally add/drop a higher percentage than this (this is examined further in Chapter 8). These nodes would be forced to employ the traditional O-E-O architecture. While the cost savings of using an OADM at a node that drops over 50% of the wavelengths may not be that significant, it is still desirable to have the option to use an OADM at such a node. An OADM would provide more agility than an O-E-O architecture with two optical terminals and a patch-panel. Moreover, OADMs that support up to 100% add/drop allow the flexibility of using an OADM at any node without having to estimate the maximum percentage drop that will ever occur at the node.

2.4.2 Wavelength vs. Waveband Granularity

The previous section described configurability on a per-wavelength basis, where the choice of add/drop versus through can be made independently for each wavelength. Alternatively, OADMs can be fully configurable on the basis of a *waveband*. A waveband is a set of wavelengths that are treated as a single unit; either the whole waveband is added/dropped or the whole waveband transits the node. Wavebands are usually composed of wavelengths that are contiguous in the spectrum. In most implementations, the wavebands are of equal size; however, non-uniform waveband sizes may be more efficient depending on the traffic [IGKV03].

Clearly, waveband granularity is not as flexible as wavelength granularity. The chief motivation for using a waveband-based OADM is the potential for reduced cost and complexity. Waveband technology is more common in metro-core networks, where sensitivity to cost is greater.

Wavebands are most effective when many connections are being routed over the same paths in the network. Otherwise, inefficiencies can arise due to some of the bandwidth being 'stranded' in partially filled bands. Studies have shown that under reasonable traffic conditions, and through the use of intelligent algorithms, the inefficiencies resulting from wavebands are small [BuWW03]. Nevertheless, some carriers are averse to using waveband technology because of the somewhat diminished flexibility.

2.4.3 Wavelength Reuse

Another key OADM property that potentially affects network efficiency is whether or not the OADM supports *wavelength reuse*. With wavelength reuse, if a particular wavelength (i.e., optical frequency) is dropped at a node from one of the network fibers, then that same wavelength can be added at the node on the other network fiber. This is illustrated in Fig. 2.6(a). In the figure, a particular wavelength (λ_2) enters the node on the East fiber and is dropped. After being dropped, the wavelength does not continue to be routed through the OADM to the West fiber. This allows the node to add the same wavelength to the West fiber, as is shown in the figure (i.e., the add traffic is 'reusing' the same wavelength that was dropped). If the dropped wavelength had continued through the OADM, then traffic could not have been added to the West fiber on this wavelength because the two signals would interfere.

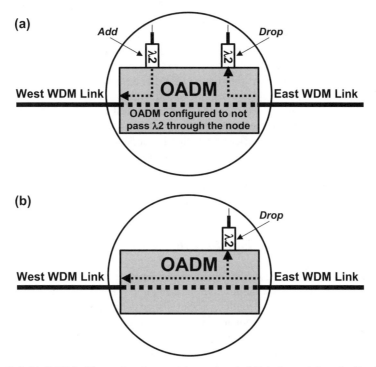

Fig. 2.6 (a) OADM with wavelength reuse. A wavelength (λ2) is dropped from the East link. The OADM is configured to not pass this wavelength through the node, such that the same wavelength can be added on the West link. (b) OADM without wavelength reuse. The wavelength that is dropped from the East link continues to be routed through the node so that the same wavelength cannot be added on the West link.

Figure 2.6(b) illustrates an OADM that does not have wavelength reuse. The wavelength entering from the East network port is dropped but also continues through the OADM to the West fiber. This clearly wastes bandwidth, as this particular wavelength cannot be used to carry useful traffic on the West fiber (i.e., it is wasted bandwidth because it is carrying traffic that has already reached its destination). If there are multiple consecutive OADMs without reuse, the signal will continue to be routed through each one of them, preventing the wavelength from being reused to carry useful traffic on all of the intermediary links. (Note that the nodes on a ring cannot all be populated by no-reuse OADMs, as the optical signals will continue to wrap around.)

Wavelength reuse is a desirable trait in OADMs to maximize the useful capacity of the network. Nevertheless, OADMs without reuse can be useful network elements and are part of the commercial offerings of several system vendors. First, no-reuse OADMs are significantly lower cost than OADMs with reuse, making them a cost-effective option for nodes that drop a small amount of traffic.

Most vendors allow just a small percentage of the wavelengths on a fiber to be added/dropped with a no-reuse OADM; e.g., a maximum of 8 add/drops from a fiber with 80 wavelengths. This is in consonance with how the no-reuse OADM should be used. Additionally, the element is ideally deployed at nodes located on lightly loaded links, so that the wasted bandwidth is inconsequential.

In many implementations, a no-reuse OADM is little more than an optical amplifier equipped with a coupler and splitter to add and drop traffic. Because of this design, it is often possible to upgrade from an optical amplifier to a no-reuse OADM without affecting the traffic already passing through the amplifier. When a system is first installed, a network site that is not generating any traffic may be equipped with just an optical amplifier. As the network grows, the site may need to source/terminate traffic; when this occurs, the optical amplifier can be upgraded to the no-reuse OADM (upgrading from an optical amplifier to an OADM *with* reuse is generally not possible without a major overhaul of the equipment).

2.4.4 Automatic Power Equalization

In addition to network efficiency, there is often another important difference between OADMs with and without reuse. In many implementations, the technology that is used to achieve wavelength reuse can also be used to automatically equalize the power levels of the individual wavelengths that are transiting the node. The wavelengths comprising a WDM signal may have originated at different nodes, such that the power levels entering a node are unequal. Unbalanced power levels also result from uneven amplifier gain across the wavelength spectrum. The need for periodic power equalization is often another factor that comes into play when determining whether a no-reuse OADM is suitable for a node.

2.4.5 Edge Configurability

OADM configurability was discussed in Section 2.4.1 with respect to flexibly allowing any wavelength to bypass or drop at the node. Here, a different type of configurability is covered.

As discussed in Section 2.3.2, one of the advantages of deploying a switch with the O-E-O architecture is that the traffic from the higher layers (e.g., IP) can be automatically directed to any of the network links at the node. This flexibility, known as *edge configurability*, is especially useful in a dynamic environment, where connections are continually set up and torn down. Edge configurability allows any connection associated with a particular transponder to be automatically routed via any of the links at the node. It is also useful for protection in the optical layer, where at the time of failure, a connection can be sent out on an alternative path using the same transponder.

Edge configurability is possible with some OADM implementations, as is covered later in the chapter. This configurability is illustrated in Fig. 2.7, where the transponder can access either the East link or the West link, depending on how the OADM is configured. Typically, fewer transponders need to be predeployed on an OADM that supports edge configurability because the transponders are not tied to a particular network link.

While having OADMs with edge flexibility is desirable, there are alternative means of achieving this configurability, as is discussed in Section 2.8.

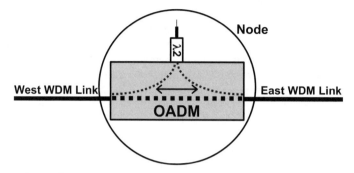

Fig. 2.7 Edge configurability in an OADM, where one transponder can access either network link. In some implementations, one transponder can access both links simultaneously to provide optical multicast.

2.4.6 Multicast

Some OADMs can be configured to support optical multicast, where a given signal is sent to multiple destinations rather than just one, and the signal replication occurs in the optical domain as opposed to in the electronic domain. This can be accomplished by implementing *drop-and-continue*, where the intermediate destination OADMs are configured to both drop a wavelength and allow that same wavelength to pass through the node. The drop-and-continue concept was illustrated in Fig. 2.6(b), where λ_2 both dropped from and passed through the OADM. (Note that any OADM not capable of wavelength reuse is performing drop-and-continue for any dropped wavelength, whether it is desired or not.)

Additionally, some OADMs support a second means of optical multicast, where a single transponder can be configured to transmit simultaneously on the East port and the West port. This launches the same signal on two different network links. (This feature is also beneficial for optical-layer protection with an active backup path, as is discussed further in Chapter 7.)

2.4.7 Slot Flexibility

The discussion regarding the benefits of slot flexibility for optical terminals (Section 2.2.1) holds for the OADM add/drop slots as well. Because of the wavelength continuity constraint in optical-bypass-enabled systems, it is often important to use a specific wavelength to carry a given connection. It is desirable to be able to insert the corresponding transponder in any add/drop slot of the OADM; i.e., colorless slots.

2.4.8 East/West Separability

East/West separability relates to the failure and repair modes of the OADM. If one side of the OADM fails, it is desirable that any add/drop traffic on that side be able to be directed to the other side for protection. It would be undesirable to have an architecture where, for example, a failure of the East link causes traffic on the West link to be shut down as well. Furthermore, if the East add/drop port fails, it would be undesirable if the process of repairing the port also requires that the West add/drop port be taken down.

2.4.9 Broadcast-and-Select and Wavelength-Selective Architectures

In this section, two common architectures for building an OADM are presented at a high level. *Broadcast-and-select* is a prevalent OADM architecture as it is suitable for optically bypassing several consecutive OADMs [BSAL02]. One common broadcast-and-select implementation is shown in Fig. 2.8. In this implementation, as the WDM signal enters from the East network fiber, some of the signal power is tapped off and directed to the East add/drop port. The dropped WDM signal is demultiplexed, and transponders are deployed only for those wavelengths that need to be dropped at the node (i.e., the entire WDM signal is 'broadcast' to the port, and transponders are deployed to each 'select' one particular wavelength). The add/drop port is a '*multi-wavelength port*', as the WDM signal is dropped there. As mentioned above, the add/drop slots are ideally colorless.

Only a relatively small portion of the signal power is tapped off on the add/drop port, say 10%, which is a major reason the architecture is suitable for optical bypass. The remainder of the optical signal power continues on through the OADM to the West fiber. In an OADM with wavelength re-use, the technology of the OADM is capable of blocking any wavelengths that have been dropped so that they do not continue to the West fiber. (Examples of such technology are covered in Section 2.5.2 and 2.5.3.) Wavelengths can be added from the West add/drop port to be multiplexed with the signals passing through the OADM (assuming there are no lambda conflicts). The operation in the West to East direction is similar.

Depending on the technology that is used, the broadcast-and-select OADM is capable of any of the properties discussed in the previous sections. However, the specific implementation shown in Fig. 2.8, with the add/drop ports being tapped off from the network fibers, does not provide edge configurability. In this figure, note that the transponders on the East add/drop port can only enter/exit from the East link; they cannot access the West link. (Other broadcast-and-select OADM implementations do provide edge configurability, however, as is discussed in Section 2.6.3.)

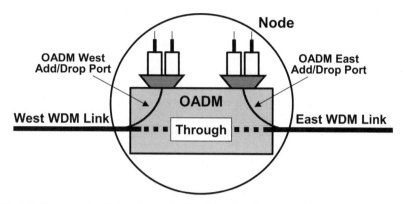

Fig. 2.8 One example of a broadcast-and-select OADM architecture. This particular implementation does not support edge configurability. The trapezoids represent general colorless or fixed add/drop ports.

An alternative to broadcast-and-select is the *wavelength-selective* architecture, an implementation of which is shown in Fig. 2.9. The WDM signal from the East network fiber is demultiplexed and the constituent wavelengths are each fed into a port on an optical switch. The switch is configured to direct some of the wavelengths to the add/drop ports on the switch; these add/drop ports are *single-wavelength* ports. The remaining wavelengths are directed to the West side, where they are multiplexed together, along with any add traffic, and sent out on the West network fiber. The operation in the West to East direction is similar.

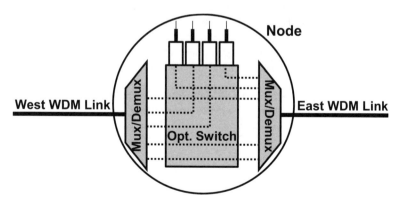

Fig. 2.9 Wavelength-selective OADM architecture.

The switch size in this wavelength-selective architecture is MxM, where M must be large enough to accommodate all of the wavelengths on the nodal fibers as well as all of the add/drop wavelengths. For example, if there are 40 wavelengths per fiber, and the OADM allows 100% drop from each fiber, then the switch must be of size 160x160. The fabric of the switch must be optical so that there is no need for O-E-O conversion in the node. MEMS technology is often used in such OADMs. (As noted in Section 2.1, the MEMS elements themselves are not wavelength-selective; however when combined with multiplexers and de-multiplexers as shown, the combination is a wavelength-selective architecture. To be more specific, the device shown in Fig. 2.9 is a WSXC.)

This type of OADM provides wavelength reuse, as dropped signals cannot also pass through the node. In fact, this architecture does not readily support drop-and-continue for optical multicast. In contrast, the broadcast-and-select architecture can support drop-and-continue simply by not blocking a wavelength that has been dropped.

The wavelength-selective OADM architecture is also compatible with any of the other properties described above (including waveband granularity [KZJP04]). As shown in the figure, edge configurability is possible because any transponder can access either network fiber by reconfiguring the switch. One potential issue with the architecture is the cost of the switch. Furthermore, for reliability reasons, the OADM is often deployed with a spare switching fabric, which adds to the cost. Another possible downside to this architecture is that it is not as suitable as the broadcast-and-select architecture for optically bypassing multiple consecutive OADMs because of higher loss on the through path [TzZT03].

2.5 Multi-Degree OADMs

The OADM network element was designed to provide optical bypass at nodes of degree two. While all nodes in a ring architecture have degree two, a significant number of nodes in interconnected-ring and mesh topologies have a degree greater than two. Table 2.1 shows the percentage of nodes of a given degree averaged over several typical United States (U.S.) mesh backbone networks. Table 2.2 and Table 2.3 show the nodal-degree percentage averaged over several U.S. metro-core networks for interconnected-ring and mesh topologies, respectively.

Table 2.1 Nodal Degree Percentage Averaged over Several Typical U.S. Backbone Networks

Nodal Degree	Percentage
2	55%
3	35%
4	7%
5	2%
6	1%

Table 2.2 Nodal Degree Percentage Averaged over Several Typical U.S. Metro-Core Networks with Interconnected-Ring Topologies

Nodal Degree	Percentage
2	85%
4	9%
6	4%
8	2%

Table 2.3 Nodal Degree Percentage Averaged over Several U.S. Metro-Core Networks with Mesh Topologies

Nodal Degree	Percentage
2	30%
3	25%
4	15%
5	15%
6	10%
7	3%
8	2%

The question is what type of equipment to deploy at nodes of degree three or higher. One thought is to continue using the optical-terminal-based O-E-O architecture at these nodes, while using OADMs at degree-two nodes. This is sometimes referred to as the 'O-E-O-at-the-hubs' architecture, where nodes of degree three or more are considered hubs. As all traffic must be regenerated at the hubs, it implies that electronic performance monitoring can be performed at the junction sites of the network, which may be advantageous for localizing faults. However, it also implies that the scalability issues imposed by O-E-O technology will still exist at a large percentage of the nodes.

Another option is to deploy OADMs, possibly in conjunction with an optical terminal, at the hubs. Figure 2.10(a) depicts a degree-three node equipped with one OADM and one optical terminal. Optical bypass is possible only for traffic transiting between the East and West links. Traffic transiting between the South and East links or the South and West links must undergo O-E-O conversion via transponders, as shown in the figure. Figure 2.10(b) depicts a degree-four node equipped with two OADMs. Optical bypass is supported between the East and South links, and between the North and West links, but not between any other link pairs. The design strategy with these quasi-optical-bypass architectures is to deploy the OADM(s) in the direction where the most transiting traffic is expected. However, if the actual traffic turns out to be very different from the forecast, then there may be an unexpectedly large amount of regeneration; i.e., the architectures of Fig. 2.10 are not 'forecast-tolerant'.

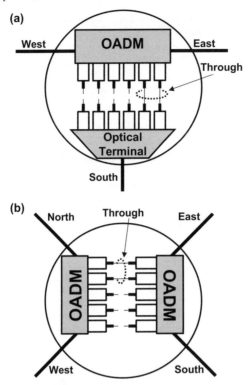

Fig. 2.10 (a) Degree-three node with one OADM and one optical terminal. (b) Degree-four node with two OADMs. In these quasi-optical-bypass architectures, some of the transponders are used for transiting traffic that crosses two different network elements at a node.

With either of the above strategies, there is potentially still a large amount of electronics needed for transiting traffic, as is explored quantitatively in Chapter 8. Another alternative is to deploy the network element known as the multi-degree OADM (OADM-MD), which extends the functionality of an OADM to higher-degree nodes. A degree-three OADM-MD, with a broadcast-and-select architecture, is shown in Fig. 2.11. With an OADM-MD, optical bypass is supported in *all* directions through the node to maximize the amount of transponders that can be eliminated. As with OADMs, transponders are needed only for the add/drop traffic. All of the properties that were discussed for the OADM are relevant for this element as well. The economic benefits of an OADM-MD relative to the 'O-E-O-at-the-hubs' and the 'OADM-only' architectures are studied in Chapter 8.

With the combination of the OADM and the OADM-MD, optical bypass can be provided in a network of arbitrary topology (subject to the maximum degree of the OADM-MD, which is discussed later in this section). For example, in Fig. 2.12(a), a degree-six OADM-MD deployed in the node at the junction of the three rings allows traffic to pass all-optically from one ring to another; the remainder of the nodes have an OADM. In the arbitrary mesh of Fig. 2.12(b), a combination of OADMs, degree-three OADM-MDs, and degree-four OADM-MDs is deployed to provide optical bypass in any direction through any node.

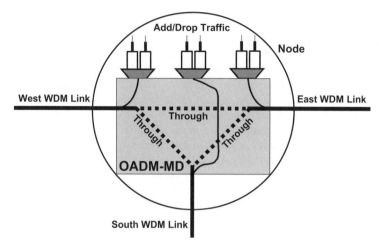

Fig. 2.11 Degree-three OADM-MD. Optical bypass is possible in all three directions through the node. Transponders are needed only for the add/drop traffic. The trapezoids represent general colorless or fixed add/drop ports. (Adapted from [Simm05]. © 2005 IEEE)

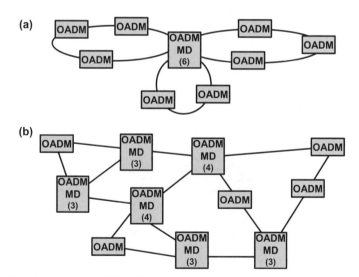

Fig. 2.12 (a) A degree-6 OADM-MD is deployed at the junction site of three rings, allowing traffic to transit all-optically between rings. The remaining nodes have OADMs. (b) In this arbitrary mesh topology, a combination of OADMs, degree-3 OADM-MDs, and degree-4 OADM-MDs are deployed according to the nodal degree. Optical bypass is supported in all directions through any node.

2.5.1 Optical Terminal to OADM to OADM-MD Upgrade Path

OADM-MDs can be part of a graceful network growth scenario. Carriers may choose to roll-out optical-bypass technology in stages in the network, where initially they deploy optical-bypass elements on just a few links to gradually grow their network. Consider deploying optical-bypass technology in just the small portion of the network shown in Fig. 2.13(a). An OADM would be deployed at Node B to allow bypass, with optical terminals deployed at Nodes A and C. As a next step, assume that the carrier wishes to extend optical bypass to the ring shown in Fig. 2.13(b). Ideally, the optical terminals at Nodes A and C are in-service upgradeable (i.e., existing traffic is not affected) to OADMs, so that all five nodes in the ring have an OADM. In the next phase, shown in Fig. 2.13(c), a link is added between Nodes A and E to enhance network connectivity. It is desirable that the OADMs at Nodes A and E be in-service upgradeable to a degree-3 OADM-MD. This element upgrade path, from optical terminal to OADM to OADM-MD, is desirable for network growth, and is supported by several commercial offerings. Furthermore, upgrading to higher-degree OADM-MDs is typically possible, up to the limit of the technology, as is discussed next.

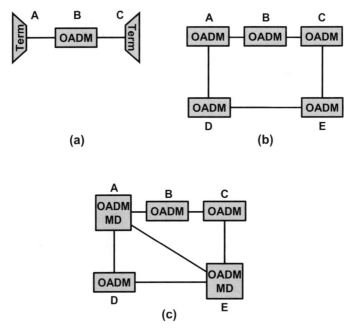

Fig. 2.13 In this network evolution, Node A is equipped with an optical terminal in (a), an OADM in (b), and a degree-three OADM-MD in (c). Ideally, these upgrades are performed in-service, without affecting any existing traffic at the node.

2.5.2 First-Generation OADM-MD Technology

The maximum degree that can be supported with an OADM-MD is very dependent on the underlying technology. The first generation of OADM-MD was built using liquid crystal wavelength blockers with a broadcast-and-select architecture, as shown in Fig. 2.14 [Tomk02, PCHN03]. The liquid crystal blockers are also called 'dynamic spectral equalizers' (DSEs) or 'dynamic channel equalizers' (DCEs); the term DSE is used here. This figure is one implementation of the OADM-MD represented in Fig. 2.11. In order to illustrate more clearly the technology in Fig. 2.14, the input network links are shown on the left of the figure and the output network links are shown on the right. The transponder receivers are shown on the left of the figure and the transponder transmitters are shown on the right.

Consider a degree-D OADM-MD, with D being three in Fig. 2.14 (for ease of discussion, D is also assumed to be the degree of the node). The signal from each incoming network fiber is initially split into a drop path and a through path. The through path is further split into D-1 paths, corresponding to the D-1 other network fibers, with a DSE along each of these through paths. (If port loopback is desired, then the through path is split into D paths.) There is typically amplification in the nodes, which can make up for the splitting loss. The DSEs can be dynamically configured to selectively pass or block any wavelength, depending on whether the wavelength is optically bypassing or being dropped at the node, respectively.

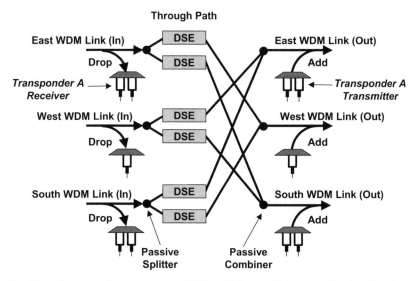

Fig. 2.14 First-generation degree-three OADM-MD built with Dynamic Spectral Equalizers (DSEs). The number of required DSEs grows quadratically with the degree of the OADM-MD. The trapezoids represent general colorless or fixed add/drop ports.

Consider a wavelength that enters from the East link that is being dropped at the node. The two DSEs associated with the incoming East link (i.e., the top two DSEs in the figure) would be configured to block this wavelength so that it is not routed to either the West or the South links. The DSE technology thus provides wavelength reuse. Next, consider a wavelength that transits the node from the East link to the South link. The upper DSE blocks the wavelength from being routed to the West link, while the second DSE allows the wavelength to pass through to the South link.

Optical multicasting on the network fibers is supported by this architecture. For example, if both DSEs associated with the incoming East link allow a wavelength to pass, then the signal is sent to both the West and South links. Furthermore, drop-and-continue is supported. For example, it is possible for a signal to drop from the East link and be transmitted to either, or both of, the West and South links.

This architecture does not support edge configurability. For example, the transponder labeled 'A' in the figure can add/drop only to/from the East link. Moreover, this architecture does not support the form of optical multiplexing where a sourced optical signal is sent out on multiple links.

A total of $D*(D-1)$ DSEs are needed in this architecture; thus, the size and cost of the switch fabric scale quadratically with the degree D. (D^2 DSEs are needed if port loopback is desired.) In practice, this architecture generally is limited to a maximum degree of four.

2.5.3 Second-Generation OADM-MD Technology

Second-generation OADM-MDs, also based on the broadcast-and-select architecture, make use of Nx1 wavelength selective switches (WSS) as shown in Fig. 2.15 [MMMT03]. For a degree-D OADM-MD, N equals $D-1$, unless port loopback is desired, in which case N equals D. This OADM-MD is functionally equivalent to the first-generation device; e.g., it supports wavelength reuse and optical multicast. As with the DSE, the WSS can be configured to selectively pass or block a particular wavelength. The difference is that one WSS operates on N incoming paths. Thus, a total of D WSSs are needed for a degree-D OADM-MD; i.e., the amount of equipment scales linearly with D, not quadratically. (The required size of the WSS increases with the degree; however, the cost of the WSS increases sublinearly with N.) The improvement in scalability is especially useful in metro-core networks where multiple rings may be interconnected at a single node. For example, a degree-eight OADM-MD deployed at the intersection of four rings allows all intra-ring and inter-ring transiting traffic to remain in the optical domain.

In a variation of this OADM-MD architecture, each of the N-way passive splitters on the input side in Fig. 2.15 is replaced by a 1xN WSS. The chief motivation for this design is to decrease the overall loss through the device, especially at high-degree nodes where the passive splitting loss is large. The disadvantages are the need for twice as many WSSs and the lack of support for multicast.

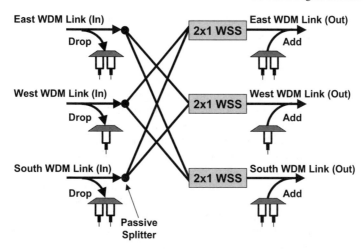

Fig. 2.15 Second-generation degree-three OADM-MD, based on Nx1 Wavelength Selective Switches (WSSs). The number of required WSSs grows linearly with the degree of the OADM-MD. The trapezoids represent general colorless or fixed add/drop ports.

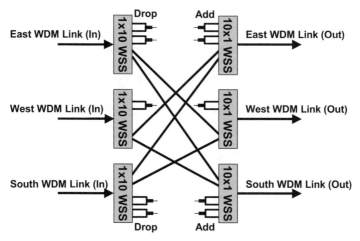

Fig. 2.16 An OADM-MD architecture based on Nx1 WSSs, where some of the WSS ports are used for single-wavelength colorless add/drop. With the configuration shown, the maximum number of add/drops per fiber is eight (but only one or two add/drops per fiber are shown).

A third OADM-MD configuration based on Nx1 WSSs is shown in Fig. 2.16. Again, the passive splitters of Fig. 2.15 are replaced by WSSs; in addition, some of the ports on the WSSs are used as single-wavelength add/drop ports. This provides colorless add/drop; i.e., any lambda can be added/dropped to/from any port. The amount of add/drop is limited by the port size of the WSS; thus, this configuration may be more appropriate for the metro-core area where the amount of add/drop at a node is more likely to be small. With the 1x10 WSSs shown in Fig. 2.16 for a degree-three node, the maximum number of add/drops is eight. This architecture does not support multicast.

(In contrast, the architecture of Fig. 2.15 has multi-wavelength add/drop terminals. As discussed in Section 2.2.1, multi-wavelength add/drop with colorless slots is also possible, depending on the technology used for the 'trapezoids'. For example, one method is to use a passive coupler/splitter for the add/drop. However, the configuration of Fig. 2.16 likely yields a lower-loss add/drop path. A second method of providing colorless terminals in Fig. 2.15 is to deploy a 1xS WSS at the drop and a Sx1 WSS at the add, where S is the maximum number of add/drop wavelengths at the node. This is closer to the configuration of Fig. 2.16, except a 1xS WSS and a 1x(D-1) splitter are used in place of one 1x(S+D-1) WSS on the input fiber, and a Sx1 and a (D-1)x1 WSS are used instead of one (S+D-1)x1 WSS on the output fiber, where D is the OADM-MD degree.)

As with the first-generation architecture, the OADM-MDs of Fig. 2.15 and Fig. 2.16 do not provide edge configurability. In fact, when the term OADM-MD was first coined, it was specifically intended to denote a device that did *not* have edge configurability. The term 'all-optical switch' is the proper classification of a higher-degree optical-bypass element that can switch all nodal traffic, including the add/drop traffic, thereby providing edge configurability. These types of switches are covered in the next section. As a historical note, all-optical switches actually existed prior to the development of the OADM-MD; however, they tended to be high-cost. Optical bypass was not introduced on a large scale at the hub nodes of carrier networks until the lower-cost OADM-MD was developed. (The industry sometimes blurs the difference between OADM-MDs and all-optical switches and uses the terms interchangeably; however, the intent was that they be distinguished based on whether or not they provide edge configurability.)

2.6 Optical Switches

The all-optical switch is one member of a broader class of switches called *optical switches*. Optical switches are also referred to as optical cross-connects. As discussed in Section 2.1, the term optical switch indicates that the switch ports operate on the granularity of a wavelength or a waveband. It does not imply that the switch supports optical bypass, nor does it imply that the switch fabric is optical. There are several flavors of optical switches as described below. For more details of the underlying optical switch technologies, see [Sala02, PaPP03, ElBa06].

2.6.1 O-E-O Optical Switch

An optical switch based on O-E-O technology is shown in Fig. 2.17(a) (it also was shown as part of Fig. 2.4). The switch fabric is electronic, and each of the switch ports is equipped with a short-reach interface to convert the incoming 1310-nm optical signal to an electronic signal. These types of switches present scalability challenges in cost, power, and heat dissipation due to the amount of electronics. Consider using such a switch to provide configurability at a degree-four O-E-O node. Assume that each fiber carries 160 wavelengths and assume that the node needs to support 50% add/drop. There needs to be a port for each wavelength on the nodal fibers as well as each add/drop wavelength. This requires a 960x960 switch, with each switch port having a short-reach interface.

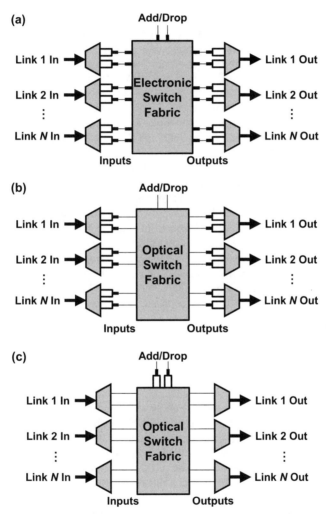

Fig. 2.17 Examples of optical-switch architectures. (a) O-E-O architecture with electronic switch fabric and electronic interfaces on all ports (Section 2.6.1). (b) Photonic switch that switches the 1310-nm optical signal (Section 2.6.2). (c) A wavelength-selective all-optical switch (Section 2.6.3).

One special type of O-E-O switch is a grooming switch, which processes the subrate signals carried on a wavelength in order to better pack the wavelengths. Grooming switches are discussed further in Chapter 6.

2.6.2 Photonic Switch

The term *photonic switch* refers to an optical switch where the switch fabric is optical so that the incoming optical signal does not have to be converted to the electrical domain.   MEMS technology is often used to build photonic switches. However, a photonic switch does not necessarily imply optical bypass through a node. For example, Fig. 2.17(b) illustrates a photonic switch that is being used to switch 1310-nm optical signals. There are WDM transponders on both the input and output fibers, and thus, optical bypass is not supported.

A carrier may choose to implement the configuration of Fig. 2.17(b) in order to isolate various technologies in the network. Because each wavelength is terminated on a WDM transponder in the node, the links are isolated from each other. This would allow, for example, the transmission technology used on each link to be supplied by different vendors. Furthermore, the switch vendor may be independent from the transmission vendor because the switch is operating on the standard 1310-nm signal as opposed to a WDM-compatible signal.

This same vendor independence is provided by the O-E-O switch of Fig. 2.17(a); however, the photonic switch configuration of Fig. 2.17(b) requires significantly less electronics. In terms of port count, however, the photonic switch is no smaller than the O-E-O switch. Using the same example of a degree-four node with 160 wavelengths per fiber and 50% add/drop, the required switch size, assuming wavelength granularity, is again 960x960. The advantage is that the switch fabric is not electronic and electronic interfaces are not required on the ports.

2.6.3 All-Optical Switch

All-optical switch is the term for a switch with an optical switch fabric that is being used to support optical bypass. There have been many technologies proposed over the years for building all-optical switches, although some tended to be costly. One cost-effective all-optical switch implementation, developed in the 2003 timeframe, is based on the same Nx1 WSS technology as the second-generation OADM-MD. Because of the scalability of the WSS-based architecture, it is possible to pass the add/drop traffic through the switch fabric, as shown in Fig. 2.18, rather than having the add/drop ports be simple taps off of the network fibers. In contrast to the OADM-MD of Fig. 2.15 (and Fig. 2.16), the all-optical switch of Fig. 2.18 allows any transponder to access any network link. The tradeoff is that the all-optical switch has greater complexity and cost. The degree of the element shown is six, with three of the six ports designated for local access (when the add/drop traffic passes through the switch fabric in this type of architecture, the ports are referred to as local access ports). Thus, six 5x1 WSSs are needed.

The edge flexibility of an all-optical switch provides similar benefits as described in Section 2.4.5 for the OADM. It can flexibly route dynamic traffic, as any transponder can access any of the network links at the node. It supports optical-layer protection, where a single transponder is used for either the working or protect paths, with the switch toggling between paths at the time of a failure. Because the transponders are not tied to a particular network link, typically fewer transponders are predeployed at a node with an all-optical switch as compared to an OADM-MD.

The architecture of Fig. 2.18 is considered broadcast-and-select (even though it contains wavelength-selective switches) because the passive splitters on the input link broadcast the signal to multiple lines and the WSSs select which wavelengths should pass through to a network link or to a local access port. As with the broadcast-and-select OADM-MD, this all-optical switch architecture supports network multicast where a signal from an incoming fiber is sent to multiple outgoing fibers. In addition, it allows an added signal to be multicast to multiple outgoing fibers; this is not supported with an OADM-MD.

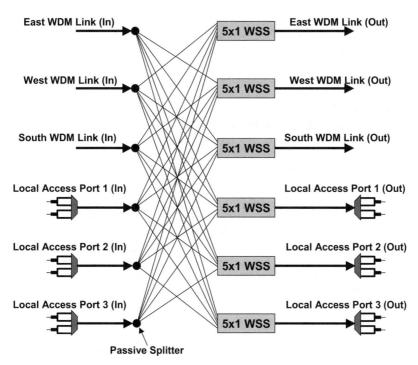

Fig. 2.18 All-optical switch with three network ports and three local access ports. This architecture provides edge configurability. The trapezoids represent general colorless or fixed local access ports.

This type of all-optical switch can be part of an element migration path, similar to the OADM-MD. One can potentially grow in-service from an optical terminal to an OADM to an all-optical switch. Note that the OADM based on this architecture would pass the add/drop traffic through the switch fabric, such that edge configurability is provided; i.e., the OADM in this migration path is composed of two network link ports, two local access ports, and four 3x1 WSSs.

With this type of all-optical switch, demultiplexing of the WDM signal typically occurs internal to the switch, so that the ports are multi-wavelength. There are D network ports, where D is the nodal degree, and L local access ports. An interesting question arises as to how large L should be relative to D. In Fig. 2.18, D and L are both three, thereby potentially supporting up to 100% drop at the node. However, because transponders can access any network port, and because the percentage of drop at a node is typically less than 100%, it is often possible to reduce the number of local access ports. However, while this strategy saves cost, it can lead to wavelength conflicts. In general, if there are L multi-wavelength local access ports, then a particular wavelength can only be dropped from L of the network ports; otherwise there would be wavelength clashes on the local access ports. This limitation becomes more of a problem when the network is heavily loaded and there are few free wavelengths on a link. While advanced routing and wavelength assignment algorithms can be used to minimize wavelength conflicts, contention may still arise if there are not enough local access ports.

Note that the DSE technology used in the first-generation OADM-MD, where the total number of ports is limited to four, could conceivably be used for an all-optical switch at degree-three nodes. Three ports would be used for network links and one port would be used for local access. This would allow transponders on the one local access port to access any of the three network links. However, this design limits the maximum add/drop at the node to 33%, which is not sufficient for many nodes. Furthermore, any given wavelength can drop from at most one network link, which would likely lead to wavelength conflicts as the network fills. Thus, the DSE technology is not favorable for the all-optical switch configuration.

Even if the number of local access ports equals the nodal degree (i.e., L equals D), wavelength conflicts may arise due to the multi-wavelength local access ports. Assume that a transponder tuned to λ_1 is in use on a local access port, where the corresponding signal is added/dropped to/from the East link. Further assume that an IP router feeds a second transponder that is plugged into the same local access port. Assume that the IP router wishes to initiate a connection via the West link and the only available wavelength (due to congestion on other links of the path) is λ_1. This connection would be blocked because there is already a transponder on this local access port using λ_1. While such a scenario is possible, its occurrence is rare and would likely only occur if the network is very heavily loaded.

Another type of all-optical switch is based on the wavelength-selective architecture, as illustrated in Fig. 2.17(c) (this is similar to the wavelength-selective OADM architecture of Fig. 2.9). A MEMS-based switch, in combination with multiplexers/demultiplexers, is often used for this application. The WDM signal entering from a fiber is demultiplexed into its constituent wavelengths, where each wavelength feeds an input port on the switch fabric. Some of the wavelengths are directed to add/drop ports while others are directed to output ports corresponding to other network fibers. In contrast to the broadcast-and-select architecture of Fig. 2.18, the add/drop ports are single-wavelength so that wavelength conflicts do not arise. However, this type of architecture does not readily support multicast, either from one input fiber to multiple output fibers or from one add port to multiple output fibers. It also tends to be more expensive than the architecture of Fig. 2.18.

Note that while the switch of Fig. 2.17(c) supports optical bypass, such that transponders are not required for the through traffic, the number of ports on the switch fabric is no smaller than that required for Fig. 2.17(a) or Fig. 2.17(b). Again, for a degree-four node supporting 50% add/drop, where each fiber carries 160 wavelengths, the switch fabric needs to be 960x960, assuming wavelength granularity. Scaling technologies such as MEMS to this size can be challenging.

2.7 Hierarchical Switches

One means of fabricating scalable optical switches is through the use of a hierarchical architecture where multiple switch granularities are supported [HSKO99, SaSi99]. Coarse switching is provided to alleviate requirements for switches with very large port counts; a limited amount of finer granularity switching is also provided. A functional illustration of a three-level hierarchical switch is shown in Fig. 2.19, where switching can be performed on a per-fiber basis, a per-waveband basis, or a per-wavelength basis. While the three levels are shown as distinct switches (or cross-connects), it is possible to build such a multi-granular switch with a single switching fabric, e.g., with MEMS technology [LiVe02].

If the traffic at a node is such that all the traffic entering from one fiber is directed to another fiber, then that traffic is passed through the fiber-level switch only. The switch provides configurability if the traffic pattern changes, while providing optical bypass for all traffic on the fiber. If fiber-level bypass is not appropriate for a network, then that level of the switching hierarchy can be removed.

For traffic that needs to be processed on a finer granularity than a fiber, the band-level switch demultiplexes the WDM signal into its constituent wavebands. Some of the wavebands are switched without any further demultiplexing, providing band-level bypass, whereas some of the wavebands have to be further demultiplexed into their constituent wavelengths so that individual wavelengths can be dropped or switched. When equipped with wavelength converters, whether electronic or optical, the wavelength-level switch can also be used to better pack the wavebands (i.e., waveband grooming). Changing the lambda of a wavelength allows the wavelength to be shifted from one waveband to another.

The hierarchical approach addresses the port count issue while providing more flexibility than a single-layer switch of a coarse granularity. Studies have shown a significant number of ports can be saved through the use of a multi-granularity switch, with the percentage of ports saved increasing with the level of traffic carried in the network [NoVD01, CaAQ04].

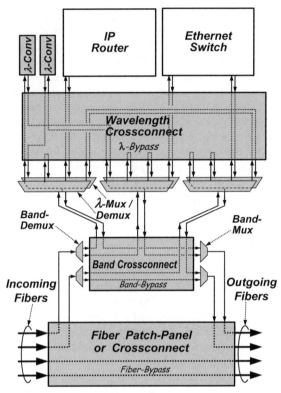

Fig. 2.19 A three-level hierarchical switch, allowing fiber-bypass, band-bypass, and wavelength-bypass. In the figure, each fiber contains two wavebands, and each waveband contains four wavelengths. Wavelength conversion is used to groom the wavebands. (Adapted from [SaSi99]. © 1999 IEEE)

2.8 Adding Edge Configurability to a Node

As discussed previously, OADM-MDs and some OADM implementations do not have built-in edge configurability. Without manual intervention, any transponder is restricted to accessing just a single network link at the node. A greater amount of automatic configurability is desirable to meet the requirements of dynamic services. Two methods for achieving the desired edge configurability with these network elements are discussed here. Whether to deploy one of these architectures, or alternatively deploy an all-optical switch that has built-in edge configurability, depends on overall nodal cost, complexity, reliability, and multicast requirements. Note that some telecommunications carriers have installed OADM-MDs in their networks due to low-cost all-optical switches not being available when the networks were deployed.

2.8.1 Adjunct Edge Switch

One option for increasing the configurability of the node is to deploy an optical switch in conjunction with the OADM-MD or OADM, as shown with a degree-three OADM-MD in Fig. 2.20. The added switch is often referred to as an '*adjunct edge switch*' or simply an '*edge switch*'. The traffic from higher layers (e.g., IP) is fed through the edge switch so that it can be automatically directed to the desired network port. Because the edge switch is only for the add/drop traffic, and not for the through traffic, the size can be smaller as compared to a switch that operates on all of the nodal wavelengths. Returning to the example of a degree-four node with 160 wavelengths per fiber and 50% add/drop at the node, the edge switch needs to be of size 640x640, as opposed to 960x960 for a 'core' switch. (Note that the IP router itself could be used to provide edge configurability by directing traffic to a different router port that is connected to the desired transponder; however, this requires more router ports, which tend to be costly.)

With the configuration of Fig. 2.20(a), the client 1310-nm optical signal is passed through the edge switch. The edge switch can be O-E-O-based or photonic; the latter is shown in the figure. The transponders are tied to a particular port, and the edge switch directs the client signal to the desired transponder. In Fig. 2.20(b), the edge switch operates on the WDM-compatible signal, and thus, must be photonic. This configuration provides more agility than Fig. 2.20(a) because the transponders can access any port. Thus, fewer transponders need to be predeployed with this configuration. There are some applications where the configuration of Fig. 2.20(b) must be used; e.g., see Section 2.10 and Section 4.5.2.

Note that deploying an edge switch at a node can be useful for other reasons, e.g., protection, as is discussed in later chapters.

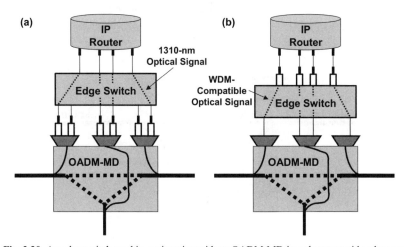

Fig. 2.20 An edge switch used in conjunction with an OADM-MD in order to provide edge configurability. In (a), the 1310-nm signal is switched; in (b), the WDM-compatible signal is switched. There is more agility with (b), so that fewer transponders need to be predeployed.

2.8.2 Flexible Transponders

Rather than incurring the cost of an edge switch, another option is to use *flexible WDM transponders,* where the output of the transponder is split into multiple paths and each path feeds into a different add port on the OADM or OADM-MD. In the reverse direction, the transponder is equipped with a switch to select a signal from one of the drop ports. This is illustrated in Fig. 2.21, where a transponder with three-way flexibility is able to access any of the network links at the degree-three node. Consider the general scenario of a degree-N node. Passively splitting the transponder output over N paths to access any of the links results in $10 \cdot \log_{10} N$ dB of optical loss, which is undesirable as N increases. One option is to have the transponder output split over M paths, where $M<N$; such a transponder is limited to accessing only M of the N nodal links. However, as proposed in [SiSa07], there is an alternative transponder architecture that uses small switches on the transmit side, so that the transponder can access any of the N nodal links while suffering no more than a nominal 3-dB optical loss. Furthermore, this transponder design allows it to simultaneously access any combination of two of the N nodal links, to support two-way multicast of an added signal, or to support optical-layer protection with an active backup path as is discussed further in Chapter 7.

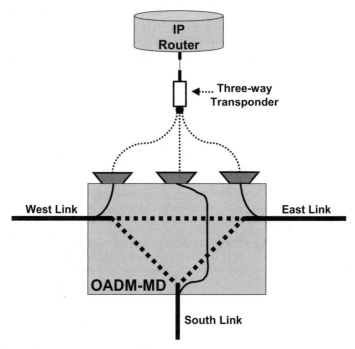

Fig. 2.21 A three-way transponder is able to access any of the network links at the degree-three node. It is desirable to use an optical backplane to simplify the cabling.

One advantage of the flexible transponder solution is that fewer transponders need to be deployed at a node as compared to an OADM-MD with regular transponders. There is one pool of transponders that can access any of the N nodal links, rather than N pools of transponders, each associated with a particular link. One possible drawback of the N-way flexible transponder is that one transponder occupies N add/drop slots. However, not all transponders at the node have to be equipped with the flexible architecture; regular transponders can still be used for the non-dynamic traffic.

2.9 Optical Reach

Network elements capable of allowing transiting traffic to remain in the optical domain, such as the OADM, OADM-MD, and the all-optical switch, is one requirement for optical bypass. In addition, the underlying transmission system must be compatible with a signal remaining in the optical domain as it traverses one or more nodes. An important property of a transmission system is the optical reach, which is the maximum distance an optical signal can be transmitted before it degrades to a level that requires the signal be regenerated, where regeneration occurs in the electronic domain. (All-optical regeneration, while feasible, has not been widely implemented. This is discussed further in Chapter 4.)

Consider the four-node linear network in Fig. 2.22. Nodes A and D are equipped with optical terminals, whereas Nodes B and C are equipped with OADMs. The distance of each of the three links is 1,000 km. Assume that the connection of interest is between Node A and Node D.

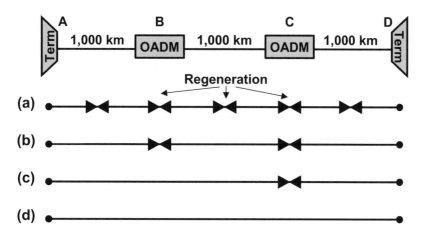

Fig. 2.22 Connection between Nodes A and D with: (a) 500-km optical reach; (b) 1,000-km optical reach; (c) 2,000-km optical reach; and (d) 3,000-km optical reach.

In Fig. 2.22(a), the optical reach is assumed to be 500 km. With this reach, not only must the connection be regenerated at Nodes B and C, it must also be regenerated at intermediate dedicated regeneration sites along the links. These sites would otherwise be equipped with just an optical amplifier, but due to the limited reach, need to regenerate all traffic that passes through them. The OADMs at Nodes B and C, while capable of optical bypass, cannot be used for that purpose in this scenario because of the limited reach. In Fig. 2.22(b), the optical reach is 1,000 km; this removes the regeneration at intermediate sites along the links, but still is not sufficient to allow the signal to optically bypass either Node B or C. With an optical reach of 2,000 km, as shown in Fig. 2.22(c), the connection can make use of the OADM at Node B to optically bypass that node, but it still must be regenerated at Node C (equivalently, it could be regenerated at Node B and optically bypass Node C). With an optical reach of 3,000 km, the connection is able to remain in the optical domain over the whole path, as shown in Fig. 2.22(d).

As this example shows, the optical reach is critical in determining how much optical bypass is achieved in a network. In legacy transmission systems based on EDFA technology, the optical reach is on the order of 500 to 600 km; with newer EDFA systems, the maximum reach is on the order of 1,000 to 1,500 km. To obtain significantly longer reach, Raman amplification is used (see Chapter 4). Raman systems generally have an optical reach in the range of 1,500 to 4,000 km, depending on the equipment vendor. Such technology is sometimes referred to as 'ultra-long-haul technology' to emphasize the extended optical reach. In typical networks, the combination of optical-bypass elements and extended optical reach may eliminate on the order of 90% of the regenerations as compared to a system with no optical bypass and 500-km reach.

Many factors go into developing a transmission system that supports extended reach, some of which are touched on here; the subject is revisited in Chapter 4. As noted above, Raman amplification, sometimes in conjunction with EDFA amplification, is generally used in such systems. It is also necessary to deal with a host of optical impairments, such as chromatic dispersion, polarization-mode dispersion, crosstalk, and four-wave mixing, all of which can degrade an optical signal (these are discussed in Chapter 4). Some of these impairments can be mitigated with special compensating equipment, while some of the problems from these impairments can be avoided by transmitting signals at a low enough power level.

Advanced transmitters and receivers, as well as robust transmission schemes, are also important for attaining extended optical reach. Furthermore, the optical network elements themselves must be compatible with extended reach. For example, they must be low loss, and not cause excessive distortion of the signal. Another key system component is advanced forward error correction (FEC). FEC allows errors picked up in transmission to be corrected at the destination. The stronger the FEC coding, the more errors can be corrected. This allows the signal to degrade further before it needs to be regenerated.

This list is not meant to be a comprehensive discussion of what goes into achieving extended optical reach. Suffice it to say it is quite an engineering accomplishment.

The desirable optical reach for a network depends on the geographic tier of where the system is deployed. In the metro-core, the longest connection paths are on the order of a few hundred kilometers. Thus, an optical reach of 500 km is sufficient to remove most, if not all, of the regeneration in the network. In regional networks, the longest connection paths are typically in the range of 1,000 to 1,500 km, requiring an optical reach of that order to eliminate regeneration. In backbone networks, the longest connection paths can be several thousand kilometers. For example, in the United States, the longest backbone connections, when including protection paths, are on the order of 8,000 km. A system with 4,000-km optical reach will eliminate much of the regeneration, but not all of it.

Note that the key factor in determining whether extended optical reach is useful in a particular network is the distribution of connection path distances. A common mistake is to focus on the link distances instead, where some may assume that extended optical reach provides no benefit unless the link distances are very long. In fact, optical bypass can be more effective in networks with high nodal density and relatively short links because a particular connection is likely to transit several intermediate nodes. Chapter 8 investigates the optimal optical reach of a network from a cost perspective.

Regarding terminology, it is perhaps not clear whether traffic that is regenerated at a node should be considered through traffic or add/drop traffic. As regeneration is typically accomplished via O-E-O means, it is usually considered add/drop traffic because it is dropping from the optical domain. This is the convention that is adopted here.

Chapter 4 discusses various regeneration architectures and strategies.

2.10 Integrating WDM Transceivers in the Client Layer

Much of the discussion so far has focused on removing the transponders for the nodal through traffic by implementing optical bypass. However, the add/drop traffic also provides an opportunity to remove some of the electronics from the node. The electronic higher layers and the optical layer typically communicate via a standard 1310-nm optical signal. In Fig. 2.23(a), the IP router is equipped with an interface to generate the 1310-nm signal, and the WDM transponder plugged into the OADM converts the 1310-nm signal to a WDM-compatible signal.

The equipment can be simplified by having a WDM-compatible transceiver deployed directly on the IP router, as shown in Fig. 2.23(b). This is referred to as *integrating* the transceivers with the IP router. (The term transceiver is used rather than transponder because the input is an electrical signal from the IP router rather than a 1310-nm optical signal; i.e., there is no transponding of an optical signal.) The IP router and OADM would then communicate via a WDM-compatible signal, which would be added to the WDM signal exiting the OADM. This eliminates the electronic interfaces between the router and the OADM. Clearly, the transceiver output must meet the specifications of the WDM transmission system for this configuration to work.

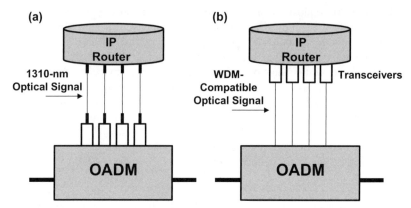

Fig. 2.23 In (a), the IP router communicates with the OADM via a standard 1310-nm optical signal. In (b), the electrical interfaces can be removed as WDM transceivers are deployed directly on the IP router.

This same configuration is possible with an IP router in combination with an OADM-MD or an all-optical switch as well. Furthermore, transceivers can be integrated with other electronic switching elements. For example, referring back to Fig. 2.17(a), the combination of a WDM transponder and an electronic interface on each switch port of the O-E-O switch can be replaced by a transceiver that is integrated with the O-E-O switch, thereby eliminating a lot of electronics. Note, however, this integrated architecture negates one of the advantages of O-E-O switches, namely the independence of the switch and transmission-system vendors. With integrated transceivers, either one vendor supplies both the switch and the transmission system, or separate vendors collaborate to ensure compatibility.

Consider the integrated-transceiver scenario where the optical network element is not capable of edge configurability. Assume that an adjunct edge switch is deployed at the node as discussed in Section 2.8.1 in order to achieve edge flexibility. The edge switch would be deployed between the transceivers and the optical network element, and would need to be capable of switching WDM-compatible signals.

2.11 Photonic Integrated Circuits

One of the motivations for removing electronics from a node is to reduce cost and physical space requirements. This led to the development of optical-bypass technology, which is now deployed in several major telecommunications networks. A more recent development is *photonic integrated circuit* (PIC) technology, used in combination with the more traditional O-E-O architecture [MDLP05, Welc06]. The idea is not to eliminate transponders but to make them lower cost and smaller through integration. For example, ten transponders may be integrated onto a single small chip. One advantage is that because the architecture is based on O-E-O at every node, there are no wavelength continuity constraints, thereby simplifying the algorithms needed to run the network. It also allows for SONET/SDH performance monitoring at every node. If there is widespread adoption of the PIC technology, this would change the direction of optical network evolution. However, it is not clear that this technology addresses operational issues such as required power, heat dissipation, reliability, and full nodal configurability.

2.12 Multi-Fiber-Pair Systems

Thus far, the discussion has implicitly assumed that each network link is populated by one fiber-pair. In this section, the multi-fiber-pair scenario is considered.

The O-E-O architecture does not change with multiple fiber-pairs per link. There is an optical terminal for every fiber-pair entering a node, and the signal from every incoming fiber is demultiplexed. In order to reduce the electronics, it may be possible to have an incoming fiber from one link be directly connected to an outgoing fiber on another link, so that fiber-bypass is achieved, though this arrangement is not readily reconfigurable.

For a network with optical bypass, there are a few options for architecting the node. First, assume that every link has N fiber-pairs. One option is to deploy N copies of a network element at a node. For example, Fig. 2.24 shows a degree-two node with two fiber-pairs per link. The node is equipped with two OADMs. Optical bypass is supported for traffic that bypasses the node using fiber-pair 1 or fiber-pair 2. However, traffic routed on fiber-pair 1 of one link and fiber-pair 2 of the other link requires O-E-O conversion.

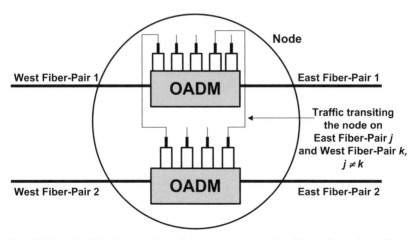

Fig. 2.24 Two OADMs deployed in parallel at a degree-two node with two fiber-pairs per link.

Alternatively, if optical bypass is desired regardless of how the traffic is routed, a degree-four OADM-MD (or all-optical switch) could be used. This may provide more flexibility than is required, however, as it allows a signal to be all-optically routed between fiber-pair 1 of a link and fiber-pair 2 of that same link. In a broadcast-and-select architecture, the OADM-MD design could be simplified so that paths from an incoming fiber are split into two paths rather than three, to allow optical bypass in just the two desired directions. For example, the OADM-MD architecture of Fig. 2.15 can be modified for a 'quasi-degree-four' node as shown in Fig. 2.25. (A similar type of simplification is possible with the all-optical switch architecture of Fig. 2.18.)

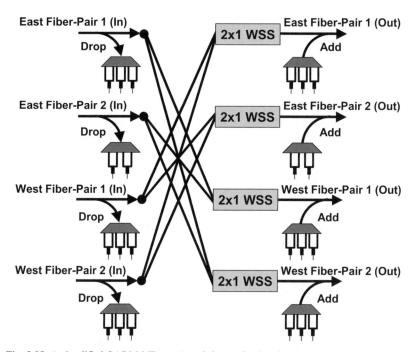

Fig. 2.25 A simplified OADM-MD at a 'quasi-degree-four' node where routing between fiber-pairs on the same link is not required. As compared to a full degree-four OADM-MD, each through path is split two ways rather than three, and there are 2x1 WSSs rather than 3x1 WSSs.

If the number of fiber-pairs on the links is unequal, due to some links being more heavily utilized, then either one large OADM-MD or all-optical switch could be deployed at a node to provide full optical bypass, or some combination of optical bypass and O-E-O could be used. For example, if a degree-two node has one link with two fiber-pairs and one link with one fiber-pair, a degree-three OADM-MD would provide optical bypass in all directions. Again, this is actually more flexibility than is typically needed as it provides an all-optical path between the two fiber-pairs that are on the same link; the OADM-MD can be simplified, analogous to what is shown in Fig. 2.25. Another option is to deploy an OADM in combination with an optical terminal, in which case the network planning process should favor routing the through traffic on the fiber-pairs interconnected by the OADM.

When considering the maximum desired size of an OADM-MD or all-optical switch it is important to take into account the possibility of multiple fiber-pairs. Thus, the maximum desired sizes of the various network elements may be larger than what is indicated by simply looking at nodal-degree distribution.

Chapter 3

Routing Algorithms

Telecommunications networks are generally so large and complex that manually designing a network in a reasonable amount of time is prohibitively difficult. Network designers primarily rely on automated algorithms to determine, for example, how to route traffic through the network, how to protect the traffic, and how to bundle the traffic into wavelengths. In systems with optical bypass, additional algorithms are needed to handle regeneration and to ensure that wavelength contention issues are minimal. The fact that networks have reached a stage where algorithms are essential in producing cost-effective and efficient network designs can be daunting. The good news is that extensive research has been done in this area and much expertise has been gained from live network deployments, resulting in the development of relatively straightforward algorithms that produce very effective network designs.

When designing network algorithms, it is important to consider the size of the problem in terms of the number of network nodes, the amount of traffic carried in the network, and the system specifications. Metro-core networks have tens of nodes whereas backbone networks may have as many as 100 nodes or more. The size of the demand set depends on whether the traffic requires grooming or not. First, consider a backbone network. If all of the traffic is at the wavelength line-rate (no grooming needed), there are typically a few hundred to a couple of thousand demands in the network. If all of the traffic is subrate, such that grooming is needed, there could be tens of thousands of demands. The number of demands in a metro-core network is significantly lower than this. A key system parameter is the number of wavelengths per fiber. Metro-core WDM networks generally have no more than 40 wavelengths per fiber, whereas backbone networks typically have 80 to 160 wavelengths per fiber. Any algorithms used in the network planning process must be scalable under these conditions.

The run-time of the network planning algorithms is very important. In a highly dynamic real-time environment, a new connection may need to be established in less than one second. The process of planning the route of the connection and de-

J.M. Simmons, *Optical Network Design and Planning*,
DOI: 10.1007/978-0-387-76476-4_3, © Springer Science+Business Media, LLC 2008

termining which network resources should be used to carry it ideally should be completed in less than ~250 milliseconds to allow time for the network to be configured appropriately to carry the connection. Furthermore, real-time design may be implemented in a distributed manner at the network nodes, where processing and memory capabilities may be limited.

In long-term network planning, design time is not as critical, although it is still important. In some long-term planning exercises, a large number of demands (e.g., thousands of subrate demands) may need to be processed at once. Furthermore, if working with a 'greenfield' network (i.e., a completely new network), numerous design scenarios typically are considered. For example, the process may include comparing different network topologies, different line-rates, or different protection strategies. Due to the number of designs that need to be run, and the often short time allocated to the network design process (especially when the design exercise is performed by a system vendor in response to a carrier request), it is desirable that the planning process for each scenario take no more than a couple of minutes.

To produce a network design that is optimal relative to a set of metrics is typically computationally complex and would require an inordinate amount of time for realistic networks. Thus, many steps in the planning process rely on heuristics, which experience has shown produce very good, though not always optimal, results, and which run in a reasonable amount of time.

A brief overview of the network planning process can be found in [Simm06]. This chapter focuses on the routing component of the algorithms, whereas Chapters 4 and 5 discuss regeneration and wavelength assignment, respectively. The initial emphasis is on treating these three components as sequential steps in the planning process; however, performing routing, regeneration, and wavelength assignment in a single step is covered in Chapter 5.

Routing is the process of selecting a path through the network for a traffic demand, where there are typically many possible paths to get from the demand source to the demand destination. It is important to take into account several factors when selecting a route. First, cost is a key consideration. The selected route should require adding minimal cost to the network when possible. The path distance and number of links in the path may also be relevant, as these are indicators of the bandwidth occupied by the path; these factors also may affect the reliability of the connection. However, this does not imply that demands should always be routed over the shortest possible path or the path with the fewest links. In fact, such a strategy may lead to inefficient designs. It is also necessary to consider the total potential capacity of the network, where a particular path may be chosen because it leaves the network in a better state to accommodate future traffic.

Several routing strategies are discussed in this chapter, with a focus on relatively straightforward methodologies that are effective in network planning for practical optical networks. The chapter is not meant to be a review of all known routing algorithms and strategies. Much of the chapter is equally applicable to a network with optical bypass as to a network based on O-E-O technology.

The first few sections of the chapter specifically examine routing a demand with one source and one destination over a single path. Section 3.1 introduces shortest path routing algorithms and Section 3.2 covers how such algorithms are used depending on the underlying technology of the network. Sections 3.3 and 3.4 describe some effective routing strategies that take into account network cost and network utilization. Section 3.5 considers the more detailed network modeling that may be needed when routing demands in real-time, where only the equipment that is already deployed in the network can be used. Routing with protection is covered in Section 3.6, where two or more diverse paths are required for a demand to enable recovery from a failure. Section 3.7 is relevant to planning scenarios where multiple demand requests are processed at one time. Assuming that the routing policy is adaptive, where the selected path is dependent on the state of the network, then the order in which the demands are considered is important. Various effective ordering strategies are presented in this section. Section 3.8 discusses multicast traffic, where there is one source and multiple destinations. The final section in the chapter discusses some of the challenges of real-time network planning when the current state of the network is not fully known.

3.1 Shortest Path Algorithms

Most routing strategies incorporate some type of shortest path algorithm to determine which path minimizes a particular metric. A general discussion of shortest path algorithms can be found in [CoLR90]. In the shortest path algorithms discussed here, it is assumed that the metric for an end-to-end path is the sum of the metrics of the links comprising the path. Any such additive metric can be used, depending on the goal of the routing process. For example, to find the path with the shortest geographic distance, each link is assigned a metric equal to its own distance. As another example, assume that each link is assigned a metric of unity. The shortest path algorithm then finds the path that traverses the fewest hops. (The term *hop* is often used to refer to each link in a path.) As a third example, assume that each network link has a certain probability of being available (i.e., not failed), and assume that each link in the network fails independently such that the availability of a path is the product of the availabilities of each link in the path (ignoring node failures). The link metric can be chosen to be the negative of the logarithm of the link availability, where the logarithm function is used in general to convert a multiplicative metric to an additive metric. Higher link availability corresponds to a smaller link metric; thus, running the shortest path algorithm with this metric produces the path with the highest availability. As this last example demonstrates, the metric may be unrelated to distance; thus, the term 'shortest path' is in general a misnomer. Nevertheless, this term will be used here to represent the path that minimizes the desired metric.

One very well known shortest path algorithm is the Dijkstra algorithm, where the inputs to the algorithm are the network topology and the source and the destination. This is a 'greedy' algorithm that is guaranteed to find the shortest path from source to destination, assuming a path exists. Greedy algorithms proceed by choosing the optimal option at each step without considering future steps. In the case of the Dijkstra algorithm, this strategy produces the optimal overall result (in general, however, greedy algorithms do not always yield the optimal solution).

Another shortest path algorithm is the Breadth-First-Search (BFS) Shortest Path algorithm, which proceeds by considering all one-hop paths from the source, then all two-hop paths from the source, etc., until the shortest path is found from source to destination [Bhan99]. If there is a unique shortest path from source to destination, the BFS and Dijkstra algorithms produce the same result. However, if there are multiple paths that are tied for the shortest, the BFS algorithm finds the shortest path with the fewest number of hops. This can be helpful in network design because fewer hops can potentially translate into lower cost or less wavelength contention, as is discussed in the next section. The Dijkstra algorithm does not in general have this same tie-breaking property. Furthermore, the BFS algorithm works with negative link metrics, as long as there are no cycles in the network where the sum of the link metrics is negative. This is relevant for one of the graph transformations that is commonly used as part of an algorithm to find two or more diverse routes, as discussed in Section 3.6. The Dijkstra algorithm needs a small modification to be used with negative link metrics. Overall, then, the BFS shortest path algorithm is somewhat better suited for network design; code for this algorithm is provided in Appendix B.

The Dijkstra and BFS algorithms can be applied whether or not the links in the network are bi-directional (a link is bi-directional if traffic can be routed in either direction over the link). Furthermore, the algorithms work if different metrics are assigned to the two directions of a bi-directional link. However, if the network is bi-directionally symmetric, such that the traffic flow is always two-way and such that the metrics are the same for the two directions, then a shortest path from source to destination also represents, in reverse, a shortest path from destination to source. (This is also called an *undirected* network.) In this scenario, which is typical of telecommunications networks, it does not matter which node is designated as the source and which is designated as the destination.

The shortest path algorithm can be incorporated as part of a larger procedure to find the K-shortest paths. K-shortest path routines find the shortest path between the source and destination, the second shortest path, etc., until the Kth shortest path is found or until no more paths exist. Note that the paths that are found are not necessarily completely disjoint from each other; i.e., the paths may have links and/or nodes in common. Many K-shortest paths algorithms exist, e.g., [Yen71, Epps94], where the ones that find only simple paths (i.e., paths without loops) are the most relevant for network design. The code for one such K-shortest paths algorithm is provided in Appendix B (the code follows the procedure described in [HeMS03]).

A variation of the shortest-path problem arises when one or more constraints are placed on the desired path; this is known as the *constrained shortest path* (CSP) problem. Some constraints are straightforward to handle. For example, if one is searching for the shortest path subject to all links of the path having at least N wavelengths free, then prior to running a shortest path algorithm, all links with less than N free wavelengths are removed from the topology. As another example, the intermediate steps of the BFS shortest path algorithm can be readily used to determine the shortest path subject to the number of path hops being less than H, for any $H > 0$ (similar to [GuOr02]). However, more generally, the CSP problem can be difficult to solve; for example, determining the shortest path subject to the availability of the path being greater than some threshold (where it is assumed the availability is based on factors other than distance). Various heuristics have been proposed to address the CSP problem; e.g., [KoKr01, LiRa01] (the latter reference addresses the constrained K-shortest paths problem). Some heuristics have been proposed to specifically address the scenario where there is just a single constraint; this is known as the *restricted shortest path* (RSP) problem. Additionally, a simpler version of the multi-constraint problem arises when *any* path satisfying all of the constraints is desired, not necessarily the shortest path; this is known as the *multi-constrained path* (MCP) problem. An overview, including a performance comparison, of various heuristics that address the RSP and MCP problems can be found in [KKKV04].

3.2 Routing Metrics

As discussed in the previous section, a variety of metrics can be used with the shortest path algorithm. Two common search strategies are: find the path with the fewest hops and find the path with the shortest distance. With respect to minimizing network cost, the optimal routing strategy to use is dependent on the underlying system technology, as discussed next. (In this section, issues such as grooming and shared protection that also may have an impact on cost are not considered; these issues are discussed in Chapters 6 and 7, respectively.)

3.2.1 Fewest-Hops Path vs. Shortest-Distance Path

In a pure O-E-O network, a connection is electronically terminated (i.e., regenerated) at every intermediate node along the path, where the electronic terminating equipment is a major component of the path cost. Thus, searching for the path from source to destination with the fewest hops is generally favored as it minimizes the amount of required regeneration. This is illustrated in Fig. 3.1 for a connection between Nodes A and Z. Path 1 is the shortest-distance path at 900 km, but includes four hops. Path 2, though it has a distance of 1,200 km, is typically lower cost because it has only two hops and thus requires less intermediate electronic termination.

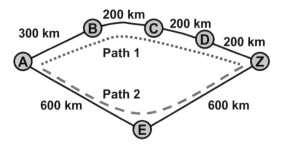

Fig. 3.1 Path 1, A-B-C-D-Z, is the shortest-distance path between Nodes A and Z, but Path 2, A-E-Z, is the fewest-hops path. In an O-E-O network, where the signal is regenerated at every intermediate node, Path 2 is typically the lower-cost path.

In networks with optical bypass, regeneration is determined by the system optical reach, which is typically based on distance. For example, an optical reach of 2,000 km indicates the connection can travel no further than 2,000 km before it needs to be regenerated. (In reality, the optical reach is determined by many factors as is discussed in Chapter 4, but for simplicity, it is usually specified in terms of a distance.) This favors searching for the shortest-distance path between the source and destination. However, with optical-bypass systems, there is a wavelength continuity constraint, such that the connection must remain on the same wavelength (i.e., lambda) as it optically bypasses nodes. Finding a wavelength that is free to carry the connection is potentially more difficult as the number of links in the path increases. This implies that the number of path hops should be considered as well.

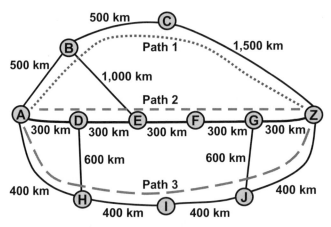

Fig. 3.2 Assume that this is an optical-bypass-enabled network with an optical reach of 2,000 km. Path 1, A-B-C-Z, has the fewest hops but requires one regeneration. Path 2, A-D-E-F-G-Z, and Path 3, A-H-I-J-Z, require no regeneration. Of these two lowest-cost paths, Path 3 is preferred because it has fewer hops.

Overall, a good strategy for optical-bypass systems is to search for a route based on distance, but of the paths that meet the minimal regeneration, favor the one with the fewest hops. This is illustrated by the three paths between Nodes A and Z shown in Fig. 3.2, where the optical reach is assumed to be 2,000 km. Path 1, with a distance of 2,500 km, has the fewest hops but requires one regeneration. Path 2, with a distance of 1,500 km, and Path 3, with a distance of 1,600 km, do not require any regeneration and are thus lower-cost paths. Of these two paths, Path 3 is generally more favorable (all other factors, e.g., link load, being equal), even though it is somewhat longer than Path 2, because it has only four hops compared to the five hops of Path 2.

3.2.2 Shortest-Distance Path vs. Minimum-Regeneration Path

While path distance is clearly related to the amount of regeneration in an optical-bypass-enabled network, it is important to note that the path with shortest physical distance is not necessarily the path with minimum regeneration. In addition to considering the distance over which a signal has traveled in determining where to regenerate, carriers generally require that any regeneration occur in a network node (i.e., the add/drop and switching locations of a network), as opposed to at an arbitrary site along a link. First, sites along a link, e.g., amplifier huts[7], may not be large enough to house regeneration equipment. Second, from a maintenance perspective, it is beneficial to limit the number of locations where regeneration equipment is deployed. (If the optical reach of the system happens to be shorter than the distance between two nodes, then there is no choice but to deploy a dedicated regeneration site along the link where all transiting traffic is regenerated. In this scenario, the dedicated regeneration site can be considered a network node.)

This additional design constraint can lead to more regeneration than would be predicted by the path length. Consider the example shown in Fig. 3.3, with a connection between Node A and Node Z, and assume that the optical reach is 2,000 km. Path 1 is the shortest path for this connection, with a length of 3,500 km. Based on the path length and the optical reach, it is expected that one regeneration would be required. However, because regeneration occurs only at nodes, two regenerations are actually required (e.g., the regenerators could be at Nodes C and E). Path 2, while longer, with a length of 3,800 km, requires just one regeneration (at Node G).

Another factor that affects the amount of regeneration is whether the network fully supports optical bypass in all directions at all nodes. There may be some nodes not fully equipped with optical-bypass equipment. For example, a degree-three node may be equipped with an OADM and an optical terminal (see Section 2.5); any transiting traffic entering via the optical terminal needs to be regenerated regardless of the distance over which it has been transmitted.

---

[7] Typically small buildings that house the optical amplifiers.

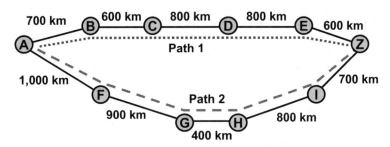

Fig. 3.3 Assume that this is an optical-bypass-enabled network with an optical reach of 2,000 km. Path 1 is 3,500 km, but requires two regenerations. Path 2 is longer at 3,800 km but requires only one regeneration. (Adapted from [Simm06]. © 2006 IEEE)

As the examples of this section illustrate, simply running a shortest path algorithm may not produce the most desirable route. Rather, it is often advantageous to generate a *set* of candidate paths, and then select a particular path to use based on other factors. For example, a particular route may be selected because of its lower cost or because it avoids a 'bottleneck' link that is already heavily loaded.

The next section discusses how to generate a good set of candidate paths, and Section 3.4 discusses how the path set is used in the routing process.

3.3 Generating a Set of Candidate Paths

Two different strategies are presented for generating a set of candidate paths. The first is a more formal strategy that uses the K-shortest paths algorithm to find minimum-cost paths. The second is a somewhat ad-hoc methodology to find a set of paths with good link diversity.

3.3.1 K-Shortest Paths Strategy

The marginal cost of adding a new demand to the network is largely a function of the number of electronic terminations (i.e., regenerations) needed to support the demand. Much of the other network costs, e.g., amplification, are incurred when the network is first installed and are amortized over the whole demand set.

Thus, roughly speaking, in an O-E-O network, paths that have the same number of hops have an equivalent cost because the number of intermediate electronic terminations is the same. To find a set of lowest-cost paths, one can run a K-shortest paths algorithm where the metric is set to unity for all links. The first N of the paths returned by the algorithm will satisfy the fewest-hops criterion, for some N where $1 \leq N \leq K$. Setting K to a number around 10 will typically ensure that all lowest-cost paths are found; i.e., N is usually less than or equal to 10 in practical optical networks.

Similarly, in an optical-bypass-enabled network, paths with the same amount of regeneration can be considered equivalent-cost paths. First, consider networks where any loopless path is shorter than the optical reach, as may be the case in a metro-core network. All paths can be considered lowest-cost because there is no need for regeneration in the network. In this scenario, one can run a K-shortest paths algorithm with a link metric of unity to generate a set of lowest-cost paths that have the fewest, or close to the fewest, number of hops.

In more general optical-bypass-enabled networks where there is regeneration, it is not quite as straightforward to find a set of paths that meet the minimum regeneration. A K-shortest paths algorithm can be run with distance as the link metric. However, as illustrated by the example of Fig. 3.3, each of the returned paths must be examined to determine the actual number of required regenerations. The ones with minimum regeneration make up a set of least-cost paths. There is no guarantee that this procedure produces a path with the lowest possible amount of regeneration. (In Chapter 4, an alternative metric that is more tied to the underlying optical system is presented that may be a better predictor of regeneration.) However, if the minimum number of regenerations found is R, and at least one of the paths found has a distance greater than $(R+1)*$[Optical Reach], then the set of minimum-regeneration paths must have been found. Practically speaking, setting K to about 10 in the K-shortest paths algorithm is sufficient to generate a good set of least-cost paths.

Alternatively, one can use a more complex topology transformation, as described in Section 3.5.1, to ensure a minimum-regeneration path is found. However, the simpler method described above is generally sufficient.

(Note: One could use the intermediate steps of the BFS algorithm to find the shortest-distance path subject to a maximum number of hops to assist in finding the minimum-regeneration path with the minimum number of hops. Again, because distance does not directly translate to regeneration, this strategy does not necessarily always succeed. Furthermore, with K large enough, such paths are generally found by the above process anyway.)

3.3.2 Bottleneck-Avoidance Strategy

While the K-shortest paths technique can generate a set of lowest-cost paths, the paths that are found may not exhibit good link diversity. For example, a link that is expected to be heavily loaded may appear in every lowest-cost path found between a particular source and destination. If the link diversity is not sufficient for a particular source/destination pair, then an alternative strategy can be used to generate a candidate path set, as described here.

The first step is to determine the links in the network that are likely to be highly loaded (i.e., the 'hot spots'). One methodology for estimating load is to perform a preliminary routing where each demand in the forecasted traffic set is routed over its fewest-hops path (O-E-O networks) or its shortest-distance path (optical-bypass-enabled networks). While a traffic forecast may not accurately predict the traffic that will actually be supported in the network, it can be used as a reasonably good estimator of which links are likely to be heavily loaded. (Alternatively, one can combine the traffic forecast with the maximum-flow method of [KaKL00] to determine the critical links.)

It is important to consider not just single links that are likely to be bottlenecks, but also sequences of consecutive links that may be heavily loaded, and find routes that avoid the whole sequence of bad links. This is illustrated in Fig. 3.4, where Links BC, CD, and DZ are assumed to be likely bottlenecks. The shortest-distance path between Nodes A and Z is Path 1, which is routed over all three of the problem links. If one were to look for a path between these nodes that avoids just Link BC, then Path 2 is the remaining shortest-distance path. This is not satis-factory as the path still traverses Links CD and DZ. It would be better to simulta-neously avoid all three bottleneck links and find Path 3.

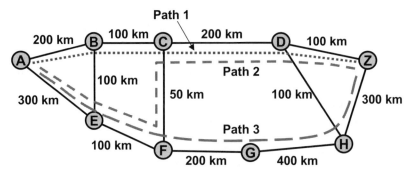

Fig. 3.4 Links BC, CD, and DZ are assumed to be bottleneck links. Path 1, A-B-C-D-Z, crosses all three of these links. If Link BC is eliminated from the topology, the resulting shortest path, Path 2, A-E-F-C-D-Z, still crosses two of the bottleneck links. All three bottleneck links must be simultaneously eliminated to yield Path 3, A-E-F-G-H-Z.

After identifying the top 10 to 20 'hot spots' in the network, the next step is to run the shortest path algorithm multiple times, where in each run, one bad link or one bad sequence of links is removed from the topology. (If a particular 'hot spot' does not appear in the lowest-cost paths for a source/destination pair, then that hot spot can be skipped. Furthermore, it is not desirable to remove *all* potentially bad links at once when running the shortest path algorithm, as the resulting set of paths may be circuitous and may shift the hot spots to other locations in the network.) This process finds paths that avoid particular bottleneck areas, if possible.

The bottleneck-avoidance strategy method can be combined with the K-shortest paths strategy such that, overall, the candidate path set includes lowest-cost paths and paths that are diverse with respect to the expected hot spots. There are clearly other methods one can devise to generate the candidate paths; however, the strat-egy described above is simple and is effective in producing designs for practical optical networks with relatively low cost and good load balancing.

3.4 Routing Strategies

The previous section discussed methods of producing a good set of candidate paths for all relevant source/destination pairs. This section considers strategies of selecting one of the paths to use for a given demand. These strategies are general such that they hold for scenarios where demand requests enter the network one at a time, as well as scenarios where there is a whole set of demands that needs to be routed but it is assumed that the demands have been ordered so that they are considered one at a time. (Ordering the demand set is covered in Section 3.7.) In either real-time or long-term planning, a particular candidate path may not be feasible as the network evolves because there is no free bandwidth on one or more of the path links. Furthermore, with real-time operation, there is the additional constraint that all necessary equipment to support the connection must already be deployed. Thus, if a particular path requires regeneration at a node, and the node does not have the requisite available equipment, the path is considered infeasible.

In optical-bypass-enabled networks, selecting a wavelength for the route is an important step of the planning process. This section focuses on selecting a route independent of wavelength assignment, where wavelength assignment is performed as a separate step later in the process. Chapter 5 considers treating both of these aspects of the planning process in a single step.

3.4.1 Fixed-Path Routing

In the strategy known as fixed-path routing, the candidate set of paths is generated prior to any demands being added to the network. For each source/destination pair, one path is chosen from the associated candidate path set. Ideally, the path is a lowest-cost path, although load balancing can also be a consideration for selecting a particular path. Whenever a demand request is received for a given source/destination pair, the selected path for that pair is used to route the demand; no other candidate paths for that source/destination pair are considered.

This is clearly a very simple strategy, with any calculations performed up-front, prior to any traffic being added. However, the performance of this strategy can be very poor, as it usually results in certain areas of the network becoming unnecessarily congested. The same path is always used for a given source/destination pair, providing no opportunity to adapt to the current network state. This often results in premature blocking of a demand even though feasible paths do exist for it. Overall, this strategy is not recommended.

3.4.2 Alternative-Path Routing

Alternative-path routing also relies on generating a candidate set of paths prior to any demands being added to the network. However, in this strategy, the candidate set is narrowed down to M paths for each source/destination pair, for some small number M. When a demand request is received for a given source/destination, one of the M paths is selected to be used for that particular demand. In practice, selecting about three paths per each source/destination pair is a good strategy, although in real-time planning, where equipment availability is an issue, it may be desirable to select somewhat more paths.

In some research regarding alternative-path routing, it is assumed that the M paths must have no links in common; however, this is unnecessarily restrictive. (Selecting paths with no links in common, however, is important for protection, as is covered in Section 3.6.) The goal is to select the M paths such that the same expected 'hot spots' do not appear in all of the paths. Furthermore, it is not necessary to pick the M paths such that no 'hot spot' appears in any of the paths; this would have the effect of simply shifting the heavy load to other links as opposed to balancing the load across the network. Ideally, the M paths are lowest-cost paths; however, in order to get enough 'hot-spot' diversity among the paths, it may be necessary to include a path that does not meet the lowest cost; e.g., one of the M paths may have an additional regeneration. (Selecting a path with a small amount of extra cost is not ideal; however, it is typically preferable to blocking a demand request due to poor load balancing.) In addition, all other factors being equal (e.g., cost, expected load), paths with shorter distance and fewer hops should be favored for inclusion in the set of M paths.

When a demand request arrives for a particular source/destination, any of the M paths can be potentially used to carry the connection, assuming the path is feasible (i.e., it has the necessary available bandwidth and equipment). Typically, the current state of the network is used in determining which of the M paths to use. A common strategy is to select the feasible path that will leave the network in the 'least-loaded' state. Assume that the most heavily loaded link in the ith path has W_i wavelengths already routed on it; the selected path is the one with the minimum W_i. If multiple paths are tied for the lowest W_i, then the load on the second most heavily loaded link in these paths is compared, and so on. (This is also known as *Least Congested Path* routing [ChYu94].) If multiple paths continue to be tied with respect to load, or if the tie is broken only when comparing links with load much less than the maximum, then one can consider other factors; e.g., in an optical-bypass-enabled network, the path with fewest hops can be used to break the tie. Furthermore, if one of the M paths has a higher cost than the other paths, then this path should not be selected unless its W_i is significantly lower than that of the other paths, or unless it is the only feasible path.

In another strategy, congestion and hops are considered together. For example, when choosing between two candidate paths, a path with H more hops is selected only if its most heavily loaded link has L fewer wavelengths routed on it, where the parameters H and L can be tuned as desired.

In real-time routing, it may also be beneficial to consider the available equipment at the nodes when selecting one of the M paths. If a particular path requires a regeneration at a node and there is very little free regeneration equipment at the node, then that path may not be favored, especially if there are other paths that have similar link loading and greater equipment availability.

The alternative-path routing strategy works very well in practice. It uses the traffic forecast to assist in generating the initial candidate path set, whereas it uses current network conditions to select one of the M paths for a particular demand. One can add a larger dynamic component to the algorithm by allowing the set of M paths to be updated periodically as the network evolves. With a good choice of paths, the network is generally fairly well loaded before all M paths for a particular source/destination pair are infeasible. When this occurs, one can revert to dynamically searching for a path, as is covered next.

(A more restrictive form of alternative-path routing is known as *fixed-alternate routing*, where the M candidate paths are considered in a fixed order, and the first such path that has available capacity is selected. This is simpler than more general alternative-path routing because it only needs to track whether a path is available or not; it does not need to track the load on every link. However, this method is not as effective at load balancing.)

3.4.3 Dynamic-Path Routing

In dynamic-path routing (also called *adaptive unconstrained routing* [MoAz98]), there is no predetermination of which paths to use for a particular source/destination combination. The path calculation is performed at the time of each demand request, based on the current state of the network. The first step is to determine if there are any links in the network with insufficient available bandwidth to carry the new demand. Any such links should be temporarily eliminated from the network topology. In addition, in real-time design with an O-E-O network, any link for which either of its endpoints does not have an available transponder should be temporarily removed from the topology, because any link in the path would require a transponder at either end for O-E-O conversion. (Recall that in O-E-O networks, transponders are plugged into optical terminals, where each optical terminal is associated with a link.)

After the topology has been pruned based on the current network state (more advanced topology transformations are discussed in Section 3.5.1), the procedure for generating a candidate set of paths can be followed; i.e., the K-shortest paths algorithm can be run and/or the bottleneck-avoidance strategy can be used where the current hot-spots in the network are systematically eliminated from the already-pruned topology. (Note that the links with no available capacity have already been eliminated, so these hot spots are the relatively heavily loaded links that still have available capacity.) This process takes tens of milliseconds to complete, so it is possible to generate a candidate set of paths every time a new demand request is received (assuming the total provisioning time is on the order of one second or more). One can use a smaller 'K' in the K-shortest paths algorithm or consider fewer hot-spots in the bottleneck-avoidance methodology in order to reduce the processing time further. Many dynamic implementations simply look for a single shortest path.

A variety of metrics can be used for dynamic routing. Typically, the metric reflects number of hops, distance, or current congestion; e.g., [BhSF01]. One method suggested in [ZTTD02] for optical-bypass-enabled networks uses a metric based on the number of wavelengths that are free on consecutive links, as this is an indicator of the likelihood of being able to assign wavelengths to the links. This was shown in [ZTTD02] to provide better performance than simply considering link congestion; however, it does involve a graph transformation in order to capture the relationship between adjacent links in the shortest paths algorithm.

After the candidate paths are generated, one path is selected for the new demand based on the current network state. (In implementations where only a single candidate path is generated, then clearly this step of choosing one of the candidate paths is not needed.) Note that in real-time planning, some of the candidate paths may be infeasible due to a lack of resources, even though some amount of topology pruning occurred up front. For example, in an optical-bypass-enabled network, a candidate path may require regeneration to occur at a node that does not have available regeneration equipment. (With optical bypass it is typically not known ahead of time whether an intermediate node in a path will require regeneration, thus the node is not pruned from the topology during the pre-processing step.) After eliminating any of the candidate paths that are infeasible, a technique such as selecting the least-loaded path, as described in Section 3.4.2, can be used to pick among the remaining paths.

The dynamic path selection methodology provides the greatest adaptability to network conditions. While this may sound good in theory, studies have shown that routing strategies that consider a more global design based on forecasted traffic can be advantageous [EMSW03]. Thus, a strictly adaptive algorithm is not necessarily ideal. Furthermore, the dynamic methodology potentially results in many different paths followed by the demands. This has the effect of decreasing the network 'interference length', which can potentially lead to more contention in the wavelength assignment process for optical-bypass-enabled networks. (The interference length is the average number of hops shared by two paths that have at least one hop in common, as defined in [BaHu96].) In addition, if the network makes use of wavebands, where groups of wavelengths are treated as a single unit, then the diversity of paths produced by a purely dynamic strategy can be detrimental from the viewpoint of efficiently packing the wavebands. Finally, the dynamic methodology is the slowest of the three routing strategies discussed and involves the most computation. It would likely not be suitable, for example, if the demand request must be provisioned in a sub-second time-frame.

Given the good results that are produced by the simpler alternative-path routing strategy of Section 3.4.2, it is often favored over a purely dynamic routing strategy. When the network is so full that none of the alternative paths are feasible, the dynamic strategy can be used instead.

3.5 Avoiding Infeasible Paths

3.5.1 Capturing the Available Equipment in the Network Model

In real-time planning, some, or even all, of the candidate paths may be infeasible due to a lack of available equipment in the appropriate nodes. As described above, the dynamic routing process first prunes out the links and nodes that would clearly be infeasible for a new demand. In a scenario where there is little available equipment, it may be necessary to perform more involved topology transformations to model the available equipment in more detail. In the examples below, it is assumed that there is available equipment at the source and destination nodes; otherwise, the demand is rejected without further analysis.

First, consider an O-E-O network with the topology shown in Fig. 3.5(a) and assume that a new demand request arrives where the source is Node A and the destination is Node Z. Node B and Node D of the network are illustrated in more detail in Fig. 3.5(b) and Fig. 3.5(c), respectively. In this example, it is assumed that pairs of transponders are interconnected via patch-cables rather than through a flexible switch (i.e., the nodal architecture is that of Fig. 2.3, not Fig. 2.4). Thus, in order for a new demand to transit Node B from Link j to Link k, ($1 \leq j, k \leq 4$), there must be an available transponder on the optical terminal for Link j, an available transponder on the optical terminal for Link k, and the two transponders must be interconnected. Assuming that Fig. 3.5(b) depicts all of the available equipment at Node B, then the only possible paths through the node for a new demand are between Links 1 and 2, Links 1 and 3, and Links 2 and 4. Similarly, the only possible paths through Node D are between Links 3 and 5 and between Links 5 and 6. It is assumed that the remaining nodes have sufficient available equipment to support any path; i.e., Node A has an available transponder on Link 1, Node Z has available transponders on both Links 4 and 6, and there are available transponders at Node C to support a path between Link 2 and Link 5.

To capture the path restrictions imposed by the limited amount of available equipment at Nodes B and D, one can perform a graph transformation where each link in the original topology becomes a node in the new topology. To be more precise, because each link shown in Fig. 3.5(a) actually represents bi-directional communication, each direction of a link becomes a node. These nodes are interconnected in the new topology only if there is equipment available in the real network to allow a new path to be routed between them. Nodes also have to be added to represent the source and destination of the new demand, i.e., Node A and Node Z, respectively.

The resulting transformed graph is illustrated in Fig. 3.6, where the node numbers in this graph correspond to the link numbers of Fig. 3.5. The single-prime nodes represent the links in the direction from the (alphabetically) lower letter to the higher letter, and the double-prime nodes represent the reverse link direction. Thus, Node 2' represents Link 2 in the original graph in the direction from Node B to Node C; Node 2" represents Link 2 in the direction from Node C to Node B.

There is no need to add a node representing Link 1" because this link enters the demand source; similarly, there is no need to add a node representing Link 4" or Link 6" because these links exit the demand destination. Note that the node representing Link 1' is connected to the nodes representing Link 2' and Link 3', but not Link 4' due to the lack of a transponder pair connecting these links (in Node B). Similarly, there is no link connecting the node representing Link 3' and the node representing Link 6'.

A shortest path algorithm is run on the transformed topology to find a feasible path, using unity as the link metric to minimize the number of O-E-O conversions. The desired path from Node A to Node Z in the transformed graph is A-1'-2'-5'-6'-Z, which corresponds to path A-B-C-D-Z in the original graph.

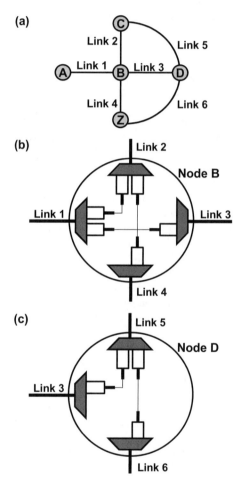

Fig. 3.5 (a) Network topology, where a new demand is requested from Node A to Node Z. (b) The available equipment at Node B. (c) The available equipment at Node D.

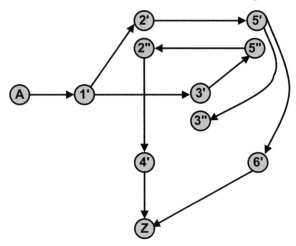

Fig. 3.6 Graph transformation to represent the available equipment in the network of Fig. 3.5. The numbered nodes correspond to the links in Fig. 3.5 with the same number, with the prime and double-prime representing the two directions of the link. Nodes A and Z are the demand endpoints.

Routing constraints such as those imposed by the available transponder pairs, where only certain directions through a node are possible, are known as *turn constraints* (this term arises from vehicular routing, where only certain turns are permissible). The graph transformation described above is one means of solving such problems, which allows a standard shortest path algorithm to be run. An alternative is to use the original graph but modify the Dijkstra algorithm or the BFS algorithm to take into account the turn constraints when building out the path from source to destination [BoUh98, SoPe02]. With this methodology, an explicit graph transformation is not required.

Note that if the nodes are equipped with switches then turn constraints do not arise, because the switch can interconnect any two transponders in the node. Additionally, if there are many available transponders at each node, then this level of modeling is not needed as typically any path through the node can be supported.

In an optical-bypass-enabled network, with edge configurability, a different graph transformation can be used in real-time planning, similar to [GeRa04]. (The methodology is described first, followed by an example.) A new topology is created, comprised of only those nodes that have available regeneration equipment, along with the source and destination of the new demand. Any two of these nodes are interconnected by a link in the new topology if a regeneration-free path with available bandwidth exists between the two nodes. Even if there are multiple regeneration-free paths between a pair of nodes, only one link is added in the new topology. A shortest path algorithm is run on this transformed topology to find a feasible path. A link metric such as [*LARGE* + *NumHops*] can be used, where *LARGE* is some number greater than any possible number of end-to-end hops, and *NumHops* is the number of hops in the minimum-hop regeneration-free path between the two nodes interconnected by the link.   This finds a minimum-regeneration feasible path of minimum number of hops, assuming one exists.

An example of such a graph transformation is shown in Fig. 3.7. The full network is shown in Fig. 3.7(a). Assume that this is an optical-bypass-enabled network with edge configurability and an optical reach of 2,000 km. Furthermore, assume that Nodes A and Z are the demand endpoints, and that only Nodes B and D are equipped with available regeneration equipment. Figure 3.7(b) illustrates the associated transformed network; Nodes C and E do not appear in the transformed graph because they do not have available regeneration equipment. Each of the links in the transformed network represents a regeneration-free path in the real network. For example, Link AD in the transformed network represents path A-E-D in the real network. Running a shortest path algorithm on the transformed graph yields the path A-D-Z, corresponding to A-E-D-Z in the true network.

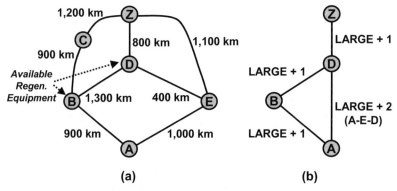

(a)   (b)

Fig. 3.7 In the network shown in (a), Nodes A and Z are assumed to be the endpoints of a new demand, and Nodes B and D are assumed to be the only nodes with available regeneration equipment. The optical reach is 2,000 km. This results in the transformed graph shown in (b). LARGE represents a number larger than any possible path hop count.

Such graph transformations as described above for O-E-O and optical-bypass-enables networks would need to be performed every time there is a new demand request to ensure that the current state of the network is taken into account, which adds to the complexity of the routing process. Ideally, sufficient equipment is predeployed in the network and the candidate paths are sufficiently diverse that one can avoid these transformations, at least until the network is heavily loaded.

3.5.2 Predeployment of Equipment

Infeasible paths due to lack of equipment can be minimized by having equipment judiciously predeployed in the network in anticipation of future demands. Predeployed equipment in large-part refers to transponders (or other regeneration equipment) and the racks for housing them. Selecting how much equipment to predeploy, as well as where to place the equipment, is strategically very important. Predeploying too little equipment leads to sub-optimal routing or excessive blocking of demand requests; predeploying too much equipment is an unnecessary expense. Clearly, the amount of equipment to predeploy depends on the underlying system characteristics; e.g., one would expect to predeploy more transponders in an O-E-O network than in an optical-bypass-enabled network. Furthermore, the different levels of configurability provided by the various system architectures mandate different levels of accuracy in the estimation process. For example, in an O-E-O architecture, one must calculate the number of transponders to predeploy on each optical terminal (i.e., a per-link estimate); in an optical-bypass architecture with edge configurability, it is necessary to estimate only the total number of transponders to predeploy at each node (i.e., a per-node estimate) [GeRa04].

Several strategies exist for estimating the amount of equipment to predeploy. A common strategy is to run simulations based on forecasted traffic, where the simulation results can be used to estimate the amount of equipment that should be predeployed at each node in order to reduce the blocking probability below a given threshold. Alternatively, one can use standard queuing theory, where each source/destination demand pair is associated with requiring particular equipment in the network. The arrival and departure processes of the demands can be modeled to estimate the required equipment to reduce the blocking probability below the desired threshold (e.g., [MoBB04]).

Another interesting strategy for optical-bypass-enabled networks, proposed in [BaLe02], involves estimating for each node the probability that any new demand will require regeneration at that node. The probabilities are determined based on the nodal position within the network and the lengths of the links feeding into the node, where being closer to the center of the network and being an endpoint of a long link increases the likelihood of regeneration at a node. This analysis can be used to assist in determining the amount of regeneration equipment to predeploy at each node.

The predeployment strategy to use may depend on the level of detail in the traffic forecast. If the forecast is very specific, then running a simulation is probably the simplest strategy to determine where to predeploy equipment. If there are only approximate models of the demand arrivals and departures, then queuing analysis can be used. If only a forecast of the total number of demands in the network is available, then the method based on nodal regeneration probabilities can be used.

In addition to intelligent predeployment, having network elements with a high degree of flexibility can be very helpful in maximizing the utility of the equipment that is deployed, as was described in Chapter 2.

3.6 Diverse Routing for Protection

The previous sections were focused on finding a single path between a source and a destination. If any of the equipment supporting a connection fails, or if the fiber over which the connection is routed is cut, the demand is brought down. Thus, it is often desirable to provide protection for a demand to improve its availability. Numerous possible protection schemes are covered in Chapter 7. Here, it is simply assumed that two paths are required from source to destination, where the two paths should be disjoint to ensure that a single failure does not bring down the demand. If one is concerned only with link failures, then the two paths can be simply link-disjoint, where nodes can be common to both paths. If one is concerned with both link and node failures, then the two paths should be both link-and-node disjoint (except for the source and destination nodes). The question is how to find the desired disjoint paths in a network.

Network designers sometimes resort to the simple strategy of first searching for a single path using a shortest path algorithm. The links in the returned path are then pruned from the topology and the shortest path algorithm is invoked a second time. (If node-disjointness is required, then the intermediate nodes from the first found path are also pruned from the topology.) If a path is found with this second invocation, then it is guaranteed to be disjoint from the first path.

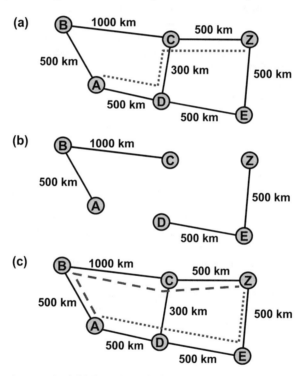

Fig. 3.8 The shortest pair of disjoint paths, with distance as the metric, is desired between Nodes A and Z. (a) The first call to the shortest path algorithm returns the path shown by the dotted line. (b) The network topology after pruning the links comprising the shortest path. The second call to the shortest path algorithm fails as no path exists between Nodes A and Z in this pruned topology. (c) The shortest pair of disjoint paths between Nodes A and Z, shown by the dotted and dashed lines.

While a simple strategy, it unfortunately fails in some circumstances. Consider the network topology shown in Fig. 3.8(a), and assume that two link-diverse paths are required from Node A to Node Z. The first invocation of the shortest path algorithm returns the path shown by the dotted line. Removing the links of this path from the topology yields the topology of Fig. 3.8(b). It is not possible to find a path between Nodes A and Z on this pruned topology, causing the strategy to fail. In fact, two diverse paths can be found in the original topology as shown in Fig. 3.8(c). This type of scenario is called a 'trap topology', where two sequential calls to the shortest path algorithm fail to find disjoint paths even though they do exist.

Even if the simple two-call strategy succeeds in finding two disjoint paths, the paths may not be optimal. In the network of Fig. 3.9(a), it is assumed O-E-O technology is used such that minimizing the number of hops is desirable. The fewest-hops path from Node A to Node Z is shown by the dotted line. The links of this path are pruned from the topology resulting in the topology shown in Fig. 3.9(b). The second call to the shortest path algorithm returns the path shown by the dashed line in this figure. While indeed disjoint, the two paths of Figs. 3.9 (a) and (b) cover a total of ten hops. However, the lowest-cost pair of disjoint paths has a total of only eight hops, as shown in Fig. 3.9(c).

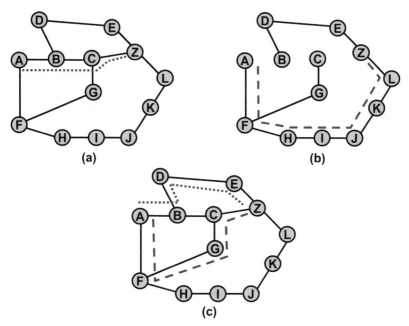

Fig. 3.9 The shortest pair of disjoint paths, with hops as the metric, is desired between Nodes A and Z. (a) The first call to the shortest path algorithm returns the path shown by the dotted line. (b) The network topology after pruning the links comprising the shortest path. The second call to the shortest path algorithm finds the path indicated by the dashed line. The total number of hops in the two paths is ten. (c) The shortest pair of disjoint paths between Nodes A and Z, shown by the dotted and dashed lines; the total number of hops in the two paths is only eight.

As these examples show, the two-call strategy may not be desirable. It is preferable to use an algorithm specifically designed for finding the optimal pair of disjoint paths, as described next.

3.6.1 Shortest Pair of Disjoint Paths

The two best known Shortest Pair of Disjoint Paths (SPDP) algorithms are the Suurballe algorithm [Suur74, SuTa84] and the Bhandari algorithm [Bhan99]. Both algorithms involve calls to a regular shortest path algorithm; however, they require different graph transformations (e.g., removing links, changing the link metrics) to ensure that the shortest pair of disjoint paths is found. The graph transformations of the Bhandari algorithm may generate links with negative metrics, which is why it requires an associated shortest path algorithm such as BFS, which can handle graphs with negative link metrics.

Both the Suurballe and the Bhandari algorithms are guaranteed to find the pair of disjoint paths between a source and destination where the sum of the metrics on the two paths is minimized (assuming that at least one pair of disjoint paths exists). As illustrated by the examples of Fig. 3.8 and Fig. 3.9, the shortest single path may not be a part of the shortest-disjoint-paths solution. The run-times of the Suurballe and Bhandari algorithms are about the same; however, the latter may be more readily extensible to other applications [Bhan99]. Appendix B provides code for the Bhandari algorithm.

The SPDP algorithms can be used to find either the shortest pair of link-disjoint paths or the shortest pair of link-and-node-disjoint paths. To illustrate the difference, Fig. 3.10(a) shows the shortest pair of link-disjoint paths between Nodes A and Z, where the paths have Node B in common; together, the paths cover seven links and 700 km. Figure 3.10(b) shows the shortest link-and-node-disjoint paths between A and Z for the same topology; these paths cover eight links and 800 km.

Furthermore, the SPDP algorithms can be modified to find the shortest *maximally* link-disjoint (and optionally node-disjoint) paths when totally disjoint paths do not exist. Consider the topology shown in Fig. 3.11, where a protected connection is required between Nodes A and Z. A pair of completely disjoint paths does not exist between these two nodes. However, the maximally disjoint pair of paths, with one common link (Link DG) and two common nodes (Nodes D and G), is shown by the dotted and dashed lines in the figure. This pair of paths minimizes the number of single points of failures for the connection.

If a demand is very susceptible to failure, or the availability requirements (i.e., the fraction of time the demand must be in-service) are very stringent, then it may be desirable to establish more than two disjoint paths for the demand. The SPDP algorithms can be extended to search for the N shortest disjoint paths between two nodes, for any N, where the N paths are mutually disjoint. In most optical networks, there are rarely more than just a small number of disjoint paths between a given source and destination (say two to four); if N is larger than this, the algorithms can be used to return the N shortest maximally disjoint paths.

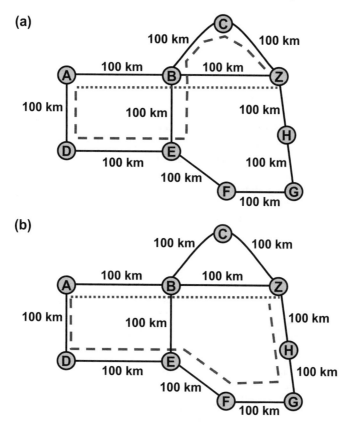

Fig. 3.10 (a) The shortest link-disjoint pair of paths is shown by the dotted and dashed lines. (b) The shortest link-and-node-disjoint pair of paths is shown by the dotted and dashed lines.

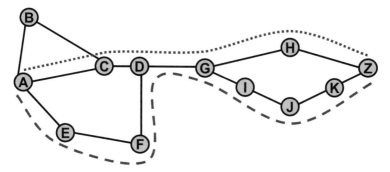

Fig. 3.11 There is no completely disjoint pair of paths between Nodes A and Z. The set of paths shown by the dotted and dashed lines represent the shortest maximally disjoint pair of paths. The paths have Nodes D and G, and the link between them, in common.

3.6.2 Shortest Pair of Disjoint Paths: Dual-Sources/Dual-Destinations

Another interesting twist to the problem of finding the shortest pair of disjoint paths arises when there are two sources and/or two destinations [Bhan99]. Consider the scenario shown in Fig. 3.12, where the source is in one regional network, the destination is in another regional network, and the two regional networks are interconnected by a backbone network. As shown in the figure, there are two nodes that serve as the gateways between each regional network and the backbone network. Each network is a separate domain, where routing occurs separately within each domain. It is desired that a protected connection be established between the source and destination. Thus, the paths in Regional Network 1 between the source and Gateways 1 and 2 should be diverse. Similarly, the paths in Regional Network 2 between the destination and Gateways 3 and 4 should be diverse.

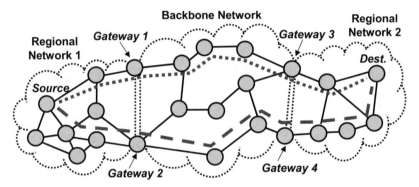

Fig. 3.12 A protected connection between the source and destination is routed from Regional Network 1, through the backbone network, to Regional Network 2. The gateways are the nodes at the boundaries of the regional networks and the backbone network. It is desirable to have diverse paths from the source to Gateways 1 and 2, diverse paths between Gateways 1 and 2 and Gateways 3 and 4, and diverse paths from Gateways 3 and 4 to the destination.

This is one application for an algorithm that finds the shortest pair of disjoint paths between one source and two destinations, or, equivalently, two sources and one destination. (A similar problem arises in backhauling subrate traffic from a node that is not equipped with a grooming switch to two nodes that are equipped with grooming switches, where the two backhaul paths should be diverse, as discussed further in Chapter 6 in Section 6.5. Backhauling refers to the general process of transporting traffic from a minor site to a major site for further distribution.)

The problem setup is illustrated in Fig. 3.13(a), where Node A is the source, and Nodes Y and Z are the two destinations. (This figure is a general illustration;

it is not related to Fig. 3.12.) Finding the shortest pair of disjoint paths is quite
simple. A 'dummy' destination node is added to the topology as shown in Fig.
3.13(b); Nodes Y and Nodes Z are connected to the dummy node via links that are
assigned a metric of zero. An SPDP algorithm is then run using Node A as the
source and the dummy node as the destination. This implicitly finds the shortest
pair of disjoint paths from Node A to Nodes Y and Z.

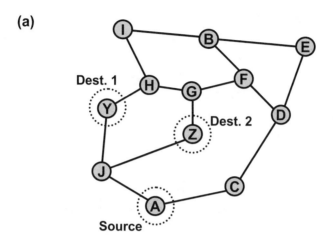

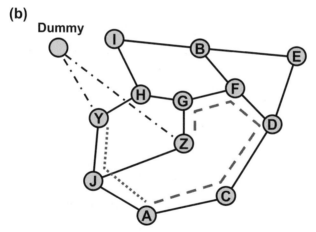

Fig. 3.13 (a) A disjoint path is desired between one source (Node A) and two destinations
(Nodes Y and Z). (b) A dummy node is added to the topology and connected to the two destinations via links that are assigned a metric of zero. An SPDP algorithm is run between Node A and
the dummy node to implicitly generate the desired disjoint paths, as shown by the dotted line and
the dashed line.

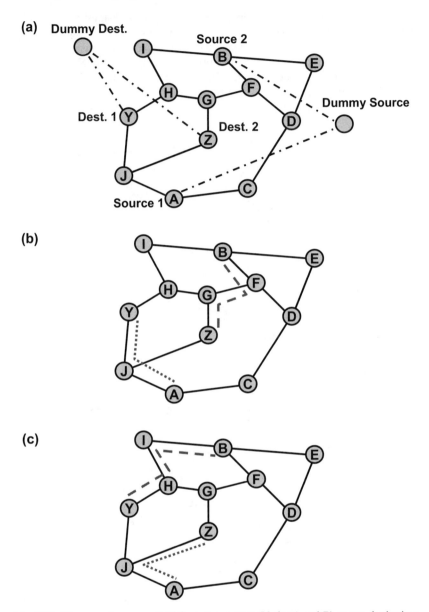

Fig. 3.14 Diverse paths are required from two sources (Nodes A and B) to two destinations (Nodes Y and Z). (a) One dummy node is connected to the two sources and one dummy node is connected to the two destinations via links that are assigned a metric of zero. (b) In this solution, one path is between Nodes A and Y and the other path is between Nodes B and Z. (c) In this solution, one path is between Nodes A and Z and the other path is between Nodes B and Y.

If disjoint paths are desired between two sources and two destinations, then both a dummy source and a dummy destination are added, and the SPDP algorithm is run between the two dummy nodes. (This application is relevant to Fig. 3.12, where it is desired to find disjoint paths in the backbone network between the two sets of gateway nodes.) This procedure is illustrated in the example network of Fig. 3.14(a) where it is assumed a shortest pair of disjoint paths from Nodes A and B to Nodes Y and Z is required. One dummy node is connected to Nodes A and B, and the other to Nodes Y and Z, via links with a metric of zero. After running the SPDP algorithm between the two dummy nodes, it is interesting that the two paths that are found may be between Nodes A and Y and between Nodes B and Z, as shown in Fig. 3.14(b); or, the two paths may be between Nodes A and Z and between Nodes B and Y, as shown in Fig. 3.14(c).

3.6.3 Shared Risk Link Groups (SRLGs)

When searching for disjoint paths for protection, it may be necessary to consider the underlying physical topology of the network in more detail. Two links that appear to be disjoint when looking at the network from the link level may actually overlap at the physical fiber level. Portions of the two links may lie in the same fiber conduit such that a failure along that conduit would simultaneously disrupt both links. Links that are part of the same failure group comprise what is known as a Shared Risk Link Group (SRLG). (There may be network resources other than links that fail as a group. Thus, the more general term is Shared Risk Group (SRG).)

A common SRLG configuration, known as the fork configuration, occurs when multiple links lie in the same conduit as they exit/enter a node. This is illustrated in Fig. 3.15. The link-level view of the network is shown in Fig. 3.15(a), where it appears Links AB, AD, and AG are mutually disjoint. The fiber-level view is shown in Fig. 3.15(b), where it is clear that these three links lie in the same conduit at Node A. Thus, it would not be desirable to have a protected demand where, for example, one path includes Link AB and the other path includes Link AG, because the common conduit is a single point of failure. To find paths that are truly diverse requires that the SPDP algorithm be modified to account for the SRLGs, as described next.

In SPDP algorithms such as the Bhandari algorithm, the first step is to find the single shortest path from source to destination, which is then used as a basis for a set of graph transformations. If the source or destination is part of an SRLG fork configuration, and one of the links included in the SRLG lies along the shortest path that is found in the first step of the SPDP, then an additional graph transformation such as the one shown in Fig. 3.16 for Node A is required [Bhan99]. A dummy node is temporarily added to the topology and the SRLG links with an endpoint of Node A are modified to have the dummy node as an endpoint. The link metrics are kept the same. Another link, with a metric of zero, is added between Node A and the dummy node. The SPDP algorithm can then proceed. Because of the presence of the added node and link and the fact that disjoint paths are required, the shortest pair of disjoint paths will not include two links from the same SRLG fork configuration.

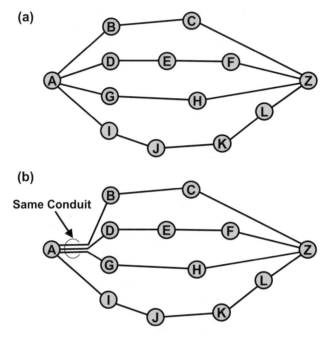

Fig. 3.15 (a) In the link-level view of the topology, Links AB, AD, and AG appear to be diverse. (b) In the fiber-level view, these three links lie in the same conduit exiting Node A, and thus are not diverse. A single cut to this section of conduit can cause all three of these links to fail.

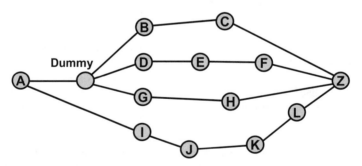

Fig. 3.16 Graph transformation performed on Fig. 3.15 to account for the SRLG extending from Node A. A dummy node is added, and each link belonging to the SRLG is modified to have this dummy node as its endpoint instead of Node A. A link is added between Node A and the dummy node, where this link is assigned a metric of zero.

In addition to links with common conduit resulting in SRLGs, certain graph transformations may produce the same effect. Consider the graph transformation for capturing available regeneration equipment in an optical-bypass-enabled network that was presented in Section 3.5.1. In this transformation, links are added to a transformed graph where the links represent regeneration-free paths in the original graph. Links in the new graph may appear to be diverse that actually are not. This effect is illustrated in Fig. 3.17. The true network is shown in Fig. 3.17(a). Assume that the optical reach is 2,000 km and assume that a protected path is desired between Nodes A and Z. Assume that the nodes with available regeneration equipment are Nodes B, D, H, J, K, L, and M. The transformed graph is shown in Fig. 3.17(b). A link is added to this transformed graph if a regeneration-free path exists in the true network between the link endpoints, and the two endpoints have available regeneration equipment. The link metrics in the transformed graph are based on the number of hops in the regeneration-free path. The next step is to find the shortest pair of diverse paths in the transformed graph, which corresponds to the minimum-regeneration diverse pair of feasible paths in the true network (i.e., feasible with respect to the available equipment).

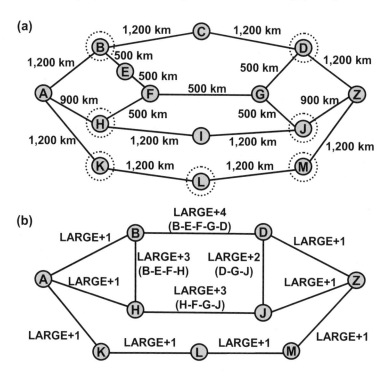

Fig. 3.17 (a) Example network, where the source and destination are A and Z, and the optical reach is 2,000 km. Only Nodes B, D, H, J, K, L, and M have available regeneration equipment. (b) Resulting transformed graph, where links represent regeneration-free paths. Links BD and HJ appear to be diverse but actually represent paths with a common link.

Links BD and HJ appear to be diverse in the transformed graph of Fig. 3.17(b). However, Link BD corresponds to path B-E-F-G-D in the real network and Link HJ corresponds to path H-F-G-J in the real network. Thus, both links correspond to paths that contain the link FG in the real network, implying that Links BD and HJ in the transformed graph comprise an SRLG. This type of SRLG formation, where only the middle portions of the links overlap, is referred to as the bridge configuration. (It is so-named because the configuration occasionally arises in actual network topologies due to a portion of the conduits of multiple links being deployed along the same bridge in order to cross a body of water.) To handle this SRLG scenario, the SPDP algorithm is run twice, each time eliminating one of the links comprising the bridge SRLG. The better result of the two runs is taken as the solution. In this example, this yields A-H-J-Z and A-K-L-M-Z in the transformed graph, corresponding to A-H-F-G-J-Z and A-K-L-M-Z in the real network.

To find the optimal pair of diverse paths in a network with multiple bridge configurations, all possible combinations have to be considered where only one link from each bridge is present. As the number of possibilities grows exponentially, this becomes intractable if the number of bridges is large. Furthermore, there are other classes of SRLGs, though not very common, for which finding an optimal routing solution is difficult, as described in [Bhan99]. In general, there are no computationally efficient algorithms that are guaranteed to find the optimal pair of disjoint paths in the presence of any type of SRLG; thus, heuristics are generally used when 'difficult' SRLGs are present in a network.

One such heuristic to handle arbitrary SRLGs is proposed in [XXQL03]. The first step in this heuristic is to find the single shortest path from source to destination; call this Path 1. The links in Path 1 are temporarily pruned from the topology. Furthermore, any link that belongs to an SRLG to which at least one of the links in Path 1 also belongs is assigned a very large metric to discourage its use. The shortest path algorithm is called again on the modified graph.

This second call may fail to find a diverse path. First, it may fail to find a path because of Path 1's links having been removed from the topology. In this case, the process restarts with a different Path 1 (one could use a K-shortest paths algorithm to generate the next path to use as Path 1). Second, it may fail because the path that is found, call it Path 2, includes a link that is a member of an SRLG that also includes a link in Path 1, in which case Path 1 and Path 2 are not diverse. (While the large metric discourages the use of such links in Path 2, it does not prevent it.) In this scenario, the most 'risky' link in Path 1 is determined; this is the link that shares the most SRLGs with the links in Path 2. The links in Path 1 are restored to the topology, with the exception of the risky link, and the process restarts; i.e., a new Path 1 is found on this reduced topology. The process continues as above, until a diverse pair of paths can be found, or until no more paths between the endpoints remain in the topology (due to risky links being sequentially removed), in which case it fails. If the procedure does find a diverse pair of paths, there is no guarantee that they are the shortest such paths; however, [XXQL03] reports achieving good results using this strategy, with reasonable run-time.

3.6.4 Routing Strategies With Protected Demands

Sections 3.3 and 3.4 covered generating candidate paths and using various routing strategies for unprotected demands (i.e., demands with a single path between the source and destination). In this section, these same topics are revisited, except for protected demands. Above, various methodologies were described to find a pair of disjoint paths between two nodes. These methodologies are used in this section to generate candidate paths. The link metric is again assumed to be unity for O-E-O networks and distance for optical-bypass-enabled networks (as noted previously, an alternative metric that is more closely tied to regeneration is discussed in Chapter 4).

As in Section 3.3, the initial step is to determine the links that are expected to be heavily loaded. This can be done by performing a preliminary design with all demands in the forecasted traffic set routed over their shortest paths, where in general the forecasted traffic will include both unprotected and protected demands. (An alternative means of estimating the critical links for protected traffic, based on maximum-flow theory, is detailed in [KaKL03]; this method can be combined with the traffic forecast to determine the links expected to be the most heavily loaded.) One can then generate a set of candidate paths for the protected demands by using the bottleneck-avoidance strategy, where the most heavily loaded links or sequence of links are systematically removed from the topology, with the SPDP algorithm run on the pruned topology. The goal is to generate a set of lowest-cost (or close to lowest-cost) disjoint path pairs that do not all contain the same expected 'bad' links. Note that a given source/destination pair may support both unprotected and protected demands. A candidate path set is independently generated for each source/destination/protection combination.

As noted in Section 3.2.2, the paths of minimum distance in an optical-bypass-enabled network do not necessarily translate to the paths of minimum regeneration. With single paths, this phenomenon is because regeneration typically must occur in network nodes as opposed to at arbitrary sites along the links (and because some nodes may not be equipped with elements that support optical bypass in all directions). This also applies to determining the regeneration locations for a pair of disjoint paths. Furthermore, the fact that regeneration is determined independently on the disjoint paths may also cause the minimum-distance pair of disjoint paths to have an extra regeneration. In Fig. 3.18, assume that a pair of link-and-node-disjoint paths is required between Nodes A and Z, and assume that the network supports optical bypass with an optical reach of 2,000 km. The shortest-distance pair of disjoint paths, shown in Fig. 3.18(a), is A-B-C-Z and A-D-E-Z. These two paths have a combined distance of 6,700 km, and require a total of three regenerations (at Nodes C, D and E). However, the minimum-regeneration pair of disjoint paths, shown in Fig. 3.18(b), is A-B-Z and A-D-C-Z, which covers a total of 7,000 km, but requires a total of only two regenerations (at Nodes B and D). As this example illustrates, each of the candidate pair of disjoint paths must be explicitly examined to determine the number of required regenerations.

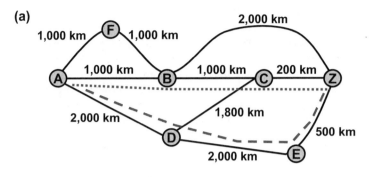

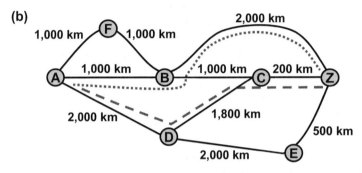

Fig. 3.18 Assume that the optical reach is 2,000 km. (a) For a protected connection from Node A to Node Z, the combination of diverse paths A-B-C-Z and A-D-E-Z is the shortest (6,700 km), but requires three regenerations. (b) The combination of diverse paths A-B-Z and A-D-C-Z is longer (7,000 km), but requires two regenerations. (Adapted from [Simm06]. © 2006 IEEE)

Any of the three routing strategies discussed in Section 3.4 – fixed-path routing, alternative-path routing, and dynamic-path routing – are applicable to protected demands. (Variations of these strategies may be used for demands that share protection bandwidth, as covered in Chapter 7.) As with unprotected demands, alternative-path routing is an effective strategy for routing protected demands in practical optical networks.

3.7 Routing Order

In real-time network planning, demand requests generally are received and processed one at a time. In long-term network planning, however, there may be a set of demands to be processed at once. If the routing strategy used is adaptive, such that the network state affects the choice of route, then the order in which the demands in the set are processed will affect how they are routed. This in turn can affect the cost of the network design, the loading in the network, and the blocking probability. Thus, some attention should be paid to the order in which the demands are processed. (Note that with global optimization techniques such as *integer linear programming*, the order of the demands is not relevant; however, such methodologies often have a very long run-time.)

One common strategy is to order the demands based on the lengths of the shortest paths for the demands, where the demands with longer paths are processed first. The idea is that demands with longer paths are harder to accommodate and thus should be handled earlier to ensure that they are assigned to optimal paths. This criterion can be combined with whether or not the demand requires protection, where protected demands are routed earlier as they require more bandwidth and there is generally less flexibility in how they can be routed.

This scheme often yields better results when combined with a round-robin strategy. Within a given 'round', the ordering is based on the required protection and the path length. However, at most one instance of a particular source/destination/protection combination is routed in each round. For example, if there are two protected demands between Node A and Node B, and three protected demands between Node A and Node C, where the A-B path is longer than the A-C path, plus one unprotected demand between Nodes A and D, then the routing order is: A-B, A-C, A-D, A-B, A-C, A-C.

Another strategy that is compatible with alternative-path routing is to order the demands based on the quality of the associated path set. If a particular source/destination/protection combination has few desirable path options, then the demands between this source and destination with this level of protection are routed earlier. For example, a particular path set may have only one path that meets the minimum cost. It is advantageous to route the associated demands earlier to better ensure that the minimum cost path can be utilized. In addition, the expected load on the links comprising the path set should be considered; the heavier the projected congestion for a particular source/destination/protection combination, the earlier it is routed.

Other ordering strategies are clearly possible. In general, no one strategy yields the best results in all network planning exercises. However, when using alternative-path routing, the routing process is so fast that multiple routing orders can be tested to determine which yields the best results. For example, routing thousands of demands with three different ordering strategies takes on the order of a couple of seconds. This is acceptable for long-term network planning.

3.8 Multicast Routing

Multicast traffic involves one source communicating with multiple destinations, where the communications is one-way. Multicast is also referred to as point-to-multipoint communications, in contrast to a point-to-point connection between a single source and single destination. (A tutorial on multicast routing can be found in [SaMu00].) The need for multicast could arise, for example, if the optical network is being used to distribute video simultaneously to multiple cities. Rather than setting up a separate unicast connection between the source and each of the destinations, where multiple copies of the signal may be transmitted on a link, a multicast tree is constructed to reduce the amount of required bandwidth. The multicast tree connects the source to each of the destinations (without any loops), such that just one copy of the signal is sent on any link. This is illustrated in Fig. 3.19, where Node Q is the source and Nodes W, X, Y, and Z are the destinations. In Fig. 3.19(a), four separate unicast connections are established between Node Q and each of the destinations. Note that four connections traverse the link between Nodes Q and R. In Fig. 3.19(b), a single multicast tree is established, as shown by the dotted line, which requires significantly less bandwidth. Nodes R and T are branching nodes of this multicast tree.

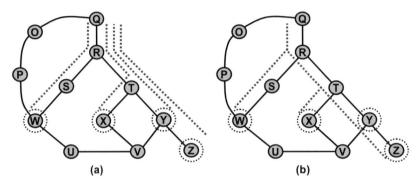

Fig. 3.19 (a) Four unicast connections are established between the source, Q, and the destinations, W, X, Y and Z. (b) One multicast connection is established between Node Q and the four destinations.

A tree that interconnects the source and all of the destinations is known as a Steiner tree (where it is assumed all links are bi-directionally symmetric; i.e., two-way links with the same metric in both directions). The weight of the tree can be considered the sum of the metrics of all links that comprise the tree. Finding the Steiner tree of minimum weight is in general a difficult problem to solve (unless the source is broadcasting to every node in the network); however, several heuristics exist to find good approximate solutions to the problem [Voss92].

The heuristic of [KoMB81] combined with the enhancement of [Waxm88] is presented here, followed by a small example to illustrate the various steps. The original network topology is referred to here as A. The first step is to create a new topology, B, composed of just the source and destination nodes. All of the nodes are fully interconnected in B, where the metric of the link connecting a pair of nodes equals the metric of the shortest path between the two nodes in A. A minimum spanning tree is found on B using an algorithm such as Prim's [CoLR90]. (A minimum spanning tree is a tree that touches all nodes in the topology where the sum of the metrics of the links comprising the tree is minimized. This is different from a minimum Steiner tree in that *all* nodes are included, and is easier to solve.) Each link in the resulting minimum spanning tree, where a link connects two nodes *i* and *j*, is expanded into the shortest path in the true topology between nodes *i* and *j* (e.g., if the shortest path in the true topology between nodes *i* and *j* has three links, then the link between nodes *i* and *j* in the new topology is replaced by the three links). Call the resulting topology *B'*. Another new topology is then formed, *C*, composed of all of the nodes in *B'*, with all of the nodes fully interconnected. The operations performed on topology *B* are repeated for topology *C*, resulting in the topology *C'*. A minimum spanning tree is then found on *C'*. Any links in the resulting tree that are not needed to get from the source to the destinations, if any, are removed, leaving the approximation to the minimum Steiner tree.

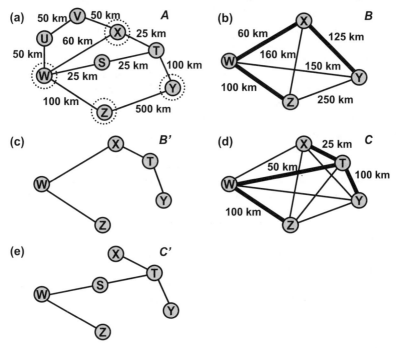

Fig. 3.20 (a) Original topology *A*, where the source and destination nodes are circled. (b) Topology *B* formed by interconnecting all source and destination nodes. (c) The minimum spanning tree on *B* expanded into paths, forming topology *B'*. (d) Topology *C* formed by interconnecting all nodes of *B'*. (e) The minimum spanning tree on *C* expanded into paths, forming topology *C'*.

This heuristic is illustrated with the small example of Fig. 3.20. Using the notation from above, the topology shown in Fig. 3.20(a) is the *A* topology. The source is Node W and the multicast destinations are Nodes X, Y, and Z. (The procedure is the same regardless of which of the four nodes is considered the source.) A fully interconnected topology composed of these four nodes is shown in Fig. 3.20(b). This is the *B* topology. The link metrics are based on distance here; e.g., the metric of Link WY, 150 km, is the shortest distance between Nodes

W and Y in the *A* topology. The minimum spanning tree on *B* is shown by the thick lines. Each link in this tree is expanded into its associated shortest path in the original topology. This produces the *B'* topology shown in Fig. 3.20(c). For example, Link XY in *B* is expanded into X-T-Y in *B'*. A new fully interconnected topology is formed with the five nodes of *B'*, as shown in Fig. 3.20(d). This is the *C* topology, where the minimum spanning tree is shown by the thick lines. The links of the tree are expanded into their associated shortest paths to form the *C'* topology, shown in Fig. 3.20(e). This topology is already a minimum spanning tree, thus, no further operations are needed, leaving *C'* as the final solution. In this example, the solution is optimal but that is not always the case.

A heuristic such as the one described can be used to generate an approximate minimum-cost multicast tree, where the link metric is set to unity for O-E-O networks and is set to link distance for optical-bypass-enabled networks. If it is known in advance the possible sets of nodes that will be involved in multicast demand requests, then one can pre-calculate the associated multicast trees. Furthermore, alternative multicast trees can be generated for each set of nodes where certain bottleneck links are avoided in each tree. The best tree to use would be selected at the time the multicast demand request arrives. If the multicast groups are unknown in advance, then the tree can be generated dynamically as the requests arrive. If it is necessary to add nodes to an existing multicast group, then greedy-like heuristics can be used to determine how to grow the tree [Waxm88].

As with unicast connections, the actual amount of regeneration in an optical-bypass multicast tree will likely depend on factors other than distance. For example, the branching points in the tree may be favored for regeneration. Refer to Fig. 3.21, which shows just the links included in the multicast tree of example Fig. 3.19, where the source is Node Q and the multicast destinations are Nodes W, X, Y, and Z. Assume that the optical reach is 2,000 km. In Fig. 3.21(a), the signal is regenerated at the furthest possible node from the source without violating the optical reach. This results in regenerations at Nodes S and T. If, however, the regeneration occurs at the branching point Node R as in Fig. 3.21(b), then no other regeneration is needed.

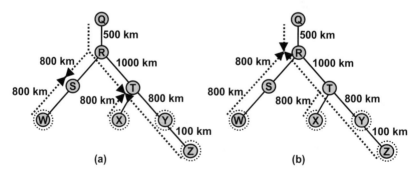

(a)   (b)

Fig. 3.21 Assume that the optical reach is 2,000 km. (a) Regeneration at both Nodes S and T. (b) Regeneration only at Node R.

An all-optical switch with multicast capability (Section 2.6.3) is very useful in a multicast tree. At a branching node where regeneration does not occur, e.g., Node T in Fig. 3.21(b), the switch can be used to all-optically multicast the signal onto the outgoing branches (the all-optical switch includes amplification such that the outgoing power level is not cut in half). Second, at a node with regeneration, e.g., Node R in the same figure, the all-optical switch can be used to multicast the regenerated signal onto both Links RS and RT. Finally, the drop-and-continue feature of the all-optical switch can be used at Node Y, where the signal drops and continues on to Node Z. (An OAMD-MD with multicast capability can perform the first and third of these functions. To allow multicast after a regeneration, a flexible regenerator card can be used, as is described in Section 4.5.2.)

3.8.1 Multicast Protection

Consider providing protection for a multicast connection such that the source remains connected to each of the destinations under a failure condition. One simplistic approach is to try to provision two disjoint multicast trees between the source and the destinations. However, because of the number of links involved in a multicast tree, it is often difficult, if not impossible, to find two completely disjoint trees.

Another approach is to find the minimum-cost ring path that traverses the source and each of the destinations [RaEl05]. The source can communicate to any destination using either the clockwise or counter-clockwise direction of the ring; thus, the source and each destination remain connected under any single failure. This idea is generalized to more complex protection topologies in [MFBG99].

A third approach is to arbitrarily order the destinations, and then sequentially find a disjoint pair of paths between the source and each destination, using an SPDP algorithm [SiSM03]. When looking for a disjoint pair of paths between the source and the $(i+1)^{st}$ destination, the links that comprise the path pairs between the source and the first i destinations are favored. The resulting sub-graph, which may not have a tree structure, provides protection against any single failure.

3.9 Routing with Inaccurate Information

The above sections on routing have presented numerous schemes for selecting a good path for each incoming demand. In long-term network planning, these strategies can be encompassed in a design tool that is run in a centralized location such as an operations center. In an on-line system, where the connection set-up time may play an important role in meeting customer requirements, an important consideration is where to locate the routing decision machinery. One strategy is to direct all incoming demand requests to a centralized controller that tracks the utilization of network resources. For example, the IETF has defined a Path Computation Element (PCE)-based architecture, where the PCE is a powerful computing platform that is capable of performing constraint-based routing [FaVA06]. In one model, the PCE is located at a centralized node or server, and all routing requests are directed to it.

While centralized routing is advantageous from the standpoint of operating from a consistent picture of the network, this approach may incur excessive signaling delays in transmitting the request to the centralized location and disseminating the set-up information to the appropriate nodes. Furthermore, if the rate of demand requests is too high, the centralized controller could become a computational bottleneck.

Decentralized route computation is also possible. For example, in the IETF model, there could be multiple active PCEs in a network, where each PCE handles the route requests from a particular network region or from a particular domain. Alternatively, multiple PCEs, one per domain, could work together to select a multi-domain end-to-end route. In an extreme decentralized approach, each node is capable of calculating a route; e.g., the source node of a new demand selects the path to use. While such a distributed route selection process may be more responsive, this approach requires state information to be maintained by each of the network nodes.

The distributed routing process may be complicated by the inaccuracy of the state information. For example, consider the scenario where all nodes are capable of selecting a route. There are a variety of factors that can result in stale state information existing at one or more nodes of a dynamic network. First, updates on the current state may be distributed throughout the network only at certain time intervals as opposed to whenever any network change occurs, in order to reduce the amount of data that is flooded in the network. There is also a delay from the time the information is initially flooded until the time it is received by all of the relevant network nodes. Furthermore, in a large network, it may be desirable to distribute only summarized information about different portions of the network as opposed to the full details regarding resource utilization. All of these factors contribute to the nodes having to select a route with less than full knowledge of the current network state.

Routing with inaccurate information may require adding a probabilistic component to the routing strategy. For example, consider a relatively heavily loaded network where a subset of the links and nodes have little or no available resources. It is assumed that the state information used in the routing process may be somewhat stale such that the non-availability of a link or node is not known with certainty. In this scenario, it may be beneficial to search for a path based on the probability of it being feasible [GuOr99]. After performing a graph transformation as described in Section 3.5.1 to better model the network resources, each link in the transformed topology can be assigned a probability that it is available to carry the new demand. This probability can be estimated using, for example, the information regarding network resources (which may be out of date) and knowledge of the rate of resources being requested in certain areas of the network based on prior history. The probability of a path being feasible can be estimated as the product of the probabilities assigned to each of the links comprising the path (resource usage on the links is unlikely to be completely independent such that the product of the probabilities is only an estimate of the probability of feasibility for the whole path). By taking the link metric to be the negative of the logarithm of its associated probability, and running a shortest path algorithm, the path with the highest estimated probability of being feasible is found.

Other methodologies for optical-bypass-enabled networks not only consider the number of wavelengths free on the links, but also consider which particular wavelengths are free to account for the wavelength continuity constraint (e.g., [MSSD03]).

Distributed routing schemes must also deal with resource contention due to concurrent demand requests being handled by multiple nodes. A path could be selected by a node that requires a particular wavelength on a given link only to find that by the time the connection establishment is initiated, a demand sourced by another node has also reserved that wavelength. One method of handling this complexity is through *contention resolution* schemes [GoLY02]; i.e., after the contention has occurred, an agreed upon rule is invoked to resolve the deadlock. For example, the GMPLS signaling specification stipulates that such contention be resolved based on the IDs of the nodes that are involved [Berg03]. Alternatively, *contention avoidance* methodologies can be implemented such that contention does not occur, or at least is minimized; this is generally preferable as it can result in smaller delays for time-critical functions such as dynamic failure restoration [GoLY02]. For example, various contention-avoidance schemes have been proposed to reduce the probability of reserving the same wavelength on the same link for two concurrent demand requests [ADHN01, OzPJ03, LiWM07]. Note that the development of intelligent reservation schemes that can be used to minimize contention without unduly blocking other demands is still an area of current research.

Chapter 4

Regeneration

Using the routing techniques of Chapter 3, a path is selected for each demand request entering the network. The next step in the planning process is selecting the regeneration sites for the demand, if any. (Chapter 5 considers techniques for accomplishing routing, regeneration, and wavelength assignment in a single step.) Regeneration 'cleans up' the optical signal, typically by reamplifying, reshaping, and retiming it; this is referred to as '3R' regeneration. As discussed in Chapter 3, paths are usually selected to minimize the amount of required regeneration, as regeneration adds to the network cost.

If the network is based on O-E-O technology, then determining the regeneration locations for a selected path is straightforward because the demand is regenerated at every intermediate node along the path. For an optical-bypass-enabled network, where it is possible to transit an intermediate node in the optical domain depending on the quality of the optical signal, determining the regeneration locations for a path may be more challenging. Numerous factors affect when an optical signal must be regenerated, including the underlying transmission technology of the system, the properties of the network elements, and the characteristics of the fiber plant on which the system is deployed. When these factors are considered together, one can estimate the nominal distance over which an optical signal can travel before requiring regeneration (i.e., the optical reach). However, while quoting the optical reach in terms of physical distance is expedient for summarizing the system performance, it is not sufficient for determining the necessary regeneration locations in an actual network. Many factors other than distance need to be considered.

Section 4.1 presents a high-level discussion of some of the optical impairments and system properties that have an impact on when a signal must be regenerated. Ultra-long-reach technology has succeeded because it is possible to mitigate many of these impairments or design a system such that their effects are negligible. In Section 4.2, the discussion focuses on one of the more important impairments, noise, which leads to a metric that can be used in the routing process instead of distance to improve the likelihood of finding minimum-regeneration paths.

J.M. Simmons, *Optical Network Design and Planning*,
DOI: 10.1007/978-0-387-76476-4_4, © Springer Science+Business Media, LLC 2008

Given that regeneration, and thus implicitly network cost, are dependent on the physical layer design, it may be beneficial to integrate algorithms for physical network design (e.g., algorithms to optimize the amplifier configuration) with the architectural design and planning tool. Some of the benefits of this approach are discussed in Section 4.3.

In Section 4.4, the discussion moves from the physical layer aspects of regeneration to the architectural facets. Several regeneration strategies are presented, where the tradeoff is between operational simplicity and cost. The strategy employed will likely affect how much regeneration is required and where a connection must be regenerated; thus, it is an important aspect of the network design.

Finally, in Section 4.5, different options for actually implementing regeneration in a node are presented. Again, there is a tradeoff of operational flexibility versus cost. While most of this chapter is relevant only to optical-bypass-enabled networks, much of this final section applies to O-E-O networks as well.

There are a few key points to be emphasized in this chapter. First, while having to encompass physical layer phenomena in a network planning tool may seem imposing, there are known methodologies for tackling this problem that have been successfully implemented in live networks. Furthermore, when optical-bypass technology is developed by system vendors, there needs to be close collaboration between the systems engineers and the network architects. If the requirements of the physical layer are so exacting that the ensuing complexities in the network planning tool are unmanageable (e.g., if the wavelengths have to be assigned in a precise order on all links), then it may be necessary to modify the technological approach. Finally, while occasional regeneration may appear to be undesirable because of its cost, it does provide an opportunity to change the wavelength on which an optical signal is carried. This can be advantageous in the wavelength assignment step of the planning process, which is covered in Chapter 5.

4.1 Factors That Affect Regeneration

4.1.1 Optical Impairments

One of the major impairments that an optical signal encounters is accumulated noise. The principal source of noise is the spontaneous emissions of optical amplifiers, which are amplified along with the signal as they propagate together through the network. (This is suitably referred to as *amplified spontaneous emissions (ASE) noise.*) The strength of the signal compared to the level of the noise is captured by the signal's *optical-signal-to-noise-ratio (OSNR)*, where signals with lower OSNR are more difficult to receive without errors.

Many other optical impairments arise from the physical properties of light propagating in a fiber. For example, the propagation speed of light within a fiber depends on the optical frequency. This causes the optical signal pulses, which have a finite spectral width, to be distorted as they propagate along a fiber (generally, the effect is that the pulses spread out in time). This phenomenon is known as *chromatic dispersion*, or simply *dispersion*. Dispersion accumulates as a linear function of the propagation distance. Higher-bit-rate signals, where pulses are closer together, are more susceptible to errors due to this effect; the amount of tolerable dispersion decreases with the square of the bit-rate.

A different type of dispersion, known as *polarization-mode dispersion* (PMD), stems from different light polarizations propagating in the fiber at different speeds. In simplistic terms, an optical signal can be locally decomposed into two orthogonal principal states of polarization, each of which propagates along the fiber at a different speed, resulting in distortion. PMD accumulates as a function of the square root of the propagation distance. As with chromatic dispersion, PMD is a larger problem as the bit-rate of the signal increases; the PMD-limited reach decreases with the square of the bit-rate.

Several nonlinear optical effects arise as a result of the fiber refractive index being dependent on the optical intensity. (The refractive index governs the speed of light propagation in a fiber.) As the optical signal power is increased, these nonlinearities become more prominent. One such nonlinearity is *self-phase modulation* (SPM), where the intensity of the light causes the phase of the optical signal to vary with time. This potentially interacts with the system dispersion to cause significant pulse distortion. *Cross-phase modulation* (XPM) is a similar effect except that it arises from the interaction of two signals, which is more likely to occur when signals are closely packed together in the spectrum. Another nonlinear effect is *four-wave mixing* (FWM). This arises when signals carried on three particularly spaced optical frequencies interact to yield a stray signal at a fourth frequency, or two frequencies interact to generate two stray signals. These stray signals can potentially interfere with desired signals near or at these frequencies.

There are various other mechanisms of fiber nonlinearities that can distort the signal. For more detailed treatment of optical impairments, see [FoTC97, BaKi02].

4.1.2 Mitigation of Optical Impairments

There are well known techniques for mitigating some of these optical impairments. For example, dispersion compensation is typically used to combat chromatic dispersion, where the level of compensation needed depends on the amount of dispersion in the transmission fiber and the dispersion tolerance of the system. Note that it is not desirable to reduce the dispersion to zero, as its presence helps reduce some of the harmful nonlinear effects [Kurt93, TkCh94, TCFG95]. In one commonly used technique, dispersion compensating fiber (DCF) having inverse dispersion relative to the transmission fiber is installed at various sites along each link. DCF is generally expensive, high loss, and provides only a static means of compensation. Further problems with DCF may result from the dispersion level of the transmission fiber not being constant across the transmission band; typically, the dispersion level of the fiber has a particular slope across the band. The DCF may not have precisely the same inverse dispersion slope, leading to different levels of residual dispersion depending on the transmission wavelength.

As networks evolve, electronic dispersion compensation (EDC) is likely to be used as an enhancement to, or a replacement of, the DCF strategy [KaSG04]. EDC can be deployed on a per-wavelength basis (e.g., as part of the WDM transponder), and can be dynamically tuned over a range of dispersion levels to better match the compensation requirements of a given connection. For example, receivers based on maximum likelihood sequence estimation (MLSE) are an active area of research as a means of combating chromatic dispersion, as well as possibly other impairments [CaCH04, ChGn06]. (MLSE operates on a sequence of bits rather than a single bit at a time, and selects the data sequence that is statistically most likely to have generated the detected signal.) In another EDC strategy, pre-compensation is used at the transmitter, based on feedback from the receiver [MORC05]. Alternatively, post-compensation can be implemented at the receiver, which may be more suitable for dynamic networking.

PMD compensation is more challenging because the level of PMD may vary as a function of time. Cost-effective adaptive PMD compensators are an active area of research. There has typically been less of a need for PMD compensation compared to dispersion compensation. However, as wavelength rates continue to increase, it may require more attention, especially on older fibers. New fiber types being developed tend to have very low PMD.

Many of the problems from nonlinear effects can be avoided by maintaining the signal power at a low enough level (but still sufficiently higher than the noise level). In addition, as mentioned above, maintaining a small amount of residual system dispersion can be effective in reducing some of the nonlinear effects.

4.1.3 Network Element Effects

In addition to impairments that accumulate due to propagation in a fiber, there are a number of potential deleterious effects that a signal may suffer when transiting an optical-bypass-enabled network element. For example, a network element may utilize optical filters to internally separate the wavelengths entering from a WDM network port. Each time a signal passes through such a filter, the bandwidth of the channel through which the signal propagates 'narrows' to some degree, distorting the signal. Another source of signal degradation is the crosstalk caused by 'leakage' within a switching element. This occurs when a small portion of the input signal power appears at outputs other than the desired output. Additionally, the optical loss of a network element may depend on the state of polarization of the signal; this is known as *polarization dependent loss (PDL)*. Since the signal polarization may vary with time, the loss may also vary over time, which is undesirable. Furthermore, the network element may contribute to dispersion, and the dispersion level may not be flat across the transmission band, making compensation more difficult.

Factors such as filter narrowing, crosstalk, PDL, and dispersion contribute to a limit on the number of network elements that can be optically bypassed before needing to regenerate the signal. However, as optical-bypass technology has matured, the performance of the network elements has significantly improved. Many commercial systems support optical bypass of up to 10 (backbone) or 16 (metro) network elements prior to requiring regeneration. With this capability, the number of network elements bypassed is not usually the limiting factor in determining where regeneration must occur, especially in backbone networks where the distance between nodes may be very long; i.e., other limiting effects 'kick-in' prior to ten nodes being traversed. (However, the network elements do have an impact on the OSNR, as is discussed in Section 4.2.1.)

4.1.4 Transmission System Design

The characteristics of the transmission system clearly influence the optical reach of the system. One of the most important system design choices is the type of amplification, where Raman technology is generally used to attain an extended optical reach. Distributed Raman amplification uses the fiber itself to amplify the optical signal, so that the rate of OSNR degradation is less steep as compared to EDFA amplification. This trend is illustrated in Fig. 4.1, which depicts the OSNR level as a function of transmission distance, in a hypothetical system, for both distributed Raman and lumped EDFA amplification (lumped indicates the amplification occurs only at the amplifier sites).

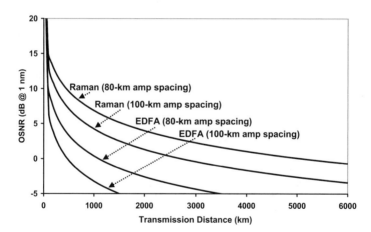

Fig. 4.1 OSNR as a function of transmission distance for distributed Raman amplification and for lumped EDFA amplification, for both 80-km and 100-km amplifier spacings. The OSNR degrades more slowly with Raman amplification. For a given amplification type, the OSNR degrades more slowly with amplifiers spaced closer together. Many carrier networks have an amplifier spacing on the order of 80 km; however, there are some carrier networks with an average amplifier spacing closer to 100 km.

The acceptable OSNR level depends on the receiver sensitivity (i.e., the minimum average optical power necessary to achieve a specified bit error rate [RaSi01]) and the desired system margin (a network is typically designed to initially perform better than the minimum acceptable level to account for degradation as the system ages). As shown in Fig. 4.1, for a desired level of OSNR at the receiver and for a given amplifier spacing, Raman amplification supports a significantly longer transmission distance. More details on Raman technology can be found in [RoSt02].

Another important design choice is the signal modulation format, which is the format used for coding the data on a lightstream. A common modulation scheme is simple on-off keying (OOK), where the presence of light indicates a '1' and the relative absence of light indicates a '0'. Two common signal formats used in conjunction with OOK are non-return-to-zero (NRZ) and return-to-zero (RZ), as shown in Fig. 4.2. The RZ format typically allows a longer optical reach. There are other formats of greater complexity (e.g., multi-level amplitude and/or phase modulation) that offer different tradeoffs in capacity, reach, complexity, and cost (e.g., see [Conr02]).

In addition to amplification and modulation, there are many other important system properties that affect optical reach, including: the spacing between channels (e.g., closer spacing reduces the reach); the initial launched powers of the optical signals (increasing the launched power increases the optical reach, up to a point; however, if the signal power is too high, the nonlinear optical impairments will have a negative impact on the reach, implying that there is an optimum value, which is system dependent); and the FEC coding strength (the stronger the FEC code, the greater its ability to detect and correct errors, which allows a longer reach; FEC was discussed in Section 2.9).

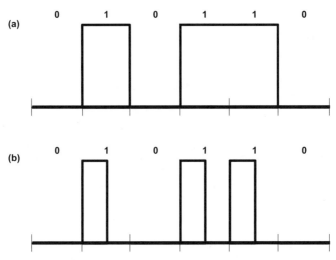

Fig. 4.2 (a) Non-Return-to-Zero (NRZ) signal format. (b) Return-to-Zero (RZ) signal format.

4.1.5 Fiber Plant Specifications

The characteristics of the physical fiber plant also have a large impact on system regeneration. Often, the optical reach of a given system depends on the fiber plant on which it is installed. For example, if the amplifier huts are spaced further apart, then the OSNR degrades more quickly leading to more frequent regeneration. This is illustrated in Fig. 4.1, where the OSNR decreases more sharply for 100-km amplifier spacing as compared to 80-km spacing, for a given amplification type.

The type of fiber in the network may have an impact as well, where the most commonly used fiber types are classified as either *non dispersion-shifted fiber* (NDSF) or *non-zero dispersion-shifted fiber* (NZ-DSF). (SMF-28® is an example of NDSF fiber; LEAF® and TrueWave® REACH LWP are examples of NZ-DSF fiber[8].) As the names imply, these classes of fiber differ in the dispersion level in the portion of the spectrum occupied by WDM systems, thereby requiring different levels of dispersion compensation and having different mitigating effects on the nonlinear optical impairments. (For example, SMF-28 fiber has roughly four times the level of dispersion as LEAF fiber in the WDM region of interest.) In addition to dispersion differences, fiber types may differ in their Raman gain efficiency, where higher efficiency may lead to longer reach.

4.1.6 System Regeneration Rules

As the above discussion indicates, there are numerous factors to take into account when determining where an optical signal needs to be regenerated. Every vendor has a different approach to the problem, such that there is no uniform set of rules for regeneration across systems. It is generally up to the individual vendor to analyze their own particular implementation and develop a set of rules that govern routing and regeneration in a network. While there are many potential aspects to consider, in some systems it is possible to come up with a minimal set of rules that are sufficient for operating a network in real-time. For example, the system rules may be along the lines of: regenerate a connection if the OSNR is below N, or if the accumulated dispersion is above D, or if the accumulated PMD is above P, or if the number of network elements optically bypassed is greater than E (where N, D, P, and E depend on the system). When determining these rules, vendors usually factor in a system margin to account for aging of the components, splicing losses (i.e., optical losses that arise when fiber cuts are repaired), and other effects. Furthermore, it is important that the rules be immune to the dynamics of the traffic in the network. Connections are constantly brought up and down in a network, either due to changing traffic patterns or due to the occurrence of, or recovery from,

---

[8] SMF-28 and LEAF are registered trademarks of Corning Incorporated; TrueWave is a registered trademark of Furukawa Electric North America, Inc.

failures; thus, it is important that the regeneration rules be independent of the number of active channels on a fiber.

While this may appear to be quite challenging, it is important to point out that there are optical-bypass-enabled networks with optical reaches over 3,000 km that have operated over a number of years using a set of relatively simple rules. It may be necessary to sacrifice a small amount of optical reach in order to come up with straightforward rules; however, as is examined more fully in Chapter 8, the marginal benefits of increasing the optical reach beyond a certain point are small anyway.

As pointed out in Section 4.1.5, the performance of a system depends on the characteristics of the fiber plant in the carrier's network. In the early stages of evaluating a system for a given carrier, prior to any equipment deployment, the exact specifications of the fiber plant may not be known. Thus, in the beginning stages of network design, one may have to rely on selecting regeneration locations based on path distance, where the optical reach in terms of a nominal distance is used. For purposes of system evaluation and cost estimation, this is generally acceptable. After the fiber spans have been fully characterized, the network planning tools can implement the more precise system regeneration rules.

4.2 Routing with Noise Figure as the Link Metric

Consider a system where the signal power levels are low enough that nonlinearities can be neglected (this is often the case in Raman-based systems). Assume that the fiber PMD is very low, such that the most important factors to consider with respect to regeneration are the OSNR, chromatic dispersion, and the number of network elements optically bypassed (due to the factors discussed in Section 4.1.3). The focus in this section is on OSNR; the other two factors are considered in Section 4.3.1 to ensure a cohesive system design.

As an optical signal propagates down a fiber link, its OSNR degrades. The *noise figure* (NF) of a link is defined as the ratio of the OSNR at the start of the link to the OSNR at the end of a link; i.e.:

$$NF_{Link} = \frac{OSNR_{Link\ Start}}{OSNR_{Link\ End}} \qquad (4.1)$$

The noise figure is always greater than or equal to unity. Low noise figure is desirable as it indicates less signal degradation. Noise figure is a quantity that can be measured in the field for each link once the amplifiers have been deployed.

Numerous factors affect the noise figure of a link. For example, the type of amplification is very important, where Raman amplification generally produces a lower noise figure than EDFA amplification. Large fiber attenuation or large splicing losses can contribute to a higher noise figure. Longer span distances (i.e., the distances between amplifier huts) also generally increase the noise figure.

OSNR is important in determining when an optical signal needs to be regenerated. Transponder receivers generally have a minimum acceptable OSNR threshold, below which the signal cannot be detected properly. Thus, the evolution of OSNR as the signal traverses an optical path is critical. Consider the two consecutive links shown in Fig. 4.3, where each link i has an associated noise figure, NF_i, and a net gain, G_i. Net gain refers to the total amplification on the link compared to the total losses. The cumulative noise figure for an optical signal traversing Link1 followed by Link 2 is given by:

$$NF_{Total} = NF_1 + \frac{(NF_2 - 1)}{G_1} \tag{4.2}$$

In most systems, the net gain on a link is unity, because the total amplification is designed to exactly cancel the total loss. In addition, typical values for the link noise figure are in the hundreds (using linear units), such that the '1' term is negligible. The formula then simplifies to:

$$NF_{Total} \approx NF_1 + NF_2 \tag{4.3}$$

Extending this formula to multiple links, the noise figure of an end-to-end path is the sum of the noise figures on each link of the path, where it is desirable to minimize this sum. Link noise figure is thus a suitable additive link metric that can be used in the shortest path algorithms of Chapter 3. Using this as a metric yields the path with least noise figure, or equivalently, the path with the highest overall OSNR (assuming other impairments are properly managed). Noise figure may be a better metric than distance in finding paths that minimize the number of required regenerations. However, it is still true that the minimum-noise-figure path may not be the minimum-regeneration path, e.g., due to regenerations occurring only at network nodes. Thus, in the process of generating candidate paths, as described in Chapter 3, each path must be evaluated to determine the actual regeneration locations, to ensure minimum-regeneration paths are chosen.

When working with the formulas for noise figure, it is important to use the correct units. The noise figure of a link is typically quoted in decibels (dB). However, the additive formula above requires that the noise figure be in linear units. The following formula is used to convert from dBs to linear units:

$$Linear\ Units = 10^{Decibel\ Units/10} \tag{4.4}$$

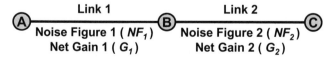

Fig. 4.3 Two consecutive links, each with their respective noise figures and net gains.

(Note: If the system is more complex such that OSNR is not the primary determinant of where to regenerate, then one can use a performance factor, which is referred to as the Q-factor, as the link metric [KTMT05].)

4.2.1 Network Element Noise Figure

In addition to each link having a measurable noise figure, the nodal network elements contribute to OSNR degradation as well. Thus, each network element, such as an OADM or an all-optical switch, has an associated noise figure. To account for this effect in the routing process, the link metric should be adjusted based on the type of equipment deployed at either end of the link. (This is simpler than modeling the network elements as additional 'links' in the topology.) One strategy is to add half of the noise figure of the elements at the endpoints to the link noise figure. (This adjustment is used only for routing purposes; it does not imply that the noise figure of the element add/drop path is half that of the through path.)

Consider Link 2 shown in Fig. 4.4, which is equipped with an OADM at one end and an OADM-MD at the other end. Assume that the noise figure of the OADM is about 16 dB and the noise figure of the OADM-MD is about 17 dB. Halving these values yields roughly 13 dB and 14 dB, respectively (subtracting 3 dB is roughly equivalent to dividing by two). These amounts should be added to the noise figure of Link 2, where the additions must be done using linear units. For example, if Link 2 has a noise figure of 25 dB, then it should be assigned a link metric of: $10^{25/10} + 10^{13/10} + 10^{14/10} = 361.3$. By adding half of the element noise figure to each link connected to the element, the full noise figure of the element is accounted for regardless of the direction in which the element is traversed. (If, instead, the full amount of the element noise figure were added to each link entering the element, then the element noise figure would be double-counted along a path. If the full amount of the element noise figure were added to just one link entering the element, then some paths may not count the element penalty at all; e.g., if the penalty of the OADM-MD were added to Link 2 only, then a path from Link 3 to Link 4 would not include any OADM-MD penalty.)

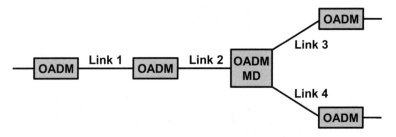

Fig. 4.4 The noise figure of each link needs to be adjusted to account for the noise figure of the network elements at either end of the link. For example, the noise figure of Link 2 is incremented by half of the noise figure of an OADM and half of the noise figure of an OADM-MD.

4.2.2 Impact of the OADM without Wavelength Reuse

There is one network element that warrants special consideration with respect to noise and regeneration: the OADM that does not have wavelength reuse (see Section 2.4.3). Such elements are typically little more than an optical amplifier with a coupler/splitter for adding/dropping traffic at the node. A simplified illustration of such an OADM is shown in Fig. 4.5. An optical signal added at this node is coupled to the light passing through the node from the network links. Assume that the added signal in the figure is carried on λ_1 and assume that it is sent out on the East link. While there is no signal at λ_1 entering the OADM from the West link, the noise in the λ_1 region of the spectrum has propagated down the fiber and undergone amplification along with the rest of the spectrum. This noise will be coupled with the signal being added on λ_1 at the node. From an OSNR perspective, it appears as if the added signal has effectively been transmitted over some distance, thus affecting where it needs to be regenerated.

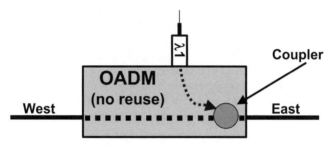

Fig. 4.5 A connection carried on $\lambda 1$ is added at a node equipped with an OADM without wavelength reuse. The added signal is combined with the WDM signal entering the node from the West link. The noise in the $\lambda 1$ region of the spectrum from the West link is added to the new connection's signal.

The effect on regeneration is illustrated more clearly in Fig. 4.6. Node B in the figure is equipped with an OADM without wavelength reuse, whereas the remaining nodes have an OADM with reuse. Assume that the nominal optical reach of the system is 2,000 km (for simplicity, we will still discuss reach in terms of a distance), and assume that a connection is established from Node B to Node Z. As the path distance from B to Z is 1,500 km, it would appear that no regeneration is required. However, the connection added at Node B has accumulated noise as if it originated at Node A. Thus, from an OSNR perspective, it is equivalent to a connection from Node A to Node Z, which covers 2,500 km. Therefore, a regeneration is required at Node C in order to clean up the signal.

This same effect does not occur with elements that have wavelength reuse because such elements are equipped with a means of blocking any wavelength along with the noise in the region of the spectrum around it. Thus, the noise is not combined with a wavelength being added at the node.

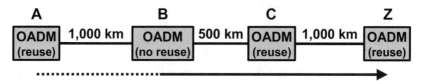

Fig. 4.6 Assume that the optical reach is 2,000 km. The OADM at Node B does not have wavelength reuse. A signal from Node B to Node Z has a level of noise as if it originated at Node A, and thus, needs to be regenerated at Node C.

Note that regenerating a connection at a node with a no-reuse OADM is not desirable. Consider the setup shown in Fig. 4.7, where Node C has a no-reuse OADM and the remaining nodes have OADMs with reuse. Assume that the optical reach is 1,000 km, and assume that a connection between Nodes A and Z is launched from Node A using wavelength λ_1. If this connection is dropped at Node C for regeneration, λ_1 will continue on through the node because of the inability of the OADM at Node C to block the wavelength. Thus, after the signal is regenerated at Node C, it must be relaunched on a different wavelength, say λ_5. This one connection will then 'burn' two wavelengths on the link between Nodes C and D (λ_1 and λ_5). Furthermore, because of noise being combined with the added signal, as described above, the signal on λ_5 will have a noise level as if it had been added at Node B. The added noise will cause the signal to require regeneration at Node D. Thus, the regeneration at Node C is not effective because one regeneration at Node D would have been sufficient for the whole end-to-end path. Additionally, the amount of add/drop supported at a no-reuse OADM is usually very limited. Using it for regeneration may prevent the node from sourcing/terminating additional traffic in the future. For these reasons, regeneration is generally not recommended at a node with a no-reuse OADM.

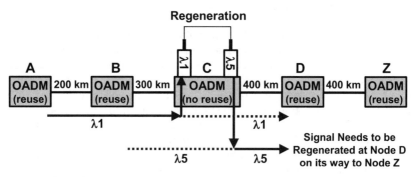

Fig. 4.7 The desired connection is between Nodes A and Z. Assume that the optical reach is 1,000 km. If the connection is regenerated at Node C, which has a no-reuse OADM, it will need to be regenerated again at Node D.

4.3 Link Engineering

Link engineering (or span engineering) is the process of designing the physical in-frastructure for a particular network. For example, this may include determining the amplifier type for each site (e.g., pure Raman or hybrid Raman/EDFA), the gain setting of each amplifier, the amount of dispersion compensation, and the lo-cations of the dispersion compensation. Algorithms specifically designed to opti-mize the system performance can be very helpful in this process. For example, *simulated annealing* [LaAa87] has proved to be one good technique for selecting the locations of the dispersion compensation modules. For a given set of amplifi-ers, proper setting of the gain and proper distribution of dispersion compensation can provide extra system margin (say on the order of 0.25 to 0.5 dB).

Clearly, the design choices in the physical layer have an impact on the network design at higher layers. It can be very advantageous to have a unified network de-sign tool that incorporates the physical-layer design. For example, consider a sys-tem that has two types of amplifiers, where the Type A amplifier provides greater gain than the Type B amplifier, but it costs more. Installing the Type A amplifier as opposed to the Type B amplifier in certain sites would reduce the noise figure of the corresponding links. However, the effect on the overall amount of regen-eration in the network may be small. By integrating physical-layer design with network planning, one can evaluate whether the extra cost of the Type A amplifier is justified by the expected reduction in regeneration. This allows better perform-ance and cost optimization of the network as a whole.

4.3.1 Cohesive System Design

In Section 4.2, a network was considered where the major factors with respect to regeneration are assumed to be the OSNR, dispersion, and the number of network elements optically bypassed. (It is assumed that the OSNR of the network ele-ments is taken into account in the system OSNR allowance.) In that section, the focus was on the OSNR, where routing is performed using noise figure as the link metric. Here, the dispersion and the network elements are considered to ensure there is a cohesive system design.

Assume that based solely on the OSNR analysis, it is anticipated that the opti-cal reach of a system is on the order of D km. It is beneficial to co-ordinate the dispersion management and the network element performance to align with this number. For example, if the OSNR forces a path to be regenerated after roughly D km, then there is no need to add dispersion compensation suitable for transmis-sion well beyond D km.

To be more concrete, assume that, based on the OSNR analysis, the optical reach is nominally on the order of 2,000 km. In addition, assume that the disper-sion tolerance of the system is 10,000 ps/nm; i.e., after accumulating this much dispersion, a signal needs to be regenerated. Assume that the fiber has a dispersion level of 15 ps/nm·km, and assume that the network elements have negligible dispersion. Assuming DCF is used for dispersion compensation, then enough DCF should be added such that after 2,000 km, the 10,000 ps/nm limit is not exceeded. The required level of DCF can be determined by solving for C in the following equation:

$$2,000\,km \cdot (15\,ps\,/\,nm \cdot km - C) = 10,000\,ps\,/\,nm \qquad (4.5)$$

This yields a C of 10 ps/nm·km, which is a guideline as to how much average dispersion compensation is needed per km of fiber.

In addition, assume that the nodes in the network are spaced such that the typical distance between nodes is 250 km. This implies that in traversing 2,000 km, it is not likely that more than seven or so network elements will be optically bypassed. This can serve as a guideline in designing the network elements, in terms of how many consecutive elements will need to be traversed optically.

As this example illustrates, by coordinating the different aspects of system design, the overall system can be made more cost effective.

4.4 Regeneration Strategies

The previous sections addressed some of the physical-layer factors that affect where regeneration is required. In this section, some of the architectural issues related to regeneration are considered.

There are several approaches to managing regeneration in a network. Three strategies are presented here, where the three differ with respect to flexibility, operational complexity, and cost. Although the above discussion has elucidated several factors other than distance that affect optical reach, the illustrative examples here will continue to refer to optical reach in terms of distance, for simplicity.

4.4.1 Islands of Transparency

In the architecture known as 'islands of transparency' [Sale98a, Sale00], a network is partitioned into multiple 'islands'. The geographic extents of the islands are such that any intra-island (loopless) path can be established without requiring regeneration. However, a regeneration is required whenever a path crosses an island boundary, regardless of the path distance.

An example of a network partitioned into three islands is shown in Fig. 4.8. Island 1 is composed of Nodes A through H; regeneration is not required for a connection between any two of these nodes. Node A is also a member of Island 2, and serves as the regeneration point for any connection routed between Island 1 and Island 2. It is assumed Node A is equipped with two OADMs: one OADM is oriented to allow traffic routed between Links AB and AC to optically bypass the node; the other OADM is oriented to allow optical bypass between Links AR and AP. Traffic between Islands 1 and 2 is dropped from one OADM and added to the other OADM, with an O-E-O regeneration occurring in between. (Refer back to Fig. 2.10(b) for an illustration of a degree-four node equipped with two OADMs.) The other nodes that fall on the boundaries between islands, i.e., Nodes H, G, and N, have similar configurations.

There are several operational advantages to this architecture. First, it completely removes the need to consider where to regenerate when establishing a new connection. Regeneration is strictly determined by the island boundaries that are crossed, if any. Second, the islands are isolated from each other, so that a carrier could potentially deploy the equipment of a different vendor in each island without having to be concerned with interoperability. This isolation is advantageous even in a single-vendor system, as it allows the carrier to upgrade the equipment of the different islands on independent time scales.

The chief disadvantage of the islands architecture is that it results in extra regeneration. In the figure, assume that a connection is required between Nodes B and P, along path B-A-P, and assume that the path between them is shorter than the optical reach. Based on distance, no regeneration is required; however, because of the island topology, the connection is regenerated at Node A. Thus, the simplicity and flexibility of the system comes with an additional capital cost.

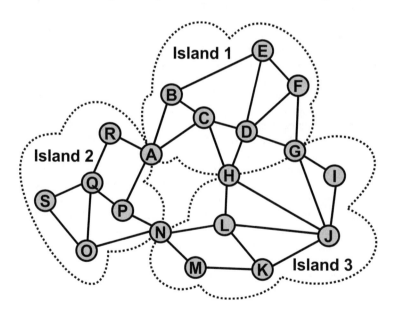

Fig. 4.8 This network is partitioned into three islands of transparency. Any traffic within an island does not need regeneration. Any traffic between islands requires O-E-O conversion.

Note that the isolation provided by the island paradigm typically exists with respect to the different geographic tiers of a network. For example, it is not common for traffic to be routed all-optically from a metro network into a backbone (or regional) network. The metro WDM system usually has coarser wavelength spacing and lower-cost components. The tolerances of the metro optics may not be stringent enough to be compatible with the system of the backbone network. Thus, the metro traffic typically undergoes O-E-O conversion prior to being carried on a backbone network regardless of the connection distance. (However, the metro and backbone tiers do not represent islands according to the definition above because regeneration may be required within either tier.)

4.4.2 Designated Regeneration Sites

A second architecture designates a subset of the nodes as regeneration sites, and allows regeneration to occur only at those sites. If an end-to-end connection is too long to be carried solely in the optical domain, then it must be routed through one or more of the regeneration sites. Either the designated regeneration sites are equipped with O-E-O equipment such that all traffic that transits them needs to be regenerated, or they have optical-bypass equipment so that regeneration occurs only when needed.

To ensure that a path is found that transits the necessary regeneration sites, one can use a graph transformation similar to the one discussed for optical-bypass-enabled networks in Section 3.5.1. (Other strategies are presented in [CMSG04, YaRa05a].) In that section, the transformation was discussed in the context of real-time planning, where only certain nodes may have available regeneration equipment. The transformed graph is comprised of only those nodes with available equipment; a link between two of these nodes is added only if a regeneration-free path exists between them in the true topology. Finding a path in this transformed graph guarantees regeneration feasibility. Restricting regeneration to certain nodes in a network is an equivalent problem, where only those nodes that are designated as regeneration sites appear in the transformed graph.

One strategy for judiciously selecting the regeneration sites is to first route all of the forecasted traffic over its shortest path. A greedy type of strategy can then be employed where nodes are sequentially picked to be regeneration sites based on the number of paths that become feasible with regeneration allowed at that site [CSGJ03]. Enough nodes are picked until all paths are feasible.

Another strategy for selecting the regeneration sites, proposed in [CSGJ03], is based on *connected dominating sets (CDSs)*. A dominating set of a graph is a subset of the nodes, S, such that all nodes not in S are directly connected to at least one of the nodes in S. The dominating set is connected if there is a path between any two nodes in S that does not pass through a node not in S. To use this methodology, the first step is to create a new topology that contains all network nodes, where two nodes are connected by a link only if there is a regeneration-free path between them in the true topology. On this new topology, a minimal CDS is found, using heuristics such as those described in [GuKh98]. (The minimal CDS is the CDS with the fewest number of nodes.) The nodes in the minimal CDS are designated as the regeneration sites. By the definition of a CDS and by virtue of how links are added to this topology, any node not in the CDS is able to reach a node in the CDS without requiring intermediate regeneration. This guarantees that for any source/destination combination, a path exists that is feasible from a regeneration standpoint.

Because this architecture limits regeneration to a subset of the nodes, extra regeneration may occur, as illustrated in Fig. 4.9. Assume that the optical reach is 1,000 km, and assume that regeneration is permitted only at Nodes B, D, and E. Assume that the nodes are equipped with optical-bypass elements, including the regeneration-capable sites. A connection between Nodes A and Z is ideally regenerated at just Node C. However, because regeneration is not permitted at Node C, the connection is regenerated at both Nodes B and E (or at Nodes B and D), resulting in an extra regeneration.

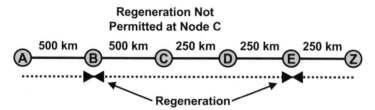

Fig. 4.9 Assume that the optical reach is 1,000 km. A connection between Nodes A and Z is ideally regenerated in just Node C. However, because it is assumed that regeneration is not permitted at this site, the connection is regenerated at both Nodes B and E instead.

Moreover, if the designated regeneration sites are equipped with O-E-O network elements (e.g., back-to-back optical terminals), then the amount of excess regeneration is likely to be significantly higher, as any connection crossing such a node needs to be regenerated. The extra regeneration cost may be partially offset by somewhat lower network element costs. Consider two adjacent degree-two network nodes where the two nodes are close enough together such that any required regeneration could equivalently occur in either node. Assume that all of the transiting traffic needs to be regenerated in one of the two nodes. In Scenario 1, assume that both nodes are equipped with OADMs and assume that they both regenerate 50% of the traffic that traverses them. In Scenario 2, one of the nodes has an OADM and regenerates none of the traffic, and the other node has two optical terminals and regenerates 100% of the transiting traffic. The total amount of regeneration is the same in either scenario. However, because two optical terminals cost less than an OADM, Scenario 2 is overall less costly. (This is an extreme example because it assumed 100% of the transiting traffic needed to be regenerated.) Setting up dedicated regeneration sites (i.e., regeneration sites with O-E-O elements) in a network may have this same effect to a degree, but the overall cost is still likely to be higher because of the extra regeneration.

A possible benefit of designating only certain nodes as regeneration sites is more streamlined equipment predeployment. With regeneration occurring at a limited number of sites, the process of predeploying equipment is more economical; i.e., with fewer pools of regeneration equipment, it is likely that less regeneration equipment needs to be predeployed across the network to reduce the blocking probability below a given threshold.

Overall, given that regeneration can be accomplished without any special equipment (i.e., the same transponders used for traffic add/drop can be used for regeneration), it is not clear if it is necessary to severely limit the number of nodes at which regeneration is supported. The biggest disadvantage is it makes the routing process more challenging and less flexible, and it likely costs more depending on the amount of extra regenerations it produces. While it may make sense to eliminate some nodes as possible regeneration sites, because the nodal offices are not large enough to house a lot of terminating equipment or because the nodes are equipped with OADMs without reuse, in general, designating only a subset of the nodes for regeneration may not be optimal.

4.4.3 Selective Regeneration

The third architecture is *selective regeneration*, which allows any (or almost any) node to perform regeneration, and regenerates a demand only when needed. The decision as to whether regeneration is needed, and if so, where to implement it, is made on a per-demand basis. This strategy is the one most commonly used in actual network deployments. Given the freedom in selecting regeneration locations, this approach yields the fewest regenerations (assuming enough available regeneration equipment is deployed at the nodes) and also allows the most flexibility when routing.

Often, there will be several options as to where the regeneration can occur for a given connection. Consider a connection between Nodes A and Z in Fig. 4.10. Assume that the optical reach is 1,000 km, and assume that regeneration is possible in any of the nodes. The minimum number of regenerations for the connection is two. As shown in the figure, there are three possible scenarios that yield two regenerations: regenerate at Nodes B and D, regenerate at Nodes C and D, or regenerate at Nodes C and E.

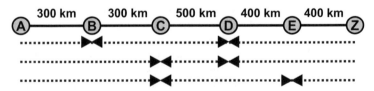

Fig. 4.10 Assume that the optical reach is 1,000 km, and assume that regeneration is permitted at any node. A connection between Nodes A and Z can be regenerated at Nodes B and D, at Nodes C and D, or at Nodes C and E.

There are several factors that should be considered when selecting one of these regeneration scenarios for the A-Z connection. In real-time planning, the amount of available equipment at each node should be considered, where regeneration is favored in the nodes that have more free equipment. Additionally, if the nodes are equipped with network elements that have a limit on the total add/drop (see Section 2.4.1), then it is important to favor regeneration at the nodes that are not close to reaching this limit. (Reaching the maximum amount of add/drop at a node could severely impact future growth, as that node will not be able to source/terminate more traffic.)

One needs to also consider the subconnections that will result from a particular regeneration scenario. The term *subconnection* is used here to refer to the portions of the connection that fall between two regeneration points or between an endpoint and a regeneration point. For example, if the connection in the figure is regenerated at Nodes C and D, the resulting subconnections are A-C, C-D, and D-Z. By aligning the newly formed subconnections with those that already exist in the network (i.e., producing subconnections with the same endpoints, on the same links), the wavelength assignment process may encounter less contention. Furthermore, in a waveband system where bands of wavelengths are treated as a single unit, creating subconnections with similar endpoints yields better packing of the wavebands.

One could also consider the system margin of the resulting subconnections when selecting where to regenerate. Regenerating the connection at Nodes C and E produces a subconnection with a length of 900 km. With the other two regeneration options, no subconnection is longer than 800 km, such that there is somewhat greater system margin. Of course, if the optical reach is specified as 1,000 km, then any of these options should work. Thus, while taking the resulting distances of each subconnection into account may produce extra margin, adding this consideration to the algorithms should not be necessary.

Note that some of the aforementioned factors may be at odds with one another. The desire to align the subconnections favors continuing to regenerate at the same nodes, but the add/drop limits of the equipment may require regeneration to be more dispersed. Given that reaching the add/drop limit at a node could restrict network growth, the limits of the network elements, if any, should dominate as the network becomes more full.

4.5 Regeneration Architectures

This chapter has thus far covered many of the physical effects, as well as the architectural strategies, that affect where a given connection must be regenerated. All of the relevant factors must be incorporated in the network planning tool to ensure that regeneration sites are selected as needed for each demand. This section examines how regeneration is actually implemented within a node.

4.5.1 Back-to-Back WDM Transponders

As has been pointed out several times already, one means of regenerating a signal is to have it exit the optical domain on one WDM transponder and re-enter the optical domain on a second WDM transponder. Figure 4.11 illustrates this architecture where the pairs of transponders used for regeneration are connected via a patch cable. The process of O-E-O conversion typically achieves full 3R regeneration, where the signal is reamplified, reshaped, and retimed. It usually provides an opportunity to change the wavelength of the optical signal as well.

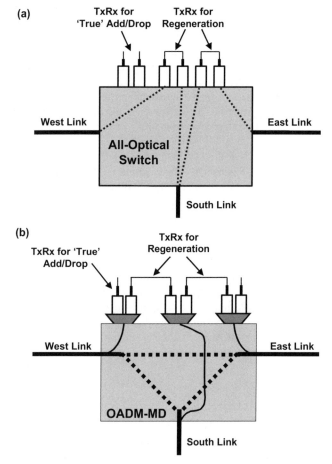

Fig. 4.11 Regeneration via back-to-back transponders (TxRx) that are interconnected by a patch cable. (a) The edge configurability provided by the all-optical switch allows any regeneration pair to access any two network links. (b) The OAMD-MD does not have edge configurability; thus, in this example, regeneration is possible only in the South/West and East/South directions. (Adapted from [Simm05]. © 2005 IEEE)

The flexibility of the back-to-back transponder architecture largely depends on the capabilities of the equipment. Consider regeneration of a connection entering from the East link and exiting on the West link. If the corresponding transponders connected to the East and West links are fully tunable, then each transponder can be independently tuned such that any (East/West) input/output wavelength combination is supported. If the transponders are not tunable, then the possible input/output wavelength combinations depend on the wavelengths of the transponders that are cabled together.

If the network element has edge configurability, as illustrated by the all-optical switch of Fig. 4.11(a), then any pair of back-to-back transponders can access any two of the network links. For example, in the figure, any pair of transponders can be used to regenerate between the East/West links, the East/South links, or the South/West links. Contrast this with Fig. 4.11(b), where the OADM-MD does not have edge configurability. With this network element, the possible regeneration directions are determined by which pairs of transponders are interconnected. Thus, in the figure, regeneration is possible between the South/West links and between the East/South links, but not between the East/West links. Manual intervention is required to rearrange the patch cables to allow for different configurations.

In Fig. 4.11, one limitation is that the transponders must be partitioned between the 'true' add/drop function and the regeneration function. (As mentioned previously, a regeneration can be considered add/drop traffic because the signal drops from the optical layer; the term '*true* add/drop' is used here to distinguish those signals that actually originate from, or terminate at, the node.) Only those transponders that are cabled together can be used for regeneration, where manual intervention is required to adjust the apportionments. To address this, one can use an adjunct edge switch, as shown in combination with an OAMD-MD in Fig. 4.12.

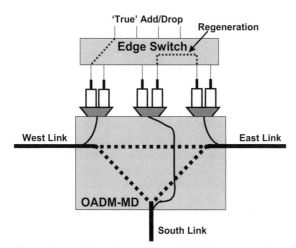

Fig. 4.12 An adjunct edge switch provides additional flexibility at the node. Any transponder can be used for either 'true' add/drop or regeneration; any regeneration direction through the node is supported. (Adapted from [Simm05]. © 2005 IEEE)

Adding the edge switch allows any transponder to be used for either 'true' add/drop or regeneration, depending on the switch configuration. While this architecture incurs the cost of the edge switch, it reduces the number of transponders that have to be predeployed at a node and/or it reduces the amount of manual intervention required to configure the node. Note that the presence of the adjunct switch also provides edge configurability for the OAMD-MD. (Using an adjunct switch with an OADM-MD to provide edge configurability was discussed in Section 2.8.1.) Thus, it allows the transponders to be used to regenerate in any direction through the node, as could already be achieved with the all-optical switch.

4.5.2 Regenerator Cards

A WDM transponder converts an incoming WDM-compatible optical signal to a 1310-nm optical signal. When two transponders are cabled together for regeneration, they communicate via the 1310-nm signal. However, for regeneration purposes, this conversion is unnecessary. A simpler device capable of 3R regeneration, referred to as a *regenerator*, is shown in the architectures of Fig. 4.13. The received optical signal on one side of the regenerator is converted to an electronic signal that directly modulates the optical transmitter on the other side, eliminating the need for short-reach interfaces (i.e., the 1310-nm interfaces). The motivation for this device is cost; the cost of one regenerator card is roughly 70% of the cost of two transponder cards.

Regenerator cards cannot be used for 'true' add/drop, as they lack the short-reach interface; thus, there is a clear division between the add/drop equipment and the regeneration equipment. The attributes of the regenerator card can have a profound impact on network operations. Consider using a regenerator card for a connection entering from the East link and exiting on the West link. It is desirable for the regenerator to allow the incoming wavelength from the East link to be different from the outgoing wavelength on the West link (the same also applies for traffic going in the reverse direction). Otherwise, wavelength conversion would be prohibited from occurring in concert with a regeneration, which is a significant restriction. Furthermore, the regenerator card is ideally fully tunable, such that any combination of incoming and outgoing wavelengths can be accommodated with a single card. If the regenerator cards are not tunable, then inventory issues become problematic if every combination of input and output wavelengths is potentially desired. (Typically, system vendors and/or carriers maintain an inventory of equipment to support new traffic or replace failed equipment. Storing thousands of different regenerator combinations would be impractical.)

Regenerator cards can be used with an all-optical switch, as shown in Fig. 4.13(a), or with an OADM-MD, as shown in Fig. 4.13(b). As with the back-to-back transponder architecture, the all-optical switch allows a regenerator to be used for regeneration in any direction through the node. In the OADM-MD, the regenerator is tied to a particular regeneration direction (e.g., East/West in the figure).

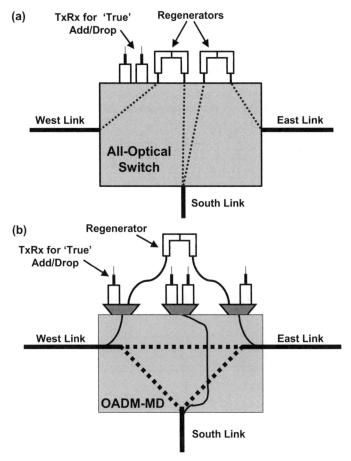

Fig. 4.13 (a) Regenerators used in conjunction with an all-optical switch. (b) Regenerator used in conjunction with an OADM-MD. In the latter, regeneration is supported only in the East/West direction. (Adapted from [Simm05]. © 2005 IEEE)

Deploying an adjunct edge switch with an OADM-MD to gain flexibility with the regenerator card is inefficient with respect to switch port utilization. The configuration is shown in Fig. 4.14. The incoming WDM signal is demultiplexed and the constituent wavelengths are fed into the edge switch. The signals that need to be regenerated are directed by the edge switch to regenerator cards, which ideally are tunable. (The signals that are truly dropping at the node are fed into transponders.) The edge switch allows wavelengths from any two network ports to be fed into a particular regenerator, thereby providing flexibility in the regeneration direction. For example, in the figure, the switch is configured to enable a regeneration in the East/South direction. However, to accomplish this, note that four ports are utilized on the edge switch for a regeneration. With the configuration shown in Fig. 4.12, only two ports are utilized on the edge switch per regeneration. Additionally, in Fig. 4.14, the edge switch must be capable of switching a WDM-compatible signal; e.g., it could be a MEMS-based switch. An electronic-based switch is not suitable for this application.

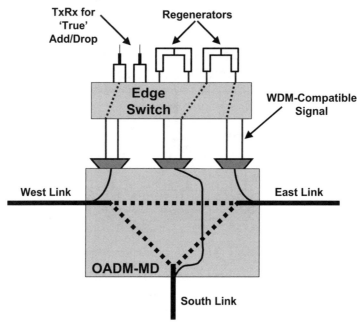

Fig. 4.14 With an edge switch used in combination with regenerator cards, four ports on the switch are utilized for each regeneration. Additionally, the edge switch must be capable of switching a WDM-compatible signal.

Rather than using an edge switch, a degree of configurability can be attained with the OADM-MD architecture through the use of flexible regenerators [SiSa07], similar to the flexible transponders that were discussed in Section 2.8.2. Refer to the flexible regenerator shown in Fig. 4.15 with a degree-four OADM-MD. One side of the regenerator is connected to the East and North links, and the other side is connected to the West and South links. Thus, this one regenerator allows regeneration to occur in either the East/West, East/South, North/West, or North/South directions; i.e., four of the six possible directions through the node are covered. Furthermore, it allows a regenerated signal to be sent out on two simultaneous links; e.g., a signal entering from the East link can be regenerated and sent out on both the West and South links. This is useful for multicast connections, as was discussed in Section 3.8.

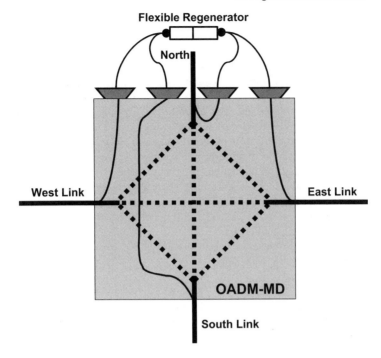

Fig. 4.15 A degree-four OADM-MD combined with a flexible regenerator that allows regeneration in either the East/West, East/South, North/West, or North/South directions. An optical backplane can be used to eliminate complex cabling. (Adapted from [SaSi06]. © 2006 IEEE)

4.5.3 All-Optical Regenerators

An all-optical regenerator card is a relatively new regeneration alternative [LLBB03, LeJC04]. Due to the scalability of optics, it is expected that the cost of all-optical regenerators will scale well with increasing line-rate. For example, there may be only a small price premium for a 40-Gb/s regenerator as compared to a 10-Gb/s regenerator. Furthermore, all-optical regenerators are expected to consume less power as compared to their electronic counterparts, which should improve nodal scalability.

All-optical regenerators may not fully replace electronic regeneration. Some all-optical regenerators provide only 2R regeneration, i.e., reamplification and reshaping, as opposed to 3R, which includes retiming as well. Thus, a combination of all-optical and electronic regeneration may be needed; the bulk of the regeneration can be performed all-optically, with electronic regenerators used intermittently to clean up the timing jitter.

The first commercial all-optical regenerators are likely to be compatible only with relatively simple modulation formats such as OOK. However, in order to meet the capacity requirements of future networks, more advanced modulation formats are needed. All-optical regeneration that is compatible with advanced modulation formats is an area of current research; e.g., [SMCS05, Mats05]. Some of the solutions that have been proposed are fiber-based, which will result in a bulky design. It is desirable that solutions that are more integratable be developed.

All-optical regeneration, at least initially, is likely to operate on a per-wavelength basis, similar to electronic regenerators; i.e., Fig. 4.13 holds for all-optical regenerators as well. It is expected that these all-optical regenerators will provide complete flexibility with respect to wavelength conversion, allowing any input/output wavelength combination. Another potential technology being researched is multi-wavelength regenerators, where a whole band of wavelengths is regenerated at once [CXBT02, HXGB05]. This potentially improves the economics of regeneration even further, although it is not clear whether such a technology can support wavelength conversion.

Overall, all-optical regeneration is a relatively new technology that is still an area of active research on many fronts.

Chapter 5

Wavelength Assignment

Wavelength assignment is an integral part of the network planning process in optical-bypass-enabled networks. Its need arises from the wavelength continuity property of optical-bypass elements, where a connection that traverses a node all-optically must enter and exit the node on the same optical frequency. Thus, the wavelengths that are in use on one link may have ramifications for the wavelengths that can be assigned on other links. Effective wavelength assignment strategies must be utilized to ensure that wavelength contention is minimized.

Wavelength assignment is tightly coupled to the routing process, as the selection of the route determines the links on which a free wavelength must be found. The two processes are often referred to together as the routing and wavelength assignment (RWA) problem. However, regeneration, when needed, is just as critical to the process. While the earliest visions of optical-bypass-enabled networks assumed that they would be completely all-optical, with no regeneration, this has not turned out to be the case in reality, especially in regional and backbone networks. The presence of regeneration has a significant impact on wavelength assignment because it allows for a change in wavelength. This is explored in Section 5.1.

The previous two chapters looked at the sequential processes of selecting a route followed by selecting regeneration locations. Wavelength assignment can be treated as the third sequential step in the planning process. With this multi-step approach, there is no guarantee that the route found will be amenable to a feasible wavelength assignment. Another option is to perform RWA as a single-step; this is more complex, but any route that is found is guaranteed to have a feasible wavelength assignment. Multi-step and single-step RWA are discussed in Sections 5.2 and 5.3, respectively.

Wavelength assignment algorithms were one of the first aspects of all-optical networks that were researched heavily. The result is an array of well-studied algorithms that represent different performance/complexity operating points. Some of the strategies that have proved effective in actual network designs are presented in Section 5.4. In scenarios where many demands are added at one time to the network, the order in which the connections are assigned wavelengths may affect the

J.M. Simmons, *Optical Network Design and Planning*,
DOI: 10.1007/978-0-387-76476-4_5, © Springer Science+Business Media, LLC 2008

performance of the wavelength assignment scheme. Effective ordering strategies are discussed in Section 5.5.

One interesting aspect of wavelength assignment relates to the two directions of a bi-directional connection. Scenarios where it may be beneficial to assign different wavelengths to the two directions are covered in Section 5.6. Another challenge encountered in some systems is the non-uniformity of the optical reach across the wavelengths in the spectrum. The ramifications are discussed in Section 5.7.

As this introduction indicates, there are numerous challenges of wavelength assignment that must be addressed. However, many studies have been performed that show the loss of network efficiency due to wavelength contention is generally small, assuming good algorithms are used; live networks exist to further bear this out. Nevertheless, the effect of wavelength contention on the performance of the network continues to be debated in the industry. Section 5.8 presents further results that demonstrate a small loss of efficiency in both a backbone network and a metro network, and offers some insight as to why this is so.

Wavelength assignment is especially important in some optical protection schemes, where the primary and backup paths may be required to be carried on the same wavelength; this is discussed further in Chapter 7. In this chapter, one can assume that any protection is client-based 1+1, where there is a primary path and a dedicated backup path, and the network client (e.g., an IP router) determines which of the two paths to use. With this type of protection, wavelengths can be assigned independently to the two paths.

It should be pointed out that the development of cost-effective all-optical wavelength converters is an active area of research. This technology would allow the wavelength of a signal to be changed without requiring O-E-O conversion, such that optical bypass would not necessarily imply wavelength continuity. The criticality of wavelength assignment in the overall network planning process would be somewhat abated with commercial deployment of this technology. However, it is unlikely that the cost of such technology will allow it to be deployed at *all* nodes on *all* wavelengths. Moreover, adding wavelength converters on every wavelength would be antithetical to the optical-bypass paradigm of reducing the amount of equipment in the network. Thus, wavelength assignment will remain an important step in the network planning process for optical-bypass-enabled networks for the foreseeable future.

Note that in O-E-O networks, wavelengths can be assigned independently on each link, and thus wavelength contention and wavelength assignment are not major issues.

5.1 Role of Regeneration in Wavelength Assignment

If a demand is carried all-optically from source to destination, then the same wavelength must be used on all links of the path (assuming all-optical wavelength conversion is not available). Consider the connection between Nodes A and Z shown in Fig. 5.1(a), which is routed on seven links. If this is an all-optical connection, then one needs to find a wavelength that is free on all seven links. In Fig. 5.1(b), this same connection is regenerated at Nodes C and E, thereby creating three subconnections: A to C, C to E, and E to Z. In this scenario, one needs to find a free wavelength for each of the subconnections, where in most cases, there is no requirement that the wavelengths be the same for each subconnection. Finding a free wavelength on a subconnection is clearly an easier problem than finding a free wavelength on the whole end-to-end connection. Thus, the presence of regeneration potentially engenders greater wavelength assignment flexibility.

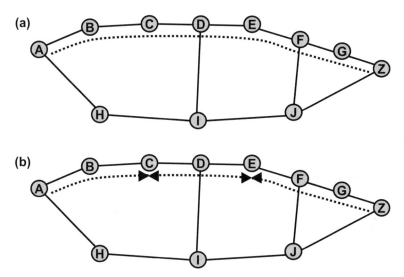

Fig. 5.1 In (a), it is necessary to find a wavelength that is free along each of the links of the path between Nodes A and Z. In (b), where regeneration occurs at Nodes C and E, it is necessary to find wavelengths that are free on each link of the subconnections; different wavelengths can be assigned to the three subconnections.

The importance of fully tunable transponders and regenerators with respect to the wavelength assignment process should be readily apparent. For example, in an architecture where two transponders are patch-cabled together for regeneration (as was shown in Fig. 4.11), fixed-tuned transponders imply that the wavelengths assigned to consecutive subconnections are determined by the wavelengths that are paired together at the regeneration node. An even more significant constriction arises in systems where the regenerator cards require that the wavelengths on the input and output sides of the regenerator (when considering either one of the two directions of a connection) be the same. Using such regenerator cards prohibits wavelength conversion at the regeneration node. With this type of equipment, all three subconnections in Fig. 5.1(b) would need to be assigned the same wavelength, which reduces the flexibility of the wavelength assignment process.

In Chapter 4, O-E-O conversion was discussed in the context of regeneration and optical reach. However, there are functions other than regeneration that require O-E-O conversion. For example, in a network with subrate traffic, it is necessary to bundle multiple connections together to utilize more fully the capacity of a wavelength. As is described in Chapter 6, this bundling process is most effective when the traffic can be groomed at various nodes in the network. The grooming process typically is accomplished in the electrical domain, thereby requiring O-E-O conversion. A second driver is shared protection, which is covered in Chapter 7. To effectively share protection bandwidth, O-E-O conversion may be required at the 'sharing' points. As these examples indicate, O-E-O conversion is not just a function of optical reach. Thus, even in small metro networks where the path distances do not warrant regeneration, not every connection may be carried all-optically end-to-end.

Any O-E-O event offers the opportunity to essentially wavelength convert 'for free' (again, assuming the equipment permits this flexibility). Wavelength assignment algorithms should take advantage of this freedom whenever possible. For simplicity, the remainder of the chapter refers to connections being broken into subconnections due to regeneration; however, keep in mind that it may be due to a factor other than optical reach, as discussed above.

5.2 Multi-Step RWA

When network planning is treated as a multi-step process, a route is selected for a connection, the connection is broken into subconnections, if necessary, and each of the subconnections is assigned a wavelength. It is possible that a feasible wavelength assignment will not be found for one or more of the subconnections, requiring some of the steps to be repeated.

Minimizing the occurrence of wavelength contention requires good routing strategies. The discussion here assumes alternative-path routing is used, possibly in combination with dynamic routing (see Section 3.4). As presented in Chapter 3, one first generates a set of candidate paths for a particular source/destination combination, where the candidate paths are chosen to minimize cost and to provide good load balancing in the network. As demands are added, the current state of the network is considered when selecting one of the candidate paths to use for a particular demand request. A good strategy is to select the least-loaded candidate path, such that the minimum number of wavelengths that are free on each link of the selected path is maximized. This does not guarantee that the *same* wavelength will be free along the various links; however, it generally improves the chances of finding a feasible wavelength assignment.

If demands are added one at a time to the network, then the algorithm can actually consider which particular wavelengths are free when selecting one of the candidate paths. Thus, the actual feasibility of the candidate path can be considered when selecting a path for the demand. When multiple demands are added at once to the network, as is often the case in long-term planning, one option is to fully process each demand individually; i.e., route, regenerate, and assign a wavelength for one demand before moving on to the next. This methodology is similar to adding demands one at a time, and allows one to consider the actual free wavelengths on a path when selecting a route. A second option, which is typically more advantageous, is to perform the routing and regeneration for all demands before the start of the wavelength assignment process. With this strategy, the wavelength assignment algorithm has full knowledge of exactly how many subconnections are routed on each link, and can use this information to better optimize the assignment process. In this methodology, considering which wavelengths are free when choosing a path for a demand is not possible; only the current load on each link, due to the demands that have already been routed, is known.

The multi-step methodology of treating routing, regeneration, and wavelength assignment separately usually performs well in practice. However, when the network is very heavily loaded, wavelength contention may occur where subconnections are created for which there are no feasible wavelength assignments. Four strategies for ameliorating the situation are discussed here.

5.2.1 Alleviating Wavelength Contention

First, in a network requiring regeneration, different regeneration sites for a connection can be considered. Consider the connection between Nodes A and Z shown in Fig. 5.2. Assume that the optical reach is 2,000 km, such that one regeneration is required for the connection. The possible locations for the regeneration are Nodes C, D, or E, as shown in Fig. 5.2 (a), (b), and (c), respectively. Each regeneration choice creates a different set of subconnections; e.g., regenerating at Node C creates the A-C and C-Z subconnections. If the initial location of the regeneration leads to wavelength contention, then a different location can be tried. For example, assume that with regeneration at Node C, a free wavelength can be found on the A-C subconnection but not on the C-Z subconnection. One can consider moving the regeneration to either Node D or Node E, to generate different subconnections. Note that moving the regeneration location(s) does not incur any additional cost.

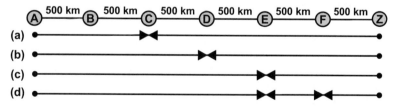

Fig. 5.2 With an optical reach of 2,000 km, one regeneration is required on the path between Nodes A and Z. The site of the regeneration determines the resulting subconnections, which can affect wavelength assignment. The regeneration is at Node C in (a), Node D in (b), and Node E in (c). In (d), a second regeneration is added to break the wavelength contention that is assumed to exist on Links EF and FZ.

If this strategy does not work, or if there are no regenerations, then one can consider using a different candidate path. In realistic networks, when wavelength assignment fails, it is typically because of a small number of heavily loaded links. Moving some demands away from these links can be enough to alleviate the wavelength contention problems. The simplest means of achieving this is to pick a different candidate path for some of the demands, where the originally selected path included one or more 'bad' links, and where the new path does not include any. Ideally, the new candidate path selected meets the minimum amount of regeneration possible for the demand so that no additional cost is incurred.

If not enough demands have candidate paths that avoid the bad links, then one can make use of dynamic routing. The links that are causing the problems in the wavelength assignment process can be temporarily eliminated from the topology before calling the shortest path algorithm. There is no guarantee that this will find a feasible path; even if a path is found, it may be very circuitous such that it requires several additional regenerations, making it undesirable.

If, after applying the above three strategies, wavelength assignment is still not successful, then one of the original candidate paths can be selected and any residual wavelength contention can be alleviated by adding in extra regeneration (the candidate path that requires the minimum amount of extra regeneration is typically selected). For example, in Fig. 5.2, assume that the wavelength contention stems from there being no common free wavelengths on Links EF and FZ. An extra regeneration can be added at Node F, as shown in Fig. 5.2(d). This breaks the interdependence between Links EF and FZ for this connection, allowing wavelengths to be assigned independently on these links. Adding in just a few extra regenerations in a network can be quite effective in alleviating wavelength contention. With very little additional cost, the network utilization can be markedly increased. This was demonstrated in [VAAD01, Simm02]. It is further explored in Section 5.8.

Note that if the network is so full that wavelength contention is causing a great deal of extra regenerations to be added, then it is probably time to add additional capacity to the network.

5.3 One-Step RWA

Rather than relying on techniques to handle infeasible assignment scenarios when they arise, it is natural to consider methodologies where routing and wavelength assignment are treated as a single problem to ensure feasibility from the start. One of the earliest advocated one-step algorithms starts with a particular wavelength and reduces the network topology to only those links on which this wavelength is available. The routing algorithm (e.g., a shortest path algorithm) is run on this pruned topology. If no path can be found, or the path is too circuitous, another wavelength is chosen and the process run through again on the correspondingly pruned topology. The process is repeated with successive wavelengths until a suitable path is found. With this combined approach, it is guaranteed that there will be a free wavelength on any route that is found. If a suitable path cannot be found after repeating the procedure for all of the wavelengths, the demand is blocked.

In a network with regeneration, using this combined routing and wavelength assignment procedure makes the problem unnecessarily more difficult because it implicitly searches for a wavelength that is free along the whole length of the path. As discussed above, it is necessary to find a free wavelength only along each sub-connection, not along the end-to-end connection. One variation is to select ahead of time where the regenerations are likely to occur along a connection, and apply the combined routing and wavelength assignment approach to each expected sub-connection individually. However, the route that is ultimately found could be somewhat circuitous and require regeneration at different sites than where was predicted, so that the process may need to be run through again. Overall, this strategy is less than ideal.

A more direct unified RWA approach is to create a transformed graph when-ever a new demand request arrives, where the transformation is similar to what was discussed in Section 3.5.1. In long-term planning, every network node ap-pears in the transformed graph; in real-time planning, only nodes with available regeneration equipment, plus the source and the destination, are added to the trans-formed graph. A link is added between a pair of nodes in the transformed graph only if there exists a regeneration-free path between the nodes in the true topol-ogy, *and* there exists a wavelength that is free along the path. (Even if there are multiple regeneration-free paths between a node pair, or multiple wavelengths free on a path, at most one link is added between a node pair.)

An example of such a transformation is shown in Fig. 5.3. The true topology is shown in Fig. 5.3(a), where the wavelengths that are assumed to still be available on a link are shown. The optical reach is assumed to be 2,000 km. Additionally, it is assumed that the transformation is being performed as part of a long-term planning exercise, so that the available regeneration equipment at a node is not a factor. The demand request is assumed to be between Nodes A and Z. The trans-formed graph is shown in Fig. 5.3(b). All of the original links appear in this graph except for Link AF, which has no available wavelengths. In addition, Links AC, AD, and BD are added because the respective associated paths, A-B-C, A-B-C-D, and B-C-D, are less than 2,000 km and have a free wavelength (i.e., on each of these paths λ_6 is free). Note that no link is added to represent the path E-F-G even though λ_4 is available on this path because the path distance is 2,500 km, which is greater than the optical reach.

In a real network with many nodes and wavelengths, creating the transformed graph can potentially be a time-consuming procedure. For each node pair, say Nodes X and Y, that possibly has a regeneration-free path between them, a search is performed to find a regeneration-free path where some wavelength is available along the whole path. To do this, one could use the one-step approach described at the beginning of this section, where the true topology is pruned down to just those links that have a particular wavelength free. A shortest path algorithm is run on the pruned topology to search for a regeneration-free path between Nodes X and Y. The process is repeated for each wavelength, until a regeneration-free path is found. (This one-step approach is better suited for generating the links in the transformed graph than it is for finding an end-to-end path; however, it still may be time consuming.) Alternatively, one can run a K-shortest paths algorithm on the true topology, where K is large enough such that any regeneration-free path between Nodes X and Y is found. The paths are then checked for a free wavelength. If a regeneration-free path with a free wavelength is found, a link is added between Nodes X and Y in the transformed graph. The algorithm should keep track of the free wavelength associated with each link in the transformed graph.

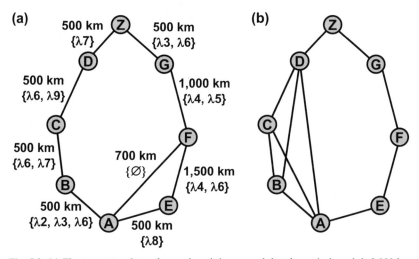

Fig. 5.3 (a) The true network topology, where it is assumed that the optical reach is 2,000 km. The wavelengths listed next to each link are the wavelengths that are assumed to be free on the link. (b) The transformed graph, where a link is added between a node pair if there is a regeneration-free path between the nodes with at least one available wavelength along the path.

To reduce the time to form the transformed graph, one can maintain a list of a few regeneration-free paths for each node pair that has at least one such regeneration-free path between them. (This is no worse than storing a set of candidate paths for alternative-path routing.) If a free wavelength can be found on any of the paths, then a link is added in the transformed graph for the corresponding node pair. This eliminates multiple calls to a shortest path routine every time a graph transformation is needed. This method is not guaranteed to find a feasible path if one exists, although in practice it usually does.

Once the transformed graph is formed, a shortest path algorithm is run from the demand source to the demand destination to find the path in the transformed graph with the fewest hops, where each hop translates to a subconnection in the true topology. If a path is found, then it is guaranteed to have the fewest feasible regenerations, and each resulting subconnection is guaranteed to have an available wavelength. In the example in the figure, path A-D-Z is found, which corresponds to the subconnections A-B-C-D and D-Z in the true topology. These subconnections are assigned λ_6 and λ_7, respectively.

One caveat with this one-step method should be noted. The path produced may include a link that is common to more than one subconnection, and where the same free wavelength is associated with the overlapping subconnections. Assuming there is just one fiber-pair per link, this would result in the same wavelength being assigned multiple times on a fiber, which is not permitted. A few strategies can be attempted to remedy the situation. Assume that there are two subconnections that overlap. A different free wavelength could be searched for on either of the subconnections. If that is not successful, one of the subconnections can be routed differently, where the subconnection path has the same endpoints, but the overlapping link is avoided. If this is also not successful, then the link associated with one of the overlapping subconnections can be removed from the transformed graph, and another search performed to find a new end-to-end path.

This type of problem is more likely to occur if the regeneration-free paths represented by the links in the transformed graph are 'meandering'. This can be minimized by using the methodology described above where a small number of fairly direct, regeneration-free paths are maintained for the nodal pairs. Only these paths are considered when forming the transformed graph. This methodology was used in the study that is reported on in Section 5.8, and no problems with overlapping subconnections were encountered.

If non-tunable transponders or regenerator cards are used, then a more complex transformation may be needed with real-time planning to ensure a feasible path is found using one-step RWA. For example, consider a transformed graph that includes only those nodes with regeneration equipment, plus the source and destination. A link is added between a pair of nodes (say A and B) in the transformed graph for *each* wavelength λ_i if there is a regeneration-free path between Nodes A and B on which λ_i is available, *and* there is an available regenerator (or transponder) at both Nodes A and B with wavelength λ_i. Call such a link in the transformed graph *ABi*. *Turn constraints* are imposed to ensure that a path can go from link *ABi* to *BCj*, for some nodes A, B and C, and some wavelengths i and j, only if there is a regenerator (or transponder-pair) at Node B that interconnects wavelengths i and j. A shortest path algorithm that enforces turn constraints is run on the transformed graph (turn constraints were discussed in Section 3.5.1.) (If the nodes are equipped with OADM-MDs rather than all-optical switches, then additional constraints must be enforced where the regenerators/transponders are tied to certain links.) The number of links in the transformed graph may be quite large if the number of wavelengths in the network is large. However, such detailed modeling is generally only needed when the network is heavily loaded and feasible paths are difficult to find. At that stage, it is expected that the number of free wavelengths and regenerators are small so that the transformed graph is not excessively large.

All of the one-step methodologies discussed in this section impose additional processing and memory burdens. When the network is not heavily loaded, the multi-step process should have little problem finding a feasible route and wavelength assignment. Thus, under these conditions, the multi-step process is favored, as it is usually faster. However, under heavy load, the one-step methodology can provide a small improvement in performance, as is investigated quantitatively in Section 5.8. Furthermore, under heavy load, the graph transformation methodology may be more manageable, as the scarcity of free wavelengths should lead to a lower density of links in the transformed graph.

5.4 Wavelength Assignment Strategies

The specific wavelength assignment strategy used affects the performance of both multi-step and one-step RWA. With multi-step RWA, a route is selected and then broken into subconnections based on where regenerations are needed. If no regeneration is needed, then the subconnection equals the whole connection; this is still referred to as a subconnection here. The wavelength assignment strategy determines the order in which wavelengths are considered when assigning a wavelength to each of the subconnections. For one-step RWA, the wavelength assignment strategy determines the order in which wavelengths are considered when forming the transformed graph that was described in Section 5.3.

With multi-step RWA, if there is no wavelength that is free along a subconnection, then one of the methods described in Section 5.2.1 is used to ease the wavelength constraints, or the corresponding demand is declared blocked. With one-step RWA, if a path from the demand source to the demand destination cannot be found in the transformed graph, the demand is declared blocked.

Wavelengths must be assigned such that the same wavelength is not used more than once on any fiber. To clarify this restriction, refer to Fig. 5.4. Link AB in Fig. 5.4(a) is populated with one fiber-pair, where one fiber carries traffic from A to B, and the other fiber carries traffic from B to A. A given wavelength can be assigned once on the A-to-B fiber and assigned once on the B-to-A fiber. Link CD in Fig. 5.4(b) is populated with three fiber-pairs. Three of the fibers carry traffic from C to D, and three from D to C. A particular wavelength can be assigned three times in each direction of the link, where each assignment is carried by a different fiber.

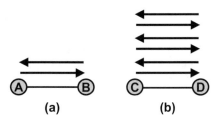

(a)   (b)

Fig. 5.4 A wavelength can be assigned to at most one subconnection per fiber. In (a), there is one fiber-pair on the link; the same wavelength can be used in both directions. In (b), there are three fiber-pairs per link; the same wavelength can be used three times in each direction.

As these two examples illustrate, it is possible that a given wavelength be used multiple times on a *link*, but it can be assigned only once per *fiber*. Of course, one exception to this rule is a system that supports bi-directional transmission of the same wavelengths on a single fiber; for wavelength assignment purposes, this can be treated like a fiber-pair.

Numerous wavelength assignment schemes have been devised over the years, where the difference in performance among the schemes is fairly small. Reference [ZaJM00] provides a good overview of the various schemes along with some performance curves. Here, the focus is on three particular schemes that have proved to be effective when preparing designs for actual carrier networks: *First-Fit*, *Most-Used*, and *Relative Capacity Loss* (RCL). The three schemes are described below, followed by a qualitative comparison. All three of these schemes are suitable for any topology. Furthermore, the schemes can be applied whether there is a single fiber-pair or multiple fiber-pairs on a link. (Note that there are schemes specifically designed for the multiple fiber-pair scenario, most notably the *Least-Loaded* scheme [KaAy98]. This is shown in [ZaJM00] to perform better than the three aforementioned schemes when there are several fiber-pairs per link. As fiber capacities have rapidly increased, however, links with several fiber-pairs are a less likely occurrence.)

5.4.1 First-Fit

First-Fit is the simplest of these three wavelength assignment schemes. Each wavelength is assigned an index from 1 to W, where W is the maximum number of wavelengths supported on a fiber. No correlation is required between the order in which a wavelength appears in the spectrum and the assigned index number. The indices remain fixed as the network evolves. Whenever wavelength assignment is needed, the search for an available wavelength proceeds in an order from the lowest index to the highest index.

It is reasonable to consider using the First-Fit indexing order to guide the network growth in a manner that may potentially improve network performance. However, due to network churn (i.e., the process of connections being established and then later torn down), the interdependence of wavelength assignment across links, and the presence of failure events, the *indexing ordering* does not guarantee the actual *assignment order* on a link. Consider the example shown in Fig. 5.5. Assume that it is desirable for wavelengths to be assigned on links such that there is a good spread across the spectrum; i.e., assign some wavelengths in the middle, some at the low end, and some at the high end of the transmission band. (The motivation for this is that some Raman amplifiers perform better when the power levels are fairly evenly distributed across the transmission band.) For simplicity, assume that there are only eight wavelengths in the system, and assume that the index order for First-Fit is: $\lambda 4, \lambda 1, \lambda 8, \lambda 5, \lambda 2, \lambda 7, \lambda 3, \lambda 6$. With a focus on the three links shown in Fig. 5.5(a), assume that eight connections are added to the network in the order shown in the figure. The figure indicates the wavelengths that are assigned to each connection, based on the indexing scheme specified above. Even though the index order is consistent with a good spread across the transmission band, the wavelengths assigned on Link CD are: $\lambda 8$, $\lambda 5$, $\lambda 7$, and $\lambda 6$. These wavelengths are all in the upper half of the spectrum as opposed to being spread across the band.

Furthermore, assume that Link CD fails; the remaining connections are shown in Fig. 5.5(b). Thus, while Link AB originally was populated with wavelengths across the band, the wavelengths remaining after the failure are: $\lambda 1$, $\lambda 2$, and $\lambda 3$. Again, the wavelengths are bunched towards one end of the band.

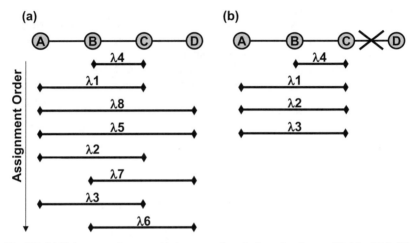

Fig. 5.5 (a) Eight connections are added one-at-a-time, in the order shown. (b) After Link CD fails, only four connections remain.

As another example of trying to use the First-Fit index to enhance performance, [HeBr06] proposes using the indexing scheme to minimize crosstalk among the wavelengths; i.e., the indices are ordered in an attempt to delay the point at which adjacent wavelengths are assigned on a link. However, because of network churn and the interdependence of wavelength assignment across links, adjacent wavelengths may need to be assigned prior to a link being half full. Furthermore, the transmission system should not be designed such that it works only if non-adjacent wavelengths are assigned on a link. An established subconnection should not fail due to any other subconnection using any other wavelength being added to the network.

Thus, the indexing strategy of First-Fit may be viewed as a short-term means of potentially providing additional system margin under certain conditions, but it should not be relied on to enforce a critical system constraint, because network growth or network churn may cause the benefits to be lost.

5.4.2 Most-Used

The second wavelength assignment scheme considered here, Most-Used, is more adaptive than First-Fit, although it requires more information. Whenever a wavelength needs to be assigned, a wavelength order is established based on the number of link-fibers on which each wavelength has already been assigned. The wavelength that has been assigned to the most link-fibers already is given the lowest index, the wavelength that has been assigned to the second-most link-fibers is given the second lowest index, etc. After the wavelengths have been indexed, the assignment procedure proceeds as in First-Fit. The motivation behind this scheme is that a wavelength that has already been assigned a lot will be more difficult to use again. Thus, if a scenario arises where a heavily used wavelength can be used, it should be assigned.

5.4.3 Relative Capacity Loss

Relative Capacity Loss (RCL), proposed in [ZhQi98], is more complex than either of the previous schemes. The idea is that wavelength assignment should take into account how much 'harm' it is doing to demands that may be added in the future; i.e., how likely will it cause wavelength contention for future demands. This scheme is more amenable to the multi-step RWA process; thus, this is the focus of the discussion.

The first step in RCL is to generate the set of possible paths in the network, e.g., based on a traffic forecast. When regeneration is present in a network, it is actually the set of possible subconnections that needs to be enumerated. If some subconnections are expected to arise more than others, as is likely to be the case, then the subconnections should be added to the list multiple times to reflect their expected relative frequency. This enumeration step is done prior to any traffic being added to the network, although the list can be updated if necessary as the network evolves.

As demand requests enter the network, a connection path is selected, and the new connection is partitioned into subconnections. Each new subconnection must then be assigned a wavelength. Consider one such new subconnection, which is denoted here by $\bar{s}$. A set S is generated that contains all forecasted subconnections that have at least one link in common with $\bar{s}$. For each subconnection s_j in S, it is determined how many wavelengths are currently available to be assigned to it. Let this quantity be represented by N_j. Note that if there are multiple fiber-pairs per link and a wavelength is available on f fiber-pairs on each link of s_j, then the quantity N_j takes this into account.

Next, for each wavelength λ_i that could possibly be assigned to $\bar{s}$, where $1 \leq i \leq W$, it is determined whether or not s_j is affected if $\bar{s}$ were assigned wavelength λ_i (i.e., if $\bar{s}$ were assigned wavelength λ_i, does that reduce N_j by one). Let I_{ij} be 1 if s_j is affected, and let I_{ij} be 0 if s_j is not affected. Then for each wavelength λ_i that is available to be assigned to $\bar{s}$, the following sum is calculated (where the sum is calculated over all s_j in S):

$$C_i = \sum_j \frac{I_{ij}}{N_j} \qquad (5.1)$$

The wavelength λ_i with the minimum C_i is selected as the wavelength to assign to $\bar{s}$. This procedure is run through whenever a wavelength needs to be assigned to a new subconnection.

This algorithm is illustrated using the example of Fig. 5.6. Assume that there are a total of four wavelengths in the system and one fiber-pair per link. The shaded boxes in the figure indicate the links on which wavelengths have already been assigned. The subconnection of interest, $\bar{s}$, extends between Nodes B and D. The set S is composed of subconnections s_1, s_2, and s_3. Subconnection s_1, which extends between Nodes A and C, has only one available wavelength, λ_2. Thus, N_1 is 1. Subconnections s_2 and s_3 each have three available wavelengths, so that N_2 and N_3 are both 3. The possible wavelengths that could be assigned to $\bar{s}$ are λ_1 and λ_2. If λ_1 were assigned to $\bar{s}$, then only s_2 and s_3 are affected, because s_1 already cannot use λ_1. Thus, C_1 is $1/3 + 1/3 = 2/3$. If λ_2 were assigned to $\bar{s}$, then only s_1 is affected; C_2 thus equals 1. C_1 is lower than C_2, resulting in λ_1 being assigned to $\bar{s}$. Even though the λ_1 assignment affects two subconnections and the λ_2 assignment would affect only one, the s_1 subconnection has fewer options, and is 'hurt' more if λ_2 were assigned to $\bar{s}$.

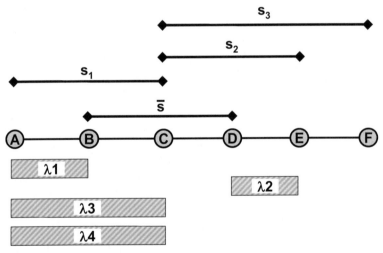

Fig. 5.6 Setup to illustrate RCL wavelength assignment, where the shaded boxes indicate the links on which a wavelength has already been assigned. $\lambda1$ is selected for $\bar{s}$ because this assignment is less 'harmful' to the remaining subconnections.

The RCL scheme is especially suitable when a whole set of demands is added at once. In this scenario, with multi-step RWA, all subconnections are known before the start of the wavelength assignment phase (thus, the step where the set of expected subconnections is enumerated based on a forecast is unnecessary). When assigning a wavelength to a subconnection and determining the relevant set S, only those subconnections that have not been assigned a wavelength yet need to be considered.

5.4.4 Qualitative Comparison

All three wavelength assignment schemes provide relatively good performance in realistic networks. For example, wavelength contention does not generally become an issue until there are at least a few links in the network with roughly 85% of the wavelengths used. RCL often performs somewhat better than the other two schemes in minimizing wavelength contention; however, it is also the most complex to implement. It is more difficult to rank the relative performance of First-Fit and Most-Used, as which performs better depends on the network topology and the traffic. In any case, the differences in performance are small. One advantage to First-Fit is that, in contrast to the other two schemes, it does not require any global knowledge to operate.

In long-term planning with a set of demands being added, and multi-step RWA being used, it is possible to try more than one scheme on the set of subconnections. For example, one could first attempt to use either First-Fit or Most-Used to assign wavelengths to the set of subconnections. If this does not result in feasible wavelength assignments for all of the subconnections, then one could restart the wavelength assignment process using RCL.

Another factor that has an impact on the success of the assignment scheme is the order in which the subconnection set is processed. This is discussed next.

5.5 Subconnection Ordering

Consider adding multiple demands to the network at once and assume that multi-step RWA is used. Assume that all routing and regeneration occurs prior to the wavelength assignment step, such that all subconnections have already been created. In the previous section, different strategies were presented for the order in which wavelengths should be considered for assignment to a particular subconnection. This section discusses the order in which the subconnections are processed.

Finding an available wavelength for a subconnection is more difficult when there are many links comprising the subconnection and when the links on which a subconnection is routed are congested. Thus, link load and number of subconnection hops are two important factors that should be considered when determining the order in which subconnections should be assigned wavelengths.

A natural first step is to determine the most heavily loaded link in the network. (It is assumed here that the heaviest load is no larger than the number of wavelengths supported on a fiber. If there are more subconnections routed on a link than there are wavelengths, then clearly a feasible assignment cannot be found. In this case, the routing process needs to be redone.) If there are multiple links tied for the heaviest load, then the link with the longest subconnections on it (in terms of hops) should be selected. Designate this as the 'worst link'. The next step is to assign wavelengths to all subconnections that are routed on the worst link, given that these are likely to be the most difficult on which to find an assignment; no conflicts can occur in this stage. (When the assignment process first starts, all wavelengths can be considered as equivalent. Thus, the wavelengths can be assigned arbitrarily to the subconnections on the worst link; any assignment could be mapped into any other.)

The next step is to order the remaining subconnections based on factors such as the load of the links traversed by the subconnection and the number of hops in the subconnection. Various heuristic ordering schemes can be devised; three examples are presented here.

In one scheme, if the load along a whole subconnection is low, then the assignment order is based only on the number of hops in the subconnection. If, however, a subconnection traverses some heavily loaded links, then the hop metric is artificially inflated to reflect the expected difficulty of finding an available wavelength for that subconnection. The subconnections are then processed for wavelength assignment in an order from the largest hop metric to the smallest hop metric.

In a second scheme, a metric is devised for each link that reflects the load, where the metric is less than unity, and the heavier the load the lower the metric. The metric for the subconnection is the product of its corresponding link metrics, thus taking into account both load and number of hops. This metric is then used to determine the subconnection assignment order, where a lower metric results in earlier assignment.

A third scheme selects a subconnection for wavelength assignment based on the number of available wavelengths that could possibly be assigned to it. The number of available wavelengths for a subconnection monotonically decreases as wavelengths are assigned to other subconnections. Thus, as the wavelength assignment process progresses, the number of available wavelengths needs to be updated for each of the remaining unassigned subconnections. At each step, the subconnection with the fewest wavelength possibilities is processed next. If there is a tie, then the subconnection with the most number of hops is selected. This is a natural ordering scheme to use with the RCL assignment algorithm because this scheme already tracks the number of wavelengths that can possibly be assigned to each remaining subconnection.

If a particular subconnection ordering does not produce a feasible assignment, then the wavelength assignment process can be restarted using a different ordering strategy. Furthermore, various combinations of subconnection ordering and wave-length assignment schemes can be considered.

Note that one of the advantages of multi-step RWA when working with a set of demands is the order in which demands are routed and the order in which subconnections are assigned wavelengths can be completely decoupled, allowing more design flexibility.

5.6 Bi-directional Wavelength Assignment

Most demands in carrier networks are bi-directional, where a connection from Node A to Node Z implies a connection from Node Z to Node A. Usually, both directions of the connection are routed over the same path and regenerated at the same sites, thereby yielding identical subconnections in the two directions. An interesting question is whether the same wavelength should be assigned to each pair of subconnections; i.e., if a particular wavelength is assigned to a subconnection extending from Node X to Node Y, should the same wavelength be assigned to the reverse subconnection extending from Node Y to Node X.

From a network management point of view, it may be most expedient to simply assign the same wavelength to the subconnection pair. However, there are scenarios where assigning different wavelengths can improve the efficiency of the network. Consider the very simple four-node topology of Fig. 5.7, and assume that the fiber capacity is just two wavelengths. Furthermore, assume that the optical reach is longer than any of the possible connection paths, such that each connection is equivalent to a subconnection.

Assume that one bi-directional connection is established between A and C, and another between A and D. In Fig. 5.7(a), λ_1 is assigned to both directions of the A-C connection, and λ_2 is assigned to both directions of the A-D connection. With this wavelength assignment, it is not possible to add a bi-directional connection between C and D without converting the wavelength at node B. There is no wavelength that is free along both links of the C-D connection. The wavelength conversion required at Node B can be accomplished via O-E-O regeneration, or, in the future, with an all-optical wavelength converter; either way, an extra cost is incurred.

In Fig. 5.7(b), different wavelengths are assigned to the two directions of connections A-C and A-D. This allows the C-D connection to be added without any need for intermediate wavelength conversion, as shown in the figure. This wavelength assignment thus provides a lower-cost solution.

While this is just a small example, this type of situation does arise in the design of real networks. Even when the number of wavelengths is large, there are scenarios where using different wavelengths in the two directions of a subconnection can result in a lower-cost network, due to less wavelength contention. It occurs most commonly in real networks when there are degree-three nodes, with a lot of bypass traffic in all three directions through the node. (It also may occur in higher-order odd-degree nodes, although this situation rarely arises in practice.)

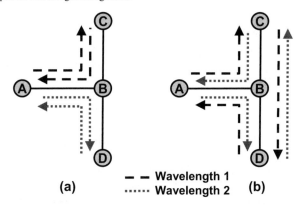

Fig. 5.7 Assume that there are only two wavelengths in this small network. In (a) the same wavelength is assigned to the two directions of each connection. If a connection between C and D is added, that connection must undergo wavelength conversion at Node B in order to avoid wavelength conflicts with the existing connections. (b) Different wavelengths are assigned to the two directions of connections A-C and A-D. The connection between C and D can be added without any wavelength conversion, as shown. (Adapted from [Simm06]. © 2006 IEEE)

Any of the wavelength assignment schemes described above can be readily modified to produce different wavelength assignments in the two directions of a subconnection. For example, consider a scheme where odd and even wavelengths are used for the two directions of a subconnection. Assume that First-Fit runs through the odd-numbered indexed wavelengths, yielding an assignment of λ_5 for a particular subconnection. Then the subconnection in the reverse direction can be assigned λ_6. If the traffic set consists of unidirectional demands as well, where traffic goes in just one direction, then more care needs to be taken because wavelengths are not always assigned in pairs.

While assigning different wavelengths to the two directions was considered above for cost reasons, there are some scenarios that *require* the wavelengths to be different. For example, there is at least one shared optical-layer protection scheme that requires different wavelengths in the two directions, as is covered in Section 7.3.1 of Chapter 7.

5.7 Wavelengths of Different Optical Reach

When assigning wavelengths to subconnections, another factor that needs to be considered in some systems is that not all wavelengths have the same optical reach. This phenomenon is often dependent on the type of fiber that is installed in the network. As discussed in Chapter 4, some amount of chromatic dispersion is desirable in a fiber as it helps to minimize the effects of nonlinear optical impairments. However, some fiber types have regions of very low dispersion that partially overlap with the portion of the spectrum used for transmission. The wavelengths that fall in this low-dispersion region suffer from greater impairments, resulting in reduced reach. These 'reduced-reach' wavelengths usually represent a small percentage of the overall wavelengths.

This effect needs to be considered when performing wavelength assignment. For example, consider an 80-wavelength system where the nominal optical reach is 2,000 km, but where 5 of the 80 wavelengths have an optical reach of only 1,000 km. It is desirable to assign these five wavelengths early in the process. Otherwise, these wavelengths may be left to the end of the assignment process, in which case extra regenerations may be needed to chop the remaining subconnections into even shorter subconnections. In a First-Fit wavelength assignment scheme, the five reduced-reach wavelengths can be assigned low indices. When looking for a free wavelength, the assignment process must check that the reach of the wavelength is suitable for the subconnection being considered. If the reach of the wavelength is too short, that wavelength is passed over. There typically is no need to proactively chop up a connection into smaller subconnections in order to utilize the reduced-reach wavelengths; there are usually enough small subconnections that are naturally formed as part of the design process.

Some carriers have a mix of fiber types in their network where just a subset of the links have fiber with low-dispersion regions. In this scenario, the wavelength assignment process must first consider the fiber type of the links on which a subconnection is routed. If a subconnection is partially on 'good' fiber and partially on 'bad' fiber, then it is up to the system engineers to come up with rules for the optical reach of the wavelengths. This also could be a factor in determining where to regenerate a connection in the scenario where regeneration is needed but it can be located at one of several nodes in the path. It may be desirable to pick the location such that the resulting subconnections are routed on homogeneous fiber types, if possible.

Note that with new builds, carriers are careful to install fiber with an appropriate level of dispersion across the transmission spectrum to avoid this problem. Thus, this issue will become less important with time.

Even if there are no dispersion issues, there still may be small differences in the optical reach of the wavelengths due to other factors. If one wants to squeeze every advantage out of an optical-bypass-enabled system, then one can use the fullest reach of each wavelength. However, there are some complications with this approach. First, it may ultimately lead to more wavelength contention. The length of the subconnection becomes the prime determinant of which wavelength is assigned to it; the orderly wavelength assignment schemes, such as the ones described in Section 5.4, are less relevant. Second, a particular connection may be regenerated differently based on the wavelength(s) that will carry it. This results in less alignment of the subconnection endpoints, which can result in more wavelength contention. Third, it makes multi-step RWA more challenging because the location of the regeneration may depend on which wavelength is assigned to it, so that these two steps must be coupled. Furthermore, as is discussed in Chapter 8, after some point, the marginal benefits of increased optical reach begin to rapidly diminish. Maximizing the reach of each individual wavelength may only result in a small cost savings, which may not justify the additional complexity. It may be more desirable to set an optical reach that almost all wavelengths can attain, with perhaps a small number of wavelengths relegated to shorter reach due to low-dispersion issues.

5.8 Wavelength Contention and Network Efficiency

Even with good wavelength assignment algorithms, wavelength contention is likely to occur when a network becomes heavily loaded. Any contention can be alleviated by adding more regeneration to reduce the wavelength dependencies among the links. However, in real-time planning, there may not be equipment available for regeneration at the desired network nodes. In addition, it is undesirable to add a significant amount of extra regeneration because it will increase the cost of the network. If wavelength contention cannot be resolved, then one or more demands may be blocked, even though there is available capacity to carry a demand.

Wavelength contention is not an issue in pure O-E-O-based networks, where wavelengths can be assigned independently on each link. This raises the question of what impact wavelength contention has on the performance of optical-bypass-enabled networks. Many studies have been performed to analyze this question. The conclusion of most of the studies is that, assuming good algorithms are used, just a small amount of wavelength conversion (whether it be accomplished with O-E-O regeneration or all-optical wavelength converters) is needed to approximate the performance of an O-E-O system. Some of these studies can be found in [SuAS96, KaAy98, VAAD01, Simm02].

Nevertheless, there is not a unanimity of opinion in the industry regarding this question. It is possible to produce studies that indicate wavelength contention has a significant negative effect on network performance. However, this is often due to the choice of algorithm. For example, some studies do not take advantage of regeneration as a chance to perform wavelength conversion; i.e., these studies unnecessarily require that the same wavelength be used end-to-end even when regeneration occurs in the middle of the path. A second example of a less-than-ideal strategy is where enough regenerations are added to an optical-bypass-enabled network in order to have *identical* efficiency to an O-E-O network. This approach may be too extreme and result in an excessive number of added regenerations. A slight reduction in network efficiency is acceptable with an optical-bypass-enabled network because the potential cost savings due to reduced electronics is still very significant.

As the topic of wavelength contention remains somewhat of a controversial issue, a small study is presented here for both a backbone network and a metro network. This study shows, again, that assuming a relatively small amount of extra regeneration can be added to the network to alleviate wavelength contention, wavelength constraints have a very small impact on the network performance. Furthermore, the discussion provides a rationale for why this is so.

5.8.1 Backbone Network Study

The network used for the backbone study is shown in Fig. 5.8. The topology is representative of U.S. backbone networks, with 60 nodes, 77 links, and an average link length of 450 km; the average nodal degree is 2.6. (Reference [Simm02] includes a similar study performed on several other backbone networks of various sizes; the results are consistent across the networks.) In the study, there was one fiber-pair per link with a maximum of 80 wavelengths per fiber. A realistic traffic set was used where roughly 20% of the nodes could be considered major nodes that generated a significant amount of the traffic; the average path length was 1,850 km. (Note that uniform all-to-all traffic is *not* a realistic traffic model.) All demands were at the line-rate such that no grooming was required; all demands were unprotected. Furthermore, the traffic was modeled as being totally dynamic, where the demands arrived one-by-one according to a Poisson process with holding times that were exponentially distributed.

The two architectures compared were a pure O-E-O network where it was assumed that the optical reach was long enough (i.e., 1,200 km) such that no regenerations were required in the middle of a link, and a pure optical-bypass-enabled network with an optical reach of 2,500 km. In both scenarios, the load on the network was increased until the desired steady-state blocking probability was reached. The study focused on the 0.1% and 1.0% blocking scenarios. (For each offered load level, several simulations were run where the system was allowed to reach steady state; the averages were obtained using the *replication/deletion* approach [LaKe91].)

Fig. 5.8 Backbone network used in the study.

The key parameter used for evaluation is the average network utilization at these blocking levels. Utilization is measured as the bandwidth-distance product of the successfully routed demands. The distance used in this calculation is the shortest possible distance for a demand source/destination pair, which is not necessarily the route taken by the demand. Thus, the utilization measure cannot be artificially increased by circuitous path routing.

Another important statistic is the average number of transponders needed per demand. The minimum is two; i.e., the transponders at the endpoints. Additionally, any regeneration along a demand counted as another two transponders.

It was assumed that equipment was available when needed. Furthermore, at a regeneration point, it was assumed that the wavelength of the incoming subconnection could be changed to any wavelength for the outgoing subconnection. Note that in many early studies of wavelength contention, limited wavelength conversion was assumed, where a given wavelength could be converted to only a small set of other wavelengths. However, in reality, tunable transponders and regenerators are generally tunable across the whole transmission band, and all-optical wavelength converters are expected to have full conversion capabilities as well; thus, this restriction is generally no longer applicable.

The O-E-O network terminated every path at every node, and thus there were no wavelength contention issues. In the optical-bypass-enabled network, wavelength contention occurred whenever a feasible wavelength assignment could not be found for a subconnection. Regenerations were judiciously added to alleviate wavelength contention; if more than one regeneration had to be added to a subconnection in order to find a feasible wavelength assignment, the associated demand was blocked. This resulted in a lower utilization than was actually possible, but moderated the number of added regenerations.

The results from the study are shown in Table 5.1. The average network utilization is normalized to 1.0 for the O-E-O architecture for both of the blocking probabilities of interest. In absolute terms, the average network utilization with 1.0% blocking was roughly 10% higher than with 0.1% blocking.

Table 5.1 Results from Backbone Network Study with 100% Dynamic Traffic

		O-E-O Network	Optical-Bypass-Enabled Network
Results at 0.1% Blocking	Normalized Average Utilization	1.0	0.98
	Average Number of Transponders Per Demand	8.42	2.60
	Average Number of Transponders Per Demand (Not counting the excess regeneration)	8.42	2.57
Results at 1.0% Blocking	Normalized Average Utilization	1.0	0.98
	Average Number of Transponders Per Demand	8.47	2.64
	Average Number of Transponders Per Demand (Not counting the excess regeneration)	8.47	2.56

Two transponder statistics are provided in the table. The first statistic listed, the average number of transponders per demand, reflects the actual transponder count in the design. The second transponder statistic indicates the number of transponders that would have been needed if no regenerations were added to alleviate wavelength contention; i.e., this is the number of transponders required simply based on optical reach. The difference in the two numbers is a measure of how much wavelength contention was encountered.

The optical-bypass-enabled network achieved 98% of the utilization of the O-E-O network, indicating that wavelength contention caused little excess blocking. (The 90% confidence intervals are on the order of ±1%.) To attain this high utilization, a relatively small amount of regeneration was added to alleviate wavelength contention. With 0.1% blocking, the average number of transponders per demand increased from 2.57 to 2.60 due to the added regeneration, which is about a 1% increase. For 1.0% blocking, the average number of transponders increased by 3% due to the added regeneration. (With a higher allowable blocking rate, the network is, on average, more heavily loaded, such that wavelength contention arises more often. This is why a greater percentage of extra regenerations were needed at 1.0% blocking as compared to 0.1% blocking.)

Even with the extra regeneration, the optical-bypass-enabled network required less than one third of the number of transponders per demand needed in the O-E-O network. (The average path distance of the successfully routed demands was about 1% longer in the optical-bypass-enabled network as compared to the O-E-O network, indicating there were not major differences with respect to fairness of the demands that were accepted. The path distance used in this calculation is the shortest possible path between the source and destination, which may be different from the path actually followed by a demand.)

5.8.2 Metro Network Study

A similar study was performed on a metro network with an interconnected-ring topology, as shown in Fig. 5.9. In this study, there was one fiber-pair per link with a maximum of 40 wavelengths per fiber. As with the backbone study, the traffic was modeled as unprotected and at the line-rate. Approximately 35% of the traffic was inter-ring, with the remainder intra-ring. The traffic was again assumed to be completely dynamic.

A pure O-E-O-based design and a pure optical-bypass-enabled design were performed for the metro network, where the latter design assumed that the optical reach was long enough to eliminate all required regeneration. The criteria for comparison are again the network utilization, as defined for the backbone study, and the average number of transponders per demand. The study focused on the 0.1% and 1.0% blocking scenarios. In the optical-bypass-enabled design, up to one regeneration could be added per demand for purposes of alleviating wavelength contention.

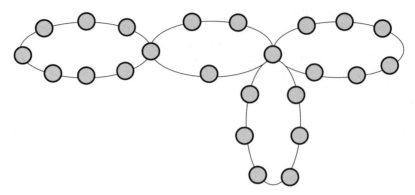

Fig. 5.9 Metro network used in the study.

The results of the study are shown in Table 5.2, where the average network utilization is normalized to 1.0 for the O-E-O designs. (In absolute terms, the average network utilization with 1.0% blocking was roughly 12% higher than with 0.1% blocking.) The optical-bypass-enabled network achieved 94% to 98% of the utilization, again indicating that wavelength contention caused little excess blocking. (The 90% confidence intervals are on the order of ±1%.) Extra regeneration resulted in a 3% increase in the average number of transponders per demand at 0.1% blocking, and a 6% increase for 1.0% blocking. The optical-bypass-enabled network required less than 40% of the number of transponders per demand needed in the O-E-O network.

(Note that, in reality, a metro network is likely to require some amount of grooming. Grooming normally occurs in the electronic domain, thereby requiring O-E-O conversion. Thus, some of the regenerations that were added to alleviate wavelength contention potentially could be needed anyway for grooming.)

Table 5.2 Results from Metro Network Study with 100% Dynamic Traffic

		O-E-O Network	Optical-Bypass-Enabled Network
Results at 0.1% Blocking	Normalized Average Utilization	1.0	0.98
	Average Number of Transponders Per Demand	5.50	2.05
	Average Number of Transponders Per Demand (Not counting the excess regeneration)	5.50	2.00
Results at 1.0% Blocking	Normalized Average Utilization	1.0	0.94
	Average Number of Transponders Per Demand	5.72	2.12
	Average Number of Transponders Per Demand (Not counting the excess regeneration)	5.72	2.00

To demonstrate that optical-bypass-enabled networks can achieve the same level of utilization as an O-E-O network by adding in more regeneration, another optical-bypass-enabled design was performed for the same metro network with 1% blocking and normalized utilization of 1.0 (instead of 0.94 as in the original design). The average number of transponders per demand increased to 2.23 (instead of 2.12 in the original design). In this particular scenario, the increase was fairly moderate. However, in general, adding the extra amount of regeneration may not be a desirable strategy because it requires more equipment and reduces some of the operating advantages afforded by optical bypass.

5.8.3 Study Conclusions

For both the backbone and metro studies, the loss in network utilization due to wavelength contention was relatively small, i.e., on the order of 5% or less. The amount of regeneration added to produce these high utilizations was also relatively small, although it was somewhat higher for the metro case where it was assumed that there were no regenerations needed based on path distances.

To provide insight into why wavelength contention is not a large problem, it is helpful to examine the number of hops in a subconnection. In the backbone network study, the average and maximum number of hops per subconnection were 3.5 and 9, respectively; in the metro study, the associated numbers were 3.3 and 9. (These statistics are the number of hops prior to dividing up subconnections due to wavelength contention.) With less than four hops in the average subconnection, it is not very difficult to find a wavelength that is free on all of the hops.

To explore this further, the backbone network study was repeated, this time with the restriction that wavelength conversion could not occur when a connection was regenerated. Thus, the same wavelength must be used on each link of the end-to-end path. The average and maximum number of hops in an end-to-end path in the backbone network study were 4.8 and 16, respectively. With this large number of hops on which to find a free wavelength, combined with no longer having the option to add regenerations to alleviate wavelength contention, the normalized average utilization with 1% blocking dropped from 0.98 to 0.72, which is a significant drop-off. This demonstrates that optical-bypass-enabled networks can perform poorly if proper design strategies are not employed.

Another factor in achieving high utilization, albeit a much more minor one, is making use of one-step RWA to find a feasible path when the optical-bypass-enabled network was heavily loaded. In the studies, a transformation similar to that described in Section 5.3 was employed when the network was heavily loaded. To speed up the transformation process, rather than searching for a set of regeneration-free paths each time a demand was added, the algorithm maintained a list of up to four possible regeneration-free paths for each node pair (in the backbone network, some node pairs had no such paths). Under heavy load, whenever a demand was added, these predetermined paths were checked for an available wavelength. If a wavelength was found, a corresponding link was added to the transformed graph. The transformed graph was then used to determine the minimum-regeneration path that had a feasible wavelength assignment. This process yielded an approximately 2% increase in the utilization of both the backbone network and the metro network, as compared to a design that always uses a multi-step RWA approach; the number of transponders needed was almost identical. (This improvement in utilization is included in the statistics shown in Table 5.1 and Table 5.2.) Thus, a small benefit can be achieved with the one-step approach. (The one-step approach would produce more significant benefits in real-time planning, where equipment must already be predeployed to be used, especially in the presence of non-tunable transponders.)

It is also worth mentioning that the First-Fit wavelength assignment algorithm was used in the studies. Thus, a very simple assignment strategy is able to produce high utilizations.

The overall conclusion is that algorithms are important in maximizing the performance of a network design. While optical-bypass-enabled networks may require more algorithms than O-E-O networks, these algorithms have already been developed and incorporated in commercial design tools; furthermore, they are not overly complex. These algorithms enable similar utilization as an O-E-O architecture, but with significantly fewer transponders. Because most system vendors provide design tools as part of their system, the burden of handling wavelength assignment is largely removed from the network operator.

Chapter 6

Grooming

As networking technology and services have evolved, one consistent characteristic is that much of the traffic requires a bit-rate less than that of a full wavelength. For example, while many backbone networks support 10-Gb/s wavelengths, most of the network demands require rates of 2.5 Gb/s or lower. Demands at the bit-rate of a wavelength are referred to as *line-rate traffic* or *wavelength services*, whereas demands at a lower bit-rate are referred to as *subrate traffic*.

With SONET/SDH framing at the physical layer, the wavelength line-rate has evolved in accordance with the SONET/SDH rate hierarchy (see Section 1.4). When used to carry subrate SONET/SDH services, the wavelength partitioning is straightforward. For example, with a SONET-based system, a line-rate of OC-N carries a maximum of N effective OC-1 (51.8 Mb/s) units. Thus, one OC-192 wavelength can carry, for example, a combination of three OC-48s and four OC-12s. Similarly, an SDH line-rate of STM-N carries a maximum of N effective STM-1 (155.5 Mb/s) units. With data-oriented services such as IP and Ethernet, the rates of the guaranteed-bandwidth traffic are more arbitrary and of finer granularity. Additionally, data services typically include bursty best-effort traffic, where there is no pre-negotiated bandwidth dedicated to carrying the traffic. Efficiently packing such bursty traffic into wavelengths is a challenge and will not be the focus of most of this chapter.

There are several options for carrying subrate traffic in a network. The most simplistic approach utilizes a full wavelength to carry a subrate demand, thereby wasting the remaining bandwidth of the wavelength. The percentage of waste can be quite large; for example, carrying one OC-3 demand in an OC-192 wavelength wastes 98% of the wavelength capacity. This is generally an untenable solution as even the large bandwidth of current transmission systems cannot support this level of inefficiency.

The more natural solution is to carry several subrate traffic demands in a single wavelength. In one such strategy, known as *end-to-end multiplexing*, subrate demands that have the same source and destination are bundled together to better fill

J.M. Simmons, *Optical Network Design and Planning*,
DOI: 10.1007/978-0-387-76476-4_6, © Springer Science+Business Media, LLC 2008

a wavelength. The demands are then routed as a single unit from source to destination. While multiplexing improves the network efficiency, it may still be inefficient if the level of traffic between node pairs is small. A more effective technique is *grooming*, where traffic bundling occurs not only at the endpoints of the demands, but also at intermediate points. Demands may ride together on the same wavelength even though the ultimate endpoints are not the same, providing opportunities for more efficient wavelength packing. The relative merits of multiplexing and grooming are discussed in Sections 6.1 and 6.2, respectively.

While grooming is an effective means of transporting subrate traffic, it can also be costly. Switches that perform grooming may be expensive and may present challenges in power consumption, heat dissipation, and physical space, particularly as the network traffic increases. This affects how such switches are architecturally deployed in a node, as is covered in Section 6.3. Furthermore, their cost may limit their deployment to only a subset of the network nodes. Selecting nodes for grooming, and strategies for effectively delivering traffic to these sites, are discussed in Sections 6.4 and 6.5, respectively.

Given a set of subrate demands, there typically is no single optimal design to groom the demands into wavelengths. For example, one design could favor minimizing cost at the expense of routing demands over very circuitous paths, whereas another design could place greater emphasis on minimizing path length and reserving capacity for future subrate demands. Such tradeoffs are explored in Section 6.6.

Similarly, there is not necessarily a single heuristic grooming algorithm that always produces the 'best' results. Rather than cover the spectrum of grooming algorithms that have been developed over time, Section 6.7 focuses on one grooming methodology that has produced cost-effective and capacity-efficient designs when applied to realistic optical networks, while maintaining a rapid run-time even with a very large number of subrate demands.

Much of this chapter is applicable to both O-E-O networks and optical-bypass-enabled networks. As grooming is typically accomplished in the electrical domain, it is advantageous to take regeneration into account when selecting grooming sites in a network with optical bypass. For example, the grooming algorithm should favor grooming a connection at sites where regeneration is required anyway. This philosophy is incorporated in the methodology of Section 6.7. In Section 6.8, a network study is performed to illustrate various grooming properties. One of the main results is that even as the amount of grooming increases, a significant amount of optical bypass is attainable, indicating that processes such as O-E-O grooming are compatible with an optical-bypass-enabled network.

As traffic has become more packet oriented as opposed to circuit oriented, grooming research has been steered towards techniques that handle bursty traffic. (With circuits, a typically fixed-rate path is established ahead of time and reserved for the duration of the connection. With packets, relatively small blocks of data are processed individually, where the network resources are utilized as needed to allow better resource sharing.) Section 6.9 provides a brief overview of some of the proposed grooming techniques, along with a discussion of where in the network such techniques may be most suitable. This section is intended to give a glimpse of how grooming may adapt to future network services, as opposed to being a definitive discourse on what carriers will actually implement.

As an aside, while this chapter addresses subrate traffic, it is also possible that a network service could require a bit-rate that is higher than what it is supported by a wavelength. For example, an IP router may generate 40-Gb/s outputs while the wavelengths carry only 10 Gb/s. This type of bit-rate mismatch is handled via *inverse multiplexing*, where the traffic is carried over more than one wavelength. Typically, all of the wavelengths supporting the traffic demand are routed over the same path. Furthermore, to minimize transmission differences, inverse multiplexing usually utilizes contiguous wavelengths in the spectrum. (One particular inverse-multiplexing scheme, which is somewhat more flexible, is known as *Virtual Concatenation* (VCAT); this allows an aggregate signal to be broken up and sent over multiple wavelengths on different paths [BCRV06].)

6.1 End-to-End Multiplexing

End-to-end multiplexing, where traffic demands with the same source and destination are packed into wavelengths, is a simple means of grouping subrate traffic to better utilize network capacity. Once the subrate demands have been grouped into a wavelength, they can be treated as if they are a single demand; i.e., routing, regeneration, and wavelength assignment can be performed on the bundle, as opposed to considering the individual demands comprising the bundle.

The network of Fig. 6.1 is used to illustrate end-to-end multiplexing. In this particular example, the services are SONET-based; however, multiplexing applies to more general service types. The line-rate is assumed to be OC-192, and the subrate demands are as shown in the figure. Each source/destination pair is grouped separately, yielding the six connections shown by the dotted lines in the figure. For example, one wavelength carries the two OC-48s between Nodes A and H, and is thus only 50% full. On average, the six wavelengths are 27% full. Note that if no multiplexing were used, such that each subrate demand is carried on a separate wavelength, the average wavelength fill-rate would be roughly 12%.

The multiplexing function can be performed with a separate multiplexer box, or with a WDM transponder with multiple client-side feeds. For example, many system vendors offer a 'quad-card', e.g., an OC-192 transponder with four OC-48 client-side feeds. The price premium for a multiplexing transponder versus a regular transponder is typically not too large, such that multiplexing is a cost-effective operation.

When multiplexing traffic together, it is important to ensure that the individual subrate demands are compatible. For example, for purposes of meeting a certain quality of service, some demands may have a requirement that they not be routed on certain links in the network. If demands are multiplexed together that each have a different set of 'forbidden' links, then it may be difficult to find a path that satisfies all of the constituent demands.

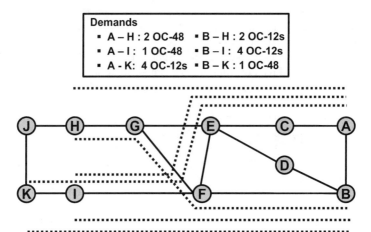

Fig. 6.1 The line-rate is OC-192 and the subrate demands are as shown. With end-to-end multiplexing, demands with the same source and destination are bundled together. Six wavelengths are required to carry the traffic in this example, as shown by the dotted lines.

As a second example, typically some demands require protection while others do not. If such demands are multiplexed together, and the multiplexed unit is protected, then all of the demands in the bundle will be protected, whether or not it is required. It may be ultimately more efficient to reserve space in a partially full protected wavelength for future protected demands rather than mix in unprotected demands.

Consider a design where multiple subrate demands are added at one time to the network. One strategy for efficiently multiplexing the traffic is to first sort the subrate demands by source/destination pair, and then by their protection level (or other quality of service parameter). Within each source/destination/protection class, the demands are then sorted in order from highest bit-rate to lowest bit-rate. Assume that the demands in a particular class will be bundled into K groups, numbered 1 through K, where each group can contain no more than a line-rate worth of traffic. Each demand, starting with the highest bit-rate, is added to the lowest numbered group that still has room for it. Each resulting group is then multiplexed onto a wavelength.

This multiplexing strategy is equivalent to the *First Fit Decreasing* bin packing methodology [GaJo79]. With SONET or SDH traffic, where each successively higher data-rate is an integral multiple of the previous one, this strategy produces the minimum number of wavelengths. With arbitrarily sized subrate demands, as is characteristic of IP guaranteed-bandwidth services, this strategy produces no more than about 20% more than the minimum number of wavelengths [Yue91].

After the groups are formed, the network planner may choose to combine groups with the same source and destination but with different protection levels, if the fill-levels of the groups are low. Again, while this may be more efficient for the current set of demands, it may be ultimately less efficient when demands are added in the future.

The efficacy of multiplexing clearly depends on how much traffic there is between each pair of nodes relative to the line-rate. If the level of traffic is low, then the wavelengths will be poorly filled, resulting in inefficient network loading. Even with high levels of traffic between node pairs, there may be inefficiently filled wavelengths. For example, assume that a node pair generates nine OC-48s in a system with an OC-192 line-rate. With two wavelengths 100% full, one wavelength will be only 25% full. Furthermore, as networks evolve, more node pairs may begin to generate traffic; the initial traffic level between these node pairs may be small, leading to poorly filled wavelengths. Additionally, if the system line-rate is increased to boost the network capacity, the average wavelength fill-rate will decrease, at least initially.

Another potential source of multiplexing inefficiency is traffic churn, where demands are periodically established and torn down. Consider two OC-192 quad-cards at a node, where four OC-48 clients feed each card; assume all eight OC-48s have the same destination. If two OC-48s on each card are torn down, the OC-192 wavelengths are only half full. One could combine the four remaining OC-48s onto a single wavelength; however, this requires either manual intervention or an adjunct edge switch in order to move two of the OC-48 clients to the other quad-card. (Note that moving the OC-48s would momentarily disrupt live traffic.)

While multiplexing is a simple and relatively cost-effective option, the resulting network efficiency may be diminished by its relative inflexibility. Network efficiency can be improved by allowing re-bundling of wavelengths to occur at intermediate nodes of a connection. The process of creating traffic bundles, breaking the bundles down at select intermediate nodes, and creating new bundles is known as grooming, which is covered next.

6.2 Grooming

Grooming attempts to form well-packed wavelengths between two particular grooming sites as opposed to between the ultimate source and destination of the subrate demands. Thus, subrate demands with different endpoints may be bundled onto the same wavelength. Furthermore, the other subrate demands with which a given subrate demand is bundled may change at various points along its path.

Grooming is illustrated in Fig. 6.2, where the network and the demands are identical to what was shown in Fig. 6.1. One possible grooming strategy is illustrated in the figure. A single wavelength is used to carry all of Node A's demands to Node E, regardless of the ultimate destination. Similarly, a single wavelength carries all of Node B's demands to Node E. It is assumed that grooming equipment is deployed at Node E, such that the wavelengths can be 'broken apart' and then reconstituted using different groupings. One wavelength produced by Node E carries all of the demands destined for Node H, regardless of the original source.

A second wavelength from Node E carries all of the demands destined for either Node I or Node K. At Node I, which is also equipped with grooming equipment, further processing occurs. The demands with a destination of Node I are dropped at this node, whereas the remaining demands are packaged into a wavelength and transmitted to Node K.

Several measures can be used to compare this grooming design with the multi-plexed design of Fig. 6.1. First, the grooming design requires five connections in contrast to the six connections required for multiplexing. Second, the groomed wavelengths are on average 75% filled, in comparison to 27% for the multiplexed wavelengths. Finally, in terms of wavelength-link units, grooming requires 9 units whereas multiplexing requires 21 units. (One wavelength utilized on one link constitutes one wavelength-link unit.) By any of these measures, grooming produces a more efficient design. Further comparisons of multiplexing and grooming efficiency are included as part of the network study in Section 6.8.

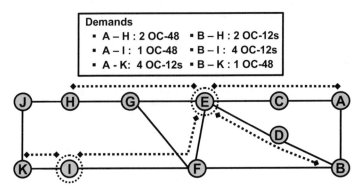

Demands
- A – H : 2 OC-48
- A – I : 1 OC-48
- A - K : 4 OC-12s
- B – H : 2 OC-12s
- B – I : 4 OC-12s
- B – K : 1 OC-48

Fig. 6.2 The network and demand set are identical to that of Fig. 6.1. Grooming is used, with intermediate grooming in Nodes E and I, to pack the wavelengths more efficiently. Five grooming connections are formed as shown by the dotted lines.

While grooming can produce a significant increase in network efficiency, the equipment needed to implement grooming is a lot more complex and costly than a multiplexing card. The required equipment depends on the type of traffic. SONET and SDH traffic demands are groomed in SONET and SDH grooming switches, respectively, where the switch granularity depends on the vendor and the application. For example, two common granularities for SONET grooming switches are OC-1 or OC-48, where the line-rate is typically OC-192. (Switches with an OC-1 granularity are often referred to as operating at the DS-3 level. A DS-3 is a 45-Mb/s signal that is commonly used to carry telephony traffic; it is mapped into a 51.8-Mb/s OC-1 signal.) The switch granularity is the smallest data-rate at which the subrate demands can be 'mixed-and-matched'. Consider a SONET switch with an OC-48 granularity and assume that four OC-12s are deliv-ered to the switch from a higher layer as a single OC-48 bundle. Because the switch granularity is coarser than an OC-12, the four OC-12s must stay bundled together; it is not possible to swap out some of the OC-12s and combine them into different OC-48 bundles. A grooming switch with OC-1 granularity, however, would allow such an operation. In general, finer switch granularity improves the network utilization, but also results in switches that are more expensive.

A very different flavor of grooming device is used for IP traffic; i.e., an IP router. Routers generally operate on the granularity of a packet or a flow, where, for simplicity, a flow can be considered a consecutive sequence of packets between the same two router endpoints. IP routers are more complex than SONET/SDH grooming switches because they also perform a per-packet or per-flow routing function. In comparison, SONET/SDH switches are circuit-based, where ports are configured for the duration of the connection.

Somewhat analogous to IP, Ethernet switches are the grooming devices at the Ethernet frame level. Ethernet switches perform less processing than IP routers, and are thus simpler devices.

Overall, grooming provides improved efficiency; however, the economic and scalability issues affect how grooming equipment is deployed both within a node and across a network. In addition to cost and physical complexity issues, grooming also requires algorithms in order to be effective. These issues are explored in the next several sections.

6.3 Grooming Node Architecture

This section examines how a grooming node should be architected, taking into account cost and scalability issues. For simplicity, the term 'grooming switch' is used in the text to represent any grooming device, including an IP router; however, the figures are labeled with 'grooming switch or router' to emphasize that the architecture applies to multiple types of grooming devices.

6.3.1 Grooming Switch at the Nodal Core

In the first nodal architecture considered here, the grooming switch serves as the 'core' network switch. In this architecture, all network traffic entering the node is processed by the grooming switch, as illustrated in Fig. 6.3. The grooming switch operates in the electrical domain, such that all of the switch ports are equipped with short-reach interfaces. (Grooming in the optical domain is touched on briefly in Section 6.9.) WDM transponders are required for all network traffic entering the node, regardless of whether the traffic is just transiting the node. This architecture is similar to the O-E-O architecture discussed in Chapter 2 (see Fig. 2.4), and has all the attendant scalability issues discussed there; e.g., cost, physical size, power, and heat dissipation.

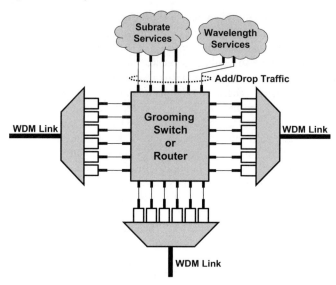

Fig. 6.3 Grooming switch or router at the nodal core. All traffic entering the node from the network, as well as all add/drop traffic, is processed by the grooming switch or router.

Furthermore, the fact that the grooming switch operates on a granularity that is typically much finer than a wavelength exacerbates the situation. Not only does the transiting traffic undergo O-E-O conversion, but it also unnecessarily 'burns' the grooming resources. Consider the case where the grooming device in a node is an IP router, and consider a wavelength that is carrying IP traffic, none of which is destined for the IP router at the node. With the architecture of Fig. 6.3, not only are transponders and router ports needed for this traffic, but also the contents of the wavelength are unnecessarily processed by the IP router. Having to process all traffic entering the node results in an excessively large IP router (as well as excess latency). Routers, and grooming devices in general, tend to be costly, such that this architecture is likely untenable as the network traffic level increases.

A further inefficiency arising from this architecture is that all nodal add/drop traffic enters the grooming switch, even the wavelength services. Such services do not require grooming because they already fill a wavelength; thus, grooming resources are wasted on such traffic.

6.3.2 Grooming Switch at the Nodal Edge

A more scalable grooming node architecture is shown in Fig. 6.4. Here, the core switch at the node is a wavelength-level switch, with the grooming switch serving as an edge switch. In the configuration shown in the figure, the wavelength-level switch enables optical bypass; i.e., it is an OADM-MD or an all-optical switch. In a degree-two node, it is an OADM. This allows any traffic transiting the node to remain in the optical domain so that no transponders or electronic switch ports are required for this traffic.

Wavelengths carrying subrate demands that are either destined for the node or being further groomed at the node are directed by the core switch to the grooming switch. Thus, the grooming switch is used only for the traffic that actually needs to be groomed at the node. As compared with Fig. 6.3, the grooming switch is appreciably smaller in this architecture, yielding a more cost effective and scalable node; e.g., greater than a 50% reduction in the number of grooming ports and the associated capacity of the switch. This is explored further in Chapter 8.

Unless otherwise stated, this is the grooming node architecture assumed in the rest of the chapter.

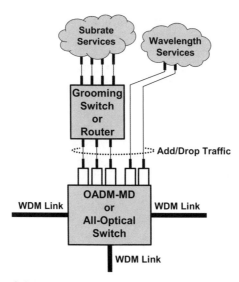

Fig. 6.4 Grooming switch or router deployed at the nodal edge with a wavelength-level switch at the core. The wavelength-level switch can provide optical bypass, as shown here. Only the subrate services that need to be groomed at the node or that need to be added/dropped at the node are processed by the grooming switch or router, yielding a more scalable architecture.

Note that the wavelength services originating at the node feed directly into the core switch, thereby avoiding the grooming switch. If the core switch is an OADM-MD as opposed to an all-optical switch, then, as discussed in Section 2.5, edge configurability is not supported. The direction in which a wavelength service can be routed is solely determined by the transponder into which it feeds, because each transponder can access only one network link. If greater routing flexibility is required with the OADM-MD architecture, then the wavelength services could be passed through the edge grooming switch so that they can be directed to different WDM transponders as required. It is not ideal to 'burn' grooming switch ports for wavelength services, but if the amount of traffic needing this flexibility is moderate, this may be acceptable. Another option is to deploy a small wavelength-level edge switch to provide edge configurability for these wavelength services. However, this would result in three separate 'boxes' in the node, which adds to the network cost and the complexity of the network management. Of course, with an all-optical switch, edge configurability is built-in, so that this issue does not arise.

Note that, in principle, the core switch could be an O-E-O switch with wavelength granularity. This still allows the transiting traffic and the add/drop wavelength services to bypass the grooming switch. Although the combination of the O-E-O wavelength-level switch and the grooming switch is more scalable than the architecture of Fig. 6.3, the amount of electronics is still likely to be an impediment to continued network growth, as was discussed in Chapter 2.

One possible enhancement to the architecture of Fig. 6.4 is to integrate the WDM transceivers on the grooming switch, as shown in Fig. 6.5. (Integrating transceivers with a general switching element was discussed in Section 2.10.) Clearly, the outputs of the transceivers must meet the technical specifications of the transmission system. With the integrated-transceiver solutions that are commercially available, either a single vendor provides both the grooming switch and the transmission system, or separate vendors collaborate to ensure compatibility.

The integration could be carried one step further such that the wavelength-level switch and the grooming switch are integrated in one box; i.e., a switch with dual switch fabrics. One fabric operates at the wavelength level and the other at the subrate level. Ideally, the switch ports are not tied to a particular fabric, so that a port can flexibly direct a wavelength to either fabric depending on its contents.

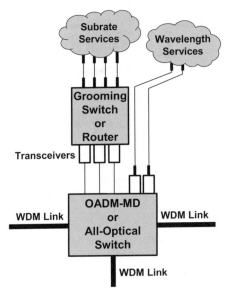

Fig. 6.5 This architecture is similar to that of Fig. 6.4 except the WDM transceivers are integrated on the grooming switch or router.

6.4 Selection of Grooming Sites

For economic reasons, some carriers choose not to deploy a grooming switch in every network node. The nodes without a grooming switch typically must *backhaul* their subrate traffic to nearby grooming nodes. If too few grooming sites are deployed in a network, there may be excessive backhauling, leading to inefficient end-to-end paths. If too many nodes have grooming switches, there are likely to be underutilized switches, resulting in unnecessary cost. From experience with actual metro-core and backbone networks, selecting about 20% to 40% of the nodes to be grooming sites produces designs that are efficient from both a cost and a network utilization perspective. (Carriers may deploy switches at more nodes, however, to provide greater flexibility and forecast tolerance.)

Selecting the nodes in which to deploy grooming switches is usually performed as part of the initial network design phase, before any traffic is provisioned in the network. The network topology and the traffic forecast are used to assist in selecting the grooming sites. Several factors should be considered in this process. First, nodes that generate a lot of subrate traffic are a natural location at which to put a grooming switch. Otherwise, there will be a large amount of traffic to backhaul to other sites, which may be inefficient. Another important factor is the geographic location of the node. Nodes near the center of the network or nodes that lie along heavily-trafficked routes are favored for grooming, as it is likely to be efficient to direct subrate traffic to these sites. Furthermore, nodes that have a high degree (i.e., several incident links) are also good candidates for grooming. Such junction sites provide a good opportunity to 'mix-and-match' the subrate traffic coming from many links so that efficiently packed wavelengths are produced.

With optical-bypass-enabled networks, another factor that should be considered is the amount of regeneration that is likely to occur at a node. Grooming is normally performed in the electronic domain, so that traffic that is groomed is automatically regenerated as well. By deploying grooming switches at sites where a large amount of regeneration may be required anyway, the overall amount of electronics in the network can be reduced further.

Each node in the network can be ranked with respect to the above criteria. Nodes that are ranked highly in two or more categories or that are ranked very highly in one category are generally good nodes to choose as grooming sites. One should also consider the proximity of the non-grooming nodes to those that do support grooming. For example, a possible goal for an optical-bypass-enabled network design may be to select the grooming nodes such that the non-grooming nodes are able to backhaul their traffic without requiring any regeneration along the backhaul path.

These criteria are used to generate an initial list of grooming sites. As part of the network design process, a few iterations can be run, where a small number of nodes are added or removed as grooming sites to check their effect on network cost and capacity (using the forecasted traffic). The results can be used to fine-tune the final selection of grooming nodes.

6.5 Backhaul Strategies

If only a subset of the network nodes are equipped with a grooming switch, then the remaining nodes with subrate traffic either use end-to-end multiplexing to carry their subrate traffic or they backhaul their subrate traffic to a grooming node. If the latter option is used, the non-grooming node is said to '*home*' on a grooming node; the grooming node is referred to here as the '*parent*' node. It is important to consider how the subrate traffic is being delivered from the higher networking layers (i.e., the client layers) to the optical network. If the traffic is packed into wavelengths on the client side without any regard to the ultimate destination, then the non-grooming node will generally send all of its subrate traffic to one particular grooming node (or two such grooming nodes for improved reliability, as discussed below). If the higher networking layer performs some grooming of its own such that the subrate traffic enters the optical network already having been grouped according to its intended parent node, then the non-grooming site may distribute the traffic to multiple grooming nodes. Additionally, the non-grooming site may choose to use end-to-end multiplexing if there is a large amount of subrate traffic that is destined for a particular destination node.

There are a few criteria that may be used to determine on which node, or nodes, a non-grooming node should home. Distance is certainly one key criterion, where the shorter the backhaul distance, the more favored a grooming node is as a parent node. The expected destination of the subrate traffic may play a role as well. If the bulk of the subrate traffic at a non-grooming node is destined for sites to the West, then selecting a parent grooming node to the West may be advantageous to produce more efficient routing. The maximum size of a grooming switch at a node may also need to be considered. If too many non-grooming nodes home on the same grooming node, the required grooming switch size may be too expensive.

Reliability is another key consideration in backhauling the subrate traffic. There are two schemes that are generally used to provide protection for the backhauled traffic. In one scheme, a non-grooming node homes on a single parent node, but the subrate traffic is routed on diverse paths to the parent node. This is illustrated in Fig. 6.6(a) where Node A is a non-grooming site that homes on Node C. After the traffic is delivered to Node C, it can be treated as if the traffic originated at that node. This is a relatively simple scheme to implement, and it does provide protection for the path between Nodes A and C. However, the traffic is vulnerable if Node C, or the grooming switch at Node C, fails.

Using this backhauling scheme, a protected end-to-end path may look as shown in Fig. 6.6(b). The path extends from Node A to Node Z, where Node Z is a non-grooming node that homes on Node E. Both Nodes C and E are points of vulnerability in this scheme.

A more robust scheme is to backhaul the traffic to diverse parent nodes, as shown in Fig. 6.7(a). Here, Node A sends traffic to both of its parent nodes, C and H, over diverse paths. Note that this is an example of where diverse routing from one source to two destinations is desired, as covered in Section 3.6.2.

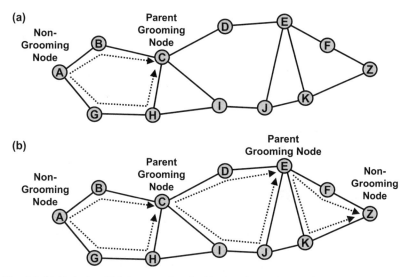

Fig. 6.6 (a) Node A, which is not equipped with grooming equipment, homes on just a single grooming node, Node C. (b) An end-to-end path from Node A to Node Z, both of which home on just a single parent grooming node, is protected against failures except at Nodes C and E.

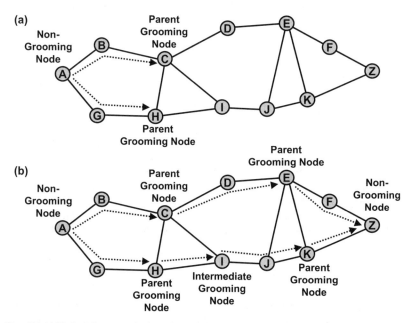

Fig. 6.7 (a) Node A homes on both Nodes C and H and delivers its subrate traffic to both parents over diverse paths. (b) An end-to-end protected path from Node A to Node Z. There are no single points of failure along the path. It is assumed that Node I is used for intermediate grooming.

The advantage of the scheme in Fig. 6.7 is that protection is provided against a grooming-node failure. The disadvantage is that the grooming costs are greater due to the redundancy. It also results in two independent end-to-end paths for the subrate traffic. For example, a protected end-to-end path between non-grooming nodes A and Z is shown in Fig. 6.7(b). Node Z homes on Nodes E and K. One path makes use of Grooming Nodes C and E, and the other path utilizes Grooming Nodes H, I, and K (it is assumed that Node I is used for intermediate grooming). This may be more difficult to manage as compared to having only one set of grooming nodes for the connection.

Protection of subrate demands is covered more fully in Chapter 7.

6.6 Grooming Tradeoffs

The process of grooming subrate traffic often presents design tradeoffs in factors such as capacity and cost. The yardstick with which a grooming design is evaluated may depend on the preferences of the carrier or may depend on the circumstances under which the design is being performed. For example, in a network that is very heavily loaded, link capacity may be the most important factor. Adding a small amount of extra grooming equipment may be justified if it results in not needing to add a second fiber-pair along a link. As another example, a carrier may issue a network planning exercise in order to evaluate the equipment costs of various system vendors. In this scenario, from the viewpoint of the vendors, producing the lowest cost design may be the most important factor.

6.6.1 Cost vs. Path Distance

The first grooming tradeoff illustrated here is between network cost and the distance over which a subrate demand is routed. Consider the network shown in Fig. 6.8, and assume that the network line-rate is OC-192, and assume that all nodes are equipped with grooming switches. Assume that there are four existing OC-48s provisioned in this network. Two OC-48s are between Nodes A and D, and are routed in a single wavelength along the path A-B-C-D; the other two OC-48s are between Nodes D and H, and are routed in a single wavelength along path D-E-F-H. Both of these wavelengths are 50% full.

Assume that a new OC-48 demand request arrives, between Nodes A and H. One option is to route this new demand along the most direct route A-G-H. This option utilizes a grooming switch port at Node A and at Node H, and utilizes a wavelength along the A-G-H path, which would be 25% full. (When counting grooming switch ports in this section, only the network-side ports are included, not the client-side ports.) If the network is O-E-O based, then there is also a regeneration required at Node G.

The second option is to carry the new OC-48 demand using the two wavelengths that have already been deployed. One wavelength carries the demand from Node A to Node D. At Node D, the traffic enters a grooming switch that directs this OC-48 to the wavelength running between Node D and Node H. This solution does not utilize any additional grooming switch ports or transponders, and is thus lower cost than the first option. Additionally, it does not require provisioning any new wavelengths, so from that viewpoint, it requires less capacity; i.e., there are a total of six wavelength-links occupied with this option as compared to eight wavelength-links with the first option.

However, if capacity is evaluated based on a finer granularity than a wavelength, e.g., an OC-48, then directly routing the new demand over A-G-H utilizes 14 OC-48-links whereas the second option utilizes 18 OC-48-links. If it is expected that there will be future subrate demands that will require the bandwidth between A and D or between D and H, then it may be desirable to directly route the OC-48 over A-G-H, with the expectation that ultimately this will result in a lower cost network.

Another factor to consider is that while the second option is less costly, it routes the new demand over a longer path and requires one intermediate grooming; thus, this option is somewhat more vulnerable to failure with respect to the new demand. Furthermore, if the traffic were IP rather than SONET, such that this option results in the traffic being processed by an intermediate IP router at Node D, then this may result in additional latency and/or jitter. (Latency is the end-to-end transmission delay of a connection; jitter is the variation in this delay over time. The latency and jitter produced by an IP router are typically much more significant than that produced by a SONET or SDH switch.)

A carrier would need to weigh these various factors to determine how the new demand should be carried.

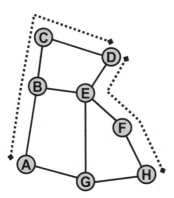

Fig. 6.8 The line-rate is OC-192. One wavelength between Node A and Node D carries two OC-48s. One wavelength between Node D and Node H also carries two OC-48s. If a new OC-48 demand is added between Node A and Node H, it can potentially be carried in the existing wavelengths, rather than establishing a new connection directly along A-G-H. This results in lower cost but a longer path that burns future capacity.

6.6.2 Cost vs. Capacity

The second grooming tradeoff considered here is between network cost and capacity. Figure 6.9 illustrates a small linear network, where it is assumed that the line-rate is OC-192. Assume that two new OC-48s are being added to the network, one between Nodes A and C and one between Nodes B and D.

One grooming option is shown in Fig. 6.9(a), where each OC-48 is simply assigned to a new wavelength. This utilizes one grooming switch port at each of Nodes A, B, C, and D. The number of utilized wavelength-links is four.

A second option is shown in Fig. 6.9(b). Here, Nodes B and C are used as intermediate grooming sites. The OC-48 from Node A is delivered to the grooming switch at Node B, which bundles it with the OC-48 originating at Node B. Both OC-48s are carried on a single wavelength to Node C, where they are delivered to the grooming switch at that node. The OC-48 destined for Node C drops at the node, whereas the remaining OC-48 is carried in a wavelength to Node D. This utilizes a total of six grooming switch ports: one at Node A, two at Node B and Node C, and one at Node D. Thus, two more ports are utilized as compared to the first option. However, the number of utilized wavelength-links is three as opposed to four, because just one wavelength needs to be provisioned on Link BC as opposed to two. If this link is heavily loaded, such that reducing the number of utilized wavelengths is important, then the second option, though more costly, may be preferred.

As mentioned in the previous section, the addition of intermediate grooming along the paths of the two OC-48s is another factor to consider. The extra grooming of Fig. 6.9(b) potentially reduces the availability of the circuits. Additionally, if the traffic were IP rather than SONET, there may be additional latency or jitter.

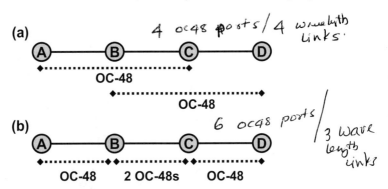

Fig. 6.9 The line-rate is OC-192. Two OC-48 demands are added, one between A and C, and one between B and D. (a) The two OC-48s are carried in separate connections between their respective endpoints. This option requires four grooming ports and utilizes two wavelengths on Link BC. (b) Nodes B and C are used to groom the traffic such that there is a single connection between B and C carrying two OC-48s. This option occupies just one wavelength on Link BC, but it requires six grooming ports.

6.6.3 Grooming Design Guidelines

To better control the grooming process, a carrier may specify certain design guidelines. For example, a limit may be imposed on the number of intermediate grooming switches through which any given demand can be processed. This may be done for reliability or latency/jitter reasons. Second, a limit may be imposed on the allowable path 'circuitousness' for any demand. For example, there could be a guideline that the end-to-end path over which a demand is routed should be no greater than $P\%$ longer than its most direct path, for some positive value P. This can be imposed for reliability reasons. Alternatively, it can be specified to serve as a guideline as to the desired balance between reducing cost in the current design at the expense of occupying bandwidth that may be needed for future demands. The higher the value of P, the greater the emphasis placed on reducing current cost. Other restrictions may be placed on the types of traffic that can be carried together in a single wavelength. There could be segregation based on service types, protection types, certain customers, etc.

It is important that the grooming algorithms be flexible enough to enforce any such rules.

6.7 Grooming Strategies

Grooming algorithms have evolved in concert with the network topology and the network traffic. Some of the earliest work on grooming specifically addressed minimizing cost in ring topologies; e.g., [SiGS98, GeRS98]. However, networks have evolved to mesh topologies, requiring grooming algorithms that work on arbitrary topologies. The algorithms must be flexible with respect to the demand granularities as well. Furthermore, the number of individual subrate demands that may need to be carried by the network can be in the tens of thousands, thereby requiring efficient grooming techniques.

One particular general grooming strategy that has produced cost-effective and wavelength-efficient designs for realistic networks is presented next. More general coverage of grooming can be found in texts such as [ZhZM05] or tutorial papers such as [DuRo02, ZhMu03, HuDu07].

6.7.1 Initial Bundling and Routing

Assume that there are multiple new subrate demands being added to the network at once. The first step is to group the new subrate demands into bundles that contain at most one wavelength worth of traffic, where all of the demands in a given bundle have the same source and destination nodes. The bin-packing scheme described in Section 6.1 in relation to end-to-end multiplexing can be used for this purpose. The algorithm must ensure that the demands that are bundled together are compatible; e.g., they cannot have conflicting service types, protection types, etc. If there is just a single new subrate demand, then it is placed in a 'bundle' by itself.

After the bundles are formed, they are routed end-to-end using the standard techniques for routing a wavelength service; e.g., alternative-path routing, as described in Section 3.4.2. (The source and destination nodes of the demands within the bundle are the endpoints of the bundle as well.) Network load can be used as one criterion for selecting which alternative path to select for a given bundle. The load on a link can be approximated by the wavelengths that have already been provisioned on the link for existing traffic and by the new bundles that have already been routed. (It is only the approximate load because the amount of grooming that will ultimately occur is not known at this point.)

Another factor that needs to be considered when selecting a path is whether the endpoints of the bundle have grooming switches. If one or both of the bundle endpoints do not have grooming capability, then the path chosen must pass through the appropriate 'parent nodes' that do have grooming switches. If none of the predetermined candidate paths pass through the desired parent nodes, then the routing can be done in steps. For example, assume that the bundle endpoints are Nodes A and Z, and assume that Node A does not have a grooming switch. The routing must go from Node A to the parent node of A, and then from the parent node of A to Node Z.

At the end of this phase of the grooming algorithm, all bundles have been routed over a tentative path. The tentative path for the bundle can be considered the baseline path for the demands in the bundle. If a different path is considered for a demand in order to improve the grooming, as described below, the new path can be compared to the baseline path to determine whether it is excessively long.

Some of the bundles that are formed may contain a full wavelength, or very close to a full wavelength, worth of traffic. No further grooming operations need to be done with these bundles. As the amount of traffic grows in the network, the number of full bundles increases as well, so that the grooming process remains scalable.

6.7.2 Grooming Operations

At this point, all bundles are routed end-to-end, similar to what would be done if the traffic were simply being multiplexed. The term *grooming connection (GC)* is used here to refer to a path that is terminated at either end on a grooming switch. This is illustrated in Fig. 6.10, where a GC extends between Nodes A and E, both of which are assumed to have grooming switches. The intermediate nodes may have grooming switches as well; however, they are not used to process this particular GC. Additionally, note that there may be regeneration along the path of a

GC, even in an optical-bypass-enabled network. In the figure, regeneration is occurring at Node C. (If a portion of a bundle's path extends from a non-grooming node to a parent grooming node, then that portion is not a GC and does not need to be considered for the GC combination operations described below.)

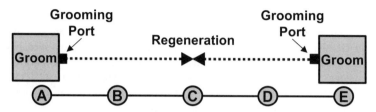

Fig. 6.10 The grooming connection (GC) represented here by the dotted line terminates on the grooming switches at Nodes A and E. A GC may need to be regenerated, as shown at Node C. The intermediate nodes, B, C, and D, may contain grooming switches as well; however, this GC is not processed by them.

Two types of GCs are distinguished here. First, the *existing GCs* encompass those grooming connections that have already been established in the network. While new subrate demands can be added to an existing GC, subject to its maximum capacity, it is assumed an existing GC cannot be rerouted, as that would disrupt existing traffic. Any existing GC that is already filled to capacity can be ignored for purposes of further grooming. Second, there are *new GCs*, formed from routing the bundles containing the new subrate demands. There is more flexibility with the new GCs: new demands can be moved into or out of a new GC; a new GC can be routed over a different path; and a new GC can be split into multiple shorter GCs. (If the grooming switch supports a 'make-before-break' feature, then moving or rearranging existing GCs may be possible without disturbing the corresponding demands. For example, if an existing GC is being rerouted, a duplicate GC is created and sent in parallel over the new route; after a short period of time, the GC on the original route is removed. With this feature, the above distinction between new and existing GCs is not necessary.)

Each GC occupies one wavelength along each hop of the GC, and utilizes a grooming switch port at either endpoint. Reducing the number of GCs can be beneficial, as it frees up capacity and switch ports, and possibly removes some regeneration equipment. In order to reduce the number of GCs, the next step is to perform various 'combination operations'. Typically, the operations proceed starting with the new GCs that have relatively low fill and that extend over several hops. In all of the operations described below, for simplicity, it is assumed that the line-rate is OC-192 and the demands are OC-48s; clearly, the operations hold for more general scenarios. In all of the examples, *it is assumed that GC 1 is a new GC* (i.e., GC 1 contains subrate demands that have not been provisioned yet), whereas the other GCs can be either new or existing. (Again, as mentioned above, if the grooming switch supports make-before-break, then GC 1 could be an existing GC. With this feature, it may be desirable to combine existing GCs due to network churn that resulted in partially full GCs.)

The first operation considered is where all of the demands in a new GC are moved into another GC, where both GCs have the same path. This simple operation, illustrated in Fig. 6.11(a), allows one GC to be removed. This type of operation often occurs after another operation 'chops' a longer GC into smaller GCs, where a resulting GC now aligns with another GC.

A similar operation is where all of the demands from one GC are moved into another GC that has the same endpoints, but a different path. This is shown in Fig. 6.11(b). GC 1 is routed along the path A-B-C-D with 2 OC-48s, and GC 2 is routed along path A-E-F-D with 2 OC-48s. The demands from GC 1 can be merged in with GC 2, such that GC 1 is removed. In performing this operation, it is necessary to check that the new path is satisfactory for the demands of GC 1.

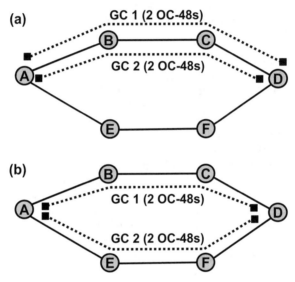

Fig. 6.11 Combining GCs with the same endpoints to allow one of the GCs to be removed. (a) The two GCs have the same path. The two OC-48s from GC 1 can be moved to GC 2. (b) The two GCs have different paths. The two OC-48s from GC 1 can be moved to GC 2, assuming this new path is satisfactory for the demands in GC 1.

It may be necessary to perform multiple operations in order to remove a GC. In Fig. 6.12, GC 1 is routed along the path A-B-C-D with 2 OC-48s, GC 2 is routed along the same path with 3 OC-48s, and GC 3 is routed along path A-E-F-D with 3 OC-48s. One OC-48 from GC 1 is moved into GC 2 and the other OC-48 is moved into GC-3, allowing GC 1 to be removed. (This assumes that the demands in GC 1 do not need to be carried in the same wavelength and do not need to be routed over the same path.)

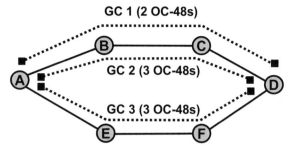

Fig. 6.12 One OC-48 from GC 1 can be moved to GC 2 and the other OC-48 can be moved to GC 3, allowing GC 1 to be removed.

In the next operation, a new GC is 'split' at an intermediate point, and the demands moved into two shorter GCs. This is illustrated in Fig. 6.13. The demands from GC 1 are placed into both GC 2 and GC 3. While the path is the same, this operation adds an intermediate grooming point for the demands from GC 1; thus, it must be verified that the maximum number of grooming points for a demand, if specified, is not violated. In a variation of this operation, GC 2 and/or GC 3 do not lie along the same path as GC 1. This both adds an extra grooming point and modifies the path. These operations can be taken a step further such that a new GC is split at two points and the demands are moved into three shorter GCs, resulting in two additional intermediate grooming points for the demands and possibly a different path.

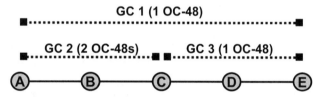

Fig. 6.13 The demands from GC 1 can be moved into both GC 2 and GC 3. This adds another grooming point for the demands in GC 1.

The operation shown in Fig. 6.14 reduces the capacity requirements, although not necessarily the switch port requirements. Figure 6.14(a) shows the original setup with GC 1 and GC 2. This requires one wavelength along Links AB and BC and two wavelengths along Links CD and DE, and one switch port at Node A, one at Node C, and two at Node E. Figure 6.14(b) shows the result of the operation. GC 1 is shrunk such that it only extends from Node A to Node C, and the demands in GC 1 are now carried in both GC 1 and GC 2. This requires one wavelength along each link, although it still requires a total of four switch ports. (It also adds another grooming point for the demands of GC 1.)

In an optical-bypass-enabled network, it is preferable to perform this operation such that regenerations are eliminated, when possible. For example, in Fig. 6.14, assume that the optical reach is 1,000 km, such that one regeneration is originally required at Node C of GC 1. By terminating GC 1 at Node C, this regeneration is removed (or, more precisely, the regeneration is occurring in concert with grooming), thus reducing the network cost. Consider performing an alternative grooming operation, where it is assumed that a half-filled GC, GC 3, extends from Node D to Node E. Assume that GC 1 is terminated at Node D instead of Node C, and its demands carried in both GC 1 and GC 3, as shown in Fig. 6.14(c). With this operation, one regeneration is still needed along GC 1, resulting in a more costly arrangement than in Fig. 6.14(b). Thus, as this example illustrates, regeneration should be a factor in selecting which grooming operations to perform.

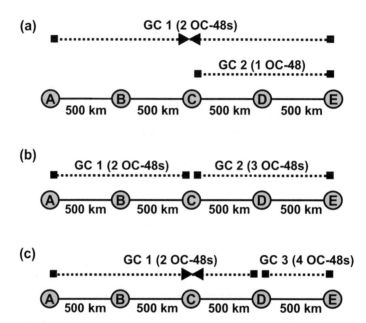

Fig. 6.14 (a) In the original setup, two wavelengths are utilized along C-D-E. (b) After GC 1 is shortened, and its demands also placed in GC 2, only one wavelength is required along C-D-E. Assuming an optical reach of 1,000 km, the regeneration at Node C is removed. (c) If GC 1 had been terminated at Node D instead of Node C, and the demands carried in both GC 1 and GC 3, one regeneration would still be required.

Another operation that saves capacity, albeit at added expense, was illustrated in Fig. 6.9 in Section 6.6.2, where the overlapping portions of two GCs are combined. This is likely to be performed only if capacity is tight on the overlapping links.

The algorithm can make several passes through the GCs to perform these various operations. It is generally preferable to perform the operations that do not change the path prior to those that do change the path. This allows the demands that ideally should be routed along a certain path to use the GCs that lie along this path, as opposed to using the GCs for a circuitously routed demand.

In scenarios where demands are added one at a time, the algorithm can be less aggressive in shifting demands to longer paths, in anticipation that future demands will be better suited to be routed along some of these links. Furthermore, it may be desirable to proactively divide a new GC into two smaller GCs, even though it provides no current benefit, so that future demands may be more efficiently groomed. These decisions can be initially guided by the traffic forecast and later by the network's traffic history.

The run-time of this grooming scheme is very manageable. For example, performing these grooming operations on 10,000 subrate demands on a 60-node network required less than 5 seconds on a 1.6 GHz PC. As an indicator of the effectiveness, the initial average GC fill-rate was 23% (this corresponds to end-to-end multiplexing); after completing the grooming operations, the average GC fill-rate was 85%. With 20,000 subrate demands on the same network, the grooming run-time increases by about 80%.

After the grooming operations are complete, the new GCs that remain can be treated like wavelength-level end-to-end demands. Regeneration sites on each GC are selected, if necessary, to break the GCs up into subconnections, as described in Chapter 4. Wavelengths are then assigned to the subconnections using the techniques of Chapter 5.

The grooming methodology described above holds for unprotected or protected demands. If the bundles that are formed contain protected traffic, then the bundle is routed end-to-end over diverse paths. When performing the various grooming operations, which may entail shifting the path of a demand, it is necessary to ensure that the working and protect paths of a given demand remain routed over diverse paths. Furthermore, depending on the protection scheme, the grooming combination operations may allow the working paths of some subrate demands to be bundled with the protection paths of other subrate demands. Protection in general is covered in more detail in Chapter 7; protection of subrate demands is specifically addressed in Section 7.10.

In real-time grooming scenarios where equipment may be limited, it may be necessary to use a graph model that captures the available equipment and available wavelengths [ZZZM03]. This is similar to the graph transformations discussed in Sections 3.5.1 and 5.3, where a graph is constructed to represent the available equipment and capacity in the network. The existing GCs with spare capacity can also be modeled in the graph. A grooming design can then be performed by running a shortest path algorithm on the graph. As discussed in the earlier sections, this type of detailed resource modeling is more helpful when the amount of available resources is very low and it becomes very difficult to find satisfactory paths where there are available grooming switches, regeneration equipment, and wavelengths. Moreover, when the amount of available resources is relatively small, the size of the transformed graph is accordingly smaller, making these modeling methods more tractable.

6.8 Grooming Network Study

To investigate various quantitative trends related to grooming, a network study was performed using the same 60-node backbone network that was shown in Fig. 5.8. The results of any grooming exercise heavily depend on the traffic distribution among nodes, the network topology, and the network line-rate. The numbers presented here are indicative of relative trends as opposed to absolute statistics that hold across all networks.

SONET-based traffic was assumed, with a line-rate of OC-192, and all demands at either the OC-48, OC-12, OC-3, or DS-3 rates. (Typically, network traffic would include some wavelength services as well; however, to focus on the grooming aspects, such services were not a part of this study.) An optical-bypass-enabled network was assumed, with an optical reach of 2,500 km. In the grooming procedure, demand paths were allowed to be up to 30% longer than their baseline (i.e., directly-routed-path) distance, or were allowed to be up to 1,000 km in length, whichever was longer.

Three different aggregate network demand scenarios were considered: a total of 1.2 Tb/s of traffic, a total of 2.5 Tb/s of traffic, and a total of 5 Tb/s of traffic. In all scenarios, 50% of the traffic required protection, which was implemented with 1+1 dedicated protection (the protection was implemented at the subrate level, as described in Section 7.10.2; two end-to-end paths were allocated for each demand). The aggregate network demand is calculated by summing the total bi-directional traffic sourced in the network. Protected demands were counted twice; e.g., a protected OC-48 demand contributed 5 Gb/s to the aggregate demand.

6.8.1 Grooming Switches at All Nodes

In the first design, a grooming switch was deployed at each node in the network (the nodal architecture was assumed to be that shown in Fig. 6.4). Figure 6.15 shows a plot of the average number of wavelengths utilized per link as a function of the allowable number of intermediate grooming points per demand, for each of the traffic scenarios. First, consider the results for the 5-Tb/s aggregate demand scenario. With no intermediate grooming, an average of 190 wavelengths were utilized per link. As expected, the number decreased as the allowable number of intermediate grooming points increased. For example, with up to five intermediate grooming points per demand, an average of 42 wavelengths were utilized per link. The average fill-rate of the resulting *grooming connections* (GCs), shown next to each data point on the graph, ranged from 24% with no intermediate grooming to 88% with up to five intermediate grooming points. Most of the grooming benefit was achieved by allowing just two intermediate grooming points per demand, which produced an average fill-rate of 76%.

With just 2.5 Tb/s of aggregate demand, more intermediate grooming was required to achieve a similar fill-rate. For example, up to three intermediate grooming points were required to achieve an average fill-rate of 75%. With 1.2 Tb/s of aggregate demand, up to five intermediate grooming points were required to achieve this fill-rate. As could be expected, lower traffic levels required more grooming to produce well-packed wavelengths.

Note that zero intermediate grooming is equivalent to end-to-end multiplexing, indicating the potential inefficiency of this scheme. With no intermediate grooming, the average fill-rates were 10%, 15%, and 24% for the 1.2-Tb/s, 2.5-Tb/s, and 5-Tb/s aggregate demand scenarios, respectively.

Given that grooming occurs in the electrical domain, it is interesting to examine the average optical-bypass percentage in the network as the amount of grooming increased. This statistic represents the percentage of wavelengths entering a node that traverse the node in the optical domain. The average optical-bypass percentage for the 5-Tb/s scenario is shown by the top curve in Fig. 6.15. The percentage of optical bypass decreased by a small amount, 75% to 70%, as the amount of grooming increased. (Though not shown in the figure, the percentage dropped from 75% to 68% for the 2.5-Tb/s scenario, and from 75% to 64% for the 1.2-Tb/s scenario.) This indicates that much of the O-E-O conversion required by the additional grooming was counterbalanced by the reduced need for regeneration due to shorter grooming connections. Thus, efficient grooming is quite compatible with an optical-bypass-enabled network.

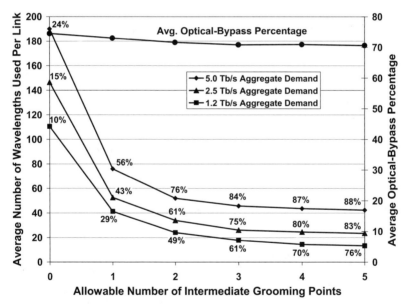

Fig. 6.15 Average link utilization (with 1.2 Tb/s, 2.5 Tb/s, and 5 Tb/s of aggregate demand) and average optical-bypass percentage as a function of the allowable number of intermediate grooming points per demand. The optical-bypass curve is for the 5-Tb/s scenario; the percentages were slightly lower for the other scenarios. The percentages specified next to the data points are the average fill-rates of the resulting Grooming Connections. All 60 network nodes were equipped with grooming switches.

One can also consider an architecture where all subrate traffic is passed through a grooming switch at *every* intermediate node (e.g., all nodes are as shown in Fig. 6.3). Thus, on each link, the wavelengths are filled as much as possible. In all three aggregate demand scenarios, the average number of utilized wavelengths per link is four to five wavelengths lower with this architecture as compared to an architecture where only five intermediate grooming points are allowed per demand. In the 5-Tb/s scenario, this reduces the average utilization by 10%, indicating that limited intermediate grooming achieves close to the optimal packing for high traffic levels. In the 1.2-Tb/s scenario, this is a 30% savings in average utilization. However, at this level of demand, the network fill-rate is low enough that efficient wavelength packing may not be critical. Thus, the extra cost of grooming at every intermediate node is likely not justified, as explored in Chapter 8.

6.8.2 Grooming Switches at a Subset of the Nodes

The study was repeated where only 15 of the 60 nodes were equipped with grooming switches. In this configuration, the 45 non-grooming nodes backhauled their unprotected traffic to a single grooming node, and their protected traffic to two diverse grooming nodes, as in Fig. 6.7. (The backhauling was not counted as intermediate grooming.) The results are shown in Fig. 6.16. After backhauling, the traffic is concentrated at a relatively small number of nodes, resulting in well-packed wavelengths even without any further grooming. For example, in the 5-Tb/s aggregate demand scenario, the *grooming connections* were 84% filled, on average, with just end-to-end multiplexing between the grooming sites. This increased to 92% with up to one intermediate grooming point allowed per demand.

The average optical-bypass percentage was roughly five percentage points lower as compared to the design where all nodes had grooming switches, due to backhauling traffic from a non-grooming node to a nearby grooming node; however, the amount of optical bypass was still high. (The curve shown holds approximately for all three aggregate demand scenarios.)

The chief motivation for concentrating the grooming at a relatively small number of nodes is cost. It is generally more cost effective to have a small number of switches where the ports are used for well-packed wavelengths rather than have many switches where the ports are used inefficiently. Moreover, the first-deployed cost of some of the switches may not be justified by the level of grooming required at the corresponding nodes. Note, however, that backhauling results in longer end-to-end path distances. In the study, the paths were roughly 10% longer when grooming switches were deployed in just 15 nodes as compared to when they were deployed in all 60 nodes. Additionally, while it did not occur in this study, concentrating the grooming in a relatively small number of nodes could lead to poor load balancing, where the links near the grooming nodes would be more heavily utilized. Furthermore, this architecture is more vulnerable to changes in the traffic; e.g., if some of the non-grooming nodes end up with a significant amount of subrate traffic, a lot of backhauling would be needed.

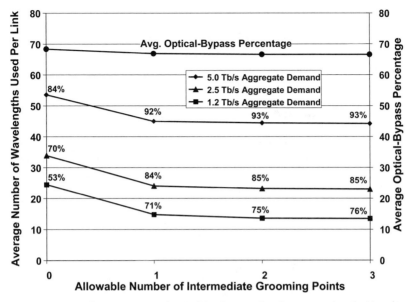

Fig. 6.16 Same as Fig. 6.15, except only 15 of the 60 network nodes were equipped with grooming switches. Less intermediate grooming was required to achieve high Grooming Connection fill-rates.

6.9 Evolving Techniques for Grooming Bursty Traffic

As networks evolve, and network services becomes more data oriented and more dynamic, the paradigm of setting up circuits for an extended period of time becomes less appropriate for some subset of the traffic. In the discussion on grooming above, it was assumed each subrate demand was allocated some fraction of a wavelength circuit and was combined with other subrate demands to better utilize the wavelength's capacity. With very dynamic traffic, a service may last for a very short period of time (e.g., milliseconds), or the service may produce data very sporadically. Establishing a circuit and dedicating bandwidth to such services can be inefficient. It may be more effective to statistically multiplex such bursty traffic onto a wavelength rather than dedicate a portion of a wavelength to each service. Whereas traditional circuit grooming slices up a wavelength by data-rate and dedicates bandwidth to individual services, more data-oriented approaches to grooming slice up a wavelength asynchronously in time and allow services to grab the wavelength as needed. While traditional electronic-based IP routers can be used for this purpose, scalability and latency issues have led to research into alternative grooming strategies for future networks.

In this section, several proposed schemes for carrying bursty traffic are briefly summarized. Most of the methods, though not all, rely on various forms of grooming in the optical domain. This discussion is meant to provide a flavor of the direction grooming research is taking; it is not intended as a predictor of what may actually be implemented in carrier networks in the future. The discussion focuses on IP traffic; however, the schemes generalize to other services as well (e.g., Ethernet).

6.9.1 Selective Randomized Load Balancing

The first approach to handling bursty traffic discussed here can be considered more of a departure from typical network routing than from traditional grooming. The *Selective Randomized Load Balancing (SRLB)* scheme [ShWi06] is based on a traffic model where the endpoints of the traffic may rapidly change, but the aggregate amount of traffic sourced/sunk at each node remains fairly constant. (This is known as the *hose* traffic model [DGGM02].) Consider a network with N nodes, where M of these nodes are selected as hub (grooming) sites. In SRLB, the traffic sourced at any node is randomly delivered to one of the M hub sites, independent of the ultimate destination (although, all traffic in a given IP flow is routed to the same hub). Traffic is groomed at the hub nodes and then sent to the destination.

The advantage of this two-phase routing approach is its smoothing effect on variable traffic, thereby making the network amenable to slow circuit switching. With the hose-traffic-model assumption, the connections between any of the N nodes to any of the M hubs are relatively constant in size even though the endpoints of the traffic may be highly variable. Additionally, because traffic passes through just one hub, the aggregate amount of required routing resources in the network is smaller as compared to a scheme that utilizes several stages of intermediate routing. Furthermore, network jitter is likely reduced because all flows are processed by just one router. The disadvantage of the scheme is the end-to-end path may be significantly longer than the shortest possible path. Additionally, the performance of the scheme is tied to the validity of the hose model.

6.9.2 Optical Flow Switching (OFS)

In *optical flow switching* [ChWM06], end-to-end wavelength connections are requested from the network to carry a data flow. Scheduling mechanisms are used to co-ordinate the assignment of resources to meet the connection requests. Once the connection is established, there is no need for further processing of the data flow within the network. It is assumed that the duration of a flow is long enough such that the process of scheduling the connection and configuring the resources prior to the flow transmission does not represent a significant inefficiency.

To better ensure that the bandwidth of the wavelength connection is well utilized, aggregation may occur in the access or metro-core portions of the network. The aggregation network could be based on a switched architecture, requiring very fast switches. Alternatively, a passive broadcast solution could be implemented, where a group of users have access to the same wavelength; a media access control (MAC) protocol is used to avoid collisions among the users sharing the wavelength. Overall, this scheme advocates fast access reconfiguration and slow core reconfiguration.

6.9.3 Optical Burst Switching (OBS)

Optical burst switching [QiYo99, ChQY04] operates on the granularity of a data burst rather than a circuit or a packet, where the data bursts are statistically multiplexed in the optical domain. The bursts of data are assembled at the edge of the network using electronic buffers. A separate control packet is sent to the destination a short time ahead of the data burst, reserving the necessary resources for the burst at each intermediate switch along the path. This allows the data burst to be immediately switched in the optical domain upon its arrival at an intermediate node, without requiring any buffering, assuming the control packet was successful in scheduling the resources. The data burst is sent without waiting for an end-to-end path to be established, thus allowing connections with a very short duration to be handled efficiently.

In one OBS implementation, *Just In Time* (JIT), the required resources are reserved for a data burst as soon as the associated control packet arrives at an intermediate node [TeRo03]. In a more recent OBS variant, *Just Enough Time* (JET), the resources are not actually configured until right before the data burst arrives; i.e., in between the control packet arrival and the data burst arrival at a node, the resources can be used for other bursts in order to improve the system efficiency [ChQY04].

One drawback to OBS, however, is the potential for contention when reserving resources. A burst may be sent partially along its path only to encounter a node where the required resources were not able to be reserved. The burst must then be sent on an alternate path or be dropped (in this context, 'dropped' indicates the burst is lost). In order to avoid a high drop rate, it may be necessary to operate the network at a relatively low level of utilization.

Contention is likely worse as the network size increases. Thus, OBS may be better suited to smaller metro-core or regional applications as opposed to backbone networks. By playing a grooming role at the edge of the backbone network, optical grooming techniques such as OBS potentially can reduce the required size of the electronic grooming boxes in the core of the network. This is explored further in Chapter 8.

6.9.4 TWIN

The *Time-Domain Wavelength Interleaved Networking* (TWIN) approach [WSGM03] can be considered a form of optical burst switching. TWIN creates optical multipoint-to-point trees to each destination node, where any node that directly communicates with the destination node is a member of the tree. Each 'destination-tree' is associated with a particular wavelength. The branching points of the tree are equipped with switches that are capable of being configured as merging devices; i.e., the signals from multiple network input ports are directed to the same output port.

The traffic source transmits a data burst to a particular destination by tuning its transmitter to the appropriate wavelength. By making use of rapidly tunable transmitters, bursts from multiple sources are multiplexed together. A MAC protocol is used to schedule the sources so that the bursts do not collide on the tree. As with OBS, TWIN may be more suitable for a regional or metro-core network as opposed to a large backbone network.

6.9.5 Lighttrail

The *lighttrail* scheme [GuCh03] takes advantage of the drop-and-continue functionality of many of the optical-bypass-enabling network elements, where a wavelength can both drop at a node and traverse the node. A wavelength connection is first created between two nodes, say Nodes A and Z. The intermediate nodes along the path are configured to allow optical bypass of the wavelength. However, the network elements (e.g., OADMs, OADM-MDs, all-optical switches) allow any of these intermediate nodes to access the wavelength as well. Essentially, a bus network is established on the wavelength between Nodes A and Z, with any two nodes along the path able to grab the bandwidth at a given time in order to communicate. This allows the wavelength to be shared in time among the nodes on the bus. A MAC protocol is used to mediate access to the wavelength.

6.9.6 Optical Packet Switching (OPS)

In the final approach discussed here, *optical packet switching*, IP routers distributed across the network continue to groom the IP traffic similar to the current routing paradigm; however, the data packets are processed all-optically rather than electronically (although the packet header may be processed with electronics) [Blum04, YaYo05]. The motivation is that grooming at a packet level is the most efficient approach for bursty traffic, but that electronic IP routers face challenges scaling to the sizes needed in future networks. By performing the bulk of the processing in optics, impediments such as the size of the switch fabric and the power consumption may be reduced. However, OPS requires advancements in technology such as optical buffers, which are very challenging.

Chapter 7

Optical Protection

Any network is subject to failures, whether it be due to fiber cuts, equipment failures, software errors, technician errors, or environmental causes. Protection against failures, by providing alternative paths or backup equipment, is a necessary component of network design. One of the key design decisions is selecting the networking layer, or layers, in which to implement protection. For example, higher-layer protocols, such as IP, typically have well defined protection mechanisms. However, these mechanisms usually operate on a relatively fine traffic granularity; as traffic levels increase, implementing failure recovery solely in these layers may be too slow. Optical protection, which operates on the granularity of a wavelength (or even a waveband or a fiber), has received growing attention, largely due to its ability to scale more gracefully with increasing traffic levels. As the wavelength bit-rate increases, e.g., from 10 Gb/s to 40 Gb/s and beyond, the amount of network traffic that can be restored by rerouting a wavelength grows accordingly.

There are numerous optical protection schemes, where the mechanisms differ in the amount of spare capacity and equipment required, the speed of protection, the number of simultaneous failures from which recovery is possible, and the operational complexity. It is possible to support a combination of protection mechanisms in a network, where the protection scheme used for a particular demand depends in large part on the associated availability requirements. Such requirements are usually specified as part of a service level agreement (SLA) between a carrier and its customers. For example, some demands may have very stringent requirements such that recovery from a failure must be almost immediate (e.g., in less than 50 milliseconds). At the other extreme, there may be demands that are contracted as best-effort, where no resources are specifically allocated for their protection.

The first four sections of this chapter describe some of the major protection classifications and their inherent tradeoffs. Specifically, these sections probe

J.M. Simmons, *Optical Network Design and Planning*,
DOI: 10.1007/978-0-387-76476-4_7, © Springer Science+Business Media, LLC 2008

dedicated vs. shared protection, client-side vs. network-side protection, ring vs. mesh protection, and failure-dependent vs. failure-independent protection. In addition to describing the basic properties of these protection classes, the discussion will address how the presence of optical bypass affects the efficacy of the various schemes.

Note that protection schemes are used for *failure recovery*, where a connection is restored after it has failed. This is in contrast to *failure repair*, which refers to actually fixing what has failed; e.g., repairing a fiber cut or replacing a failed piece of equipment. Failure recovery generally occurs on the order of seconds or less, whereas failure repair may take several hours. Because of the length of time required to repair a failure, additional failures may occur such that a network may be affected by multiple concurrent failures. This is especially true in networks of large geographic extent, or in networks deployed in hostile environments. Thus, for some demands, it may be necessary to allocate enough spare capacity such that recovery from any two (or possibly more) concurrent failures is supported. This is discussed in Section 7.5.

In Section 7.6, the relationship of optical bypass and optical protection is probed in greater depth. There are specific properties of optical-bypass-enabled networks that need to be taken into account when developing a protection scheme, most notably the sensitivity to optical amplifier transients that arise from sudden changes of the optical power level on a fiber. This favors employing protection schemes that do not require rapid turning on or off of lasers or rapid switching of wavelengths.

Sections 7.7 and 7.8 discuss two specific protection methodologies in more detail, where both are applicable to shared protection on mesh topologies. Section 7.7 describes protection based on predeployed subconnections; the scheme is notable because it avoids issues with optical amplifier transients. Section 7.8 describes protection schemes based on 'pre-cross-connected' bandwidth; these schemes are significant because of their fast recovery speed and their efficiency in terms of the required spare capacity, thereby challenging the conventional wisdom that protection schemes must trade off capacity for speed.

This is followed by a discussion of protection planning methodologies in Section 7.9. Each of the various protection types requires different design techniques and different optimizations. Rather than cover the full gamut of protection planning, this section mainly focuses on design methodologies for shared mesh protection.

Chapter 6 addressed the multiplexing and grooming of subrate demands. Section 7.10 specifically addresses some of the options for protecting such demands, where protection can be provided at the wavelength-level, at the subrate-level, or both. While wavelength-level protection can be rapid, it is not necessarily sufficient for recovering from all network failures. Subrate-level protection may be more efficient and may provide more fault coverage, but it also tends to be slow. Multilayer protection combines the approaches to achieve the benefits of both layers; however, it does present challenges in coordination of the layers.

The last section of this chapter deals with determining the location of a failure, where again the focus is on optical-bypass-enabled networks. Fault isolation is somewhat more challenging in the presence of optical bypass because per-connection performance monitoring in the electrical domain does not occur at every node.

It is necessary to clarify some of the terminology that is used in this chapter. The connection path from source to destination that is used under the condition of no failures is referred to as the *working path* or the *primary path*. The alternative path that it used after a failure occurs is referred to as the *protect path*, the *backup path,* or the *secondary path*. The network capacity that is allocated for protection is referred to here as the *protection capacity* or the *spare capacity*. In the past, the terms *protection* and *restoration* have been used to distinguish the relative speed of the recovery method, where protection mechanisms are generally pre-planned and fast, whereas restoration schemes require calculations at the time of failure and are thus relatively slow. However, because the differences between such schemes have become blurred and because there are no universally accepted definitions, the terms will be used interchangeably here.

Most of this chapter discusses recovery from a link, node, or transponder failure; more general equipment protection is not addressed. In some networks, it is assumed that node failures are so infrequent that nodal protection is not required; however, unless otherwise noted, it is assumed here that nodal protection is necessary. The major difference is that the protection mechanism looks for link-and-node-disjoint paths as opposed to just link-disjoint paths. Note that the term 'node-disjoint paths' generally refers to paths with no *intermediate* nodes in common; however, the endpoints of the paths can be the same (it is assumed that if a demand endpoint fails, the connection cannot be recovered anyway).

Finally, it should be emphasized that optical protection is a very rich topic. This chapter touches on some of the more major points, with an emphasis on optical-bypass-enabled networks. There are several books that provide more comprehensive coverage; e.g., [Grov03, VaPD04, OuMu05]. Additionally, [GeRa00, EBRL03] are good tutorial papers.

7.1 Dedicated vs. Shared Protection

One of the basic dichotomies among protection schemes is whether the protection is *dedicated* or *shared*. This dichotomy exists with protection at any network layer, not just optical protection. For ease of discourse, this section compares dedicated and shared protection in relation to link/node failures (as opposed to, e.g., equipment failures). Furthermore, it assumes the protection mechanism is path-based, where recovery is provided by moving the connection to an alternative end-to-end path; however, the dedicated vs. shared paradigm applies to more general protection schemes. (Path-based protection is discussed further in Section 7.4, along with alternative schemes.)

7.1.1 Dedicated Protection

In dedicated protection, spare resources are specifically allocated for a particular demand. If a demand is brought down by a failure, it is guaranteed that there will be available resources to recover from the failure, assuming the backup resources have not failed also.

Dedicated protection generally falls into one of two categories. In 1+1 dedicated protection, the backup path is 'active'; i.e., there are two live connections between the source and destination, and the destination is equipped with decision circuitry to select the better of the two paths. In contrast, in 1:1 dedicated protection, the backup path does not become active until after a failure has occurred on the primary path. After the failure is repaired, the connection may return to the primary path (i.e., revertive mode) or may remain on the backup path (i.e., non-revertive mode).

There are several advantages to operating dedicated protection in a 1+1 mode. First, recovery from a failure can be almost immediate. As soon as the receiver detects that the primary path has become unsatisfactory, it can switch over to using the secondary path. (There is usually a small synchronization delay due to the transmission latency of the two paths being different.) The 1:1 mode is slower, as the failure must first be detected (usually by the destination), and then the source must be notified of the failure so that it can begin to transmit over the backup path. Another advantage to 1+1 is that failures on the backup path can be detected when they occur. With 1:1, a 'silent failure' can occur on the backup path, such that the failure is not detected until the backup path is actually needed. One possible disadvantage to 1+1 protection is that it may require more equipment at the source and destination, to support two active paths.

The downside of dedicated protection, whether 1+1 or 1:1, is the large amount of spare capacity that it generally requires. In typical networks, the ratio of the dedicated backup capacity to the working capacity is often on the order of 2 to 1.

7.1.2 Shared Protection

Shared protection addresses this inefficiency by allowing the spare capacity to be used as protection resources for multiple working paths. The working paths that share protection capacity should have no links or intermediate nodes in common so that a single network failure does not bring down more than one of the paths. (If the endpoint of one working path is an intermediate node of another working path, then whether to allow the two paths to share protection resources depends on the scheme; for details, see [EBRL03].)

While sharing protection resources improves the capacity efficiency, one drawback is that contention for the resources may arise when there are multiple concurrent failures. Only one path can use the shared resources at a time, such that the other paths sharing the resources are vulnerable if a second failure occurs. Shared protection also requires greater coordination in the network so that the working paths are aware of whether the shared protection capacity is available or not. Note that shared protection usually operates in a revertive mode, such that the protection resources are released by the connection after the failure is repaired.

Shared protection is often referred to as $1:N$ protection, indicating there is one protection element for every N working elements (or, more generally, $M:N$ protection). As N increases, the protection efficiency increases; however, the vulnerability of the scheme to multiple failures also increases. One study on path-based shared protection indicated that limiting N to about five provides significant capacity savings without leaving the network too vulnerable to multiple failures [RLAC03]. Ultimately, however, for a given network failure rate, it is the required availability of a connection that determines whether shared protection is suitable, and if so, what level of sharing is acceptable.

7.1.3 Comparison of Dedicated and Shared Protection

Dedicated and shared protection are illustrated in Fig. 7.1 for two wavelength-level demands. In Fig. 7.1(a), the working paths are routed over paths A-B-C-D and A-H-I-D. Both of the paths are protected with spare capacity allocated along the path A-E-F-G-D. Note that two wavelengths of spare capacity are allocated along this path, one wavelength dedicated to each of the working paths.

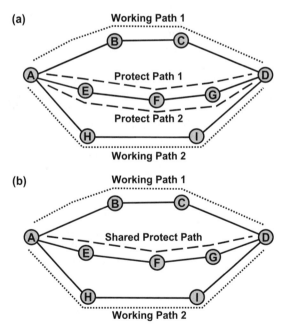

Fig. 7.1 (a) Dedicated protection where a protect path is reserved for each working path. (b) Shared protection where the two working paths share the protection resources.

Figure 7.1(b) illustrates shared protection for the same two working paths. Only one wavelength of spare capacity is allocated along A-E-F-G-D. The two working paths are link-and-node disjoint and thus can share this spare capacity. As illustrated by this small example, shared protection usually requires significantly less resources than dedicated protection. In typical networks, shared protection requires roughly 50% to 70% less spare capacity than dedicated protection.

In addition to saving capacity, shared protection may reduce the network cost, although the cost savings is usually much more significant in an O-E-O-based network than in an optical-bypass-enabled network. In Fig. 7.1, shared protection allows one protection wavelength along A-E-F-G-D to be removed. In an O-E-O network, this also removes the three regenerations that would occur along this path, providing a cost savings as well. However, with optical bypass, if the distance along path A-E-F-G-D is less than the optical reach, then no regenerations are saved by implementing shared protection. In fact, shared protection may be more expensive than dedicated protection in an optical-bypass-enabled network, as is illustrated next.

In Fig. 7.1(b), the protection capacity was shared by working paths with the same endpoints. More generally, however, working paths with different endpoints can share portions of the protection capacity, as illustrated by the A-B-C-D and A-I-J working paths in Fig. 7.2. With dedicated protection, assume that the two protect paths are A-E-F-G-D and A-E-F-H-J. Because the two protect paths overlap along A-E-F and the working paths are disjoint, a single protect wavelength along A-E-F is sufficient if shared protection is used. Assume that the network in Fig. 7.2 is optical-bypass enabled with an optical reach of 2,000 km. Then, with dedicated protection, no regeneration is required along either of the protect paths. Thus, while shared protection saves capacity compared to dedicated protection, it does not save any regenerations in this scenario. Furthermore, with shared protection, switching is required at Node F to establish the desired protect path in response to whichever working path has failed. In an optical-bypass-enabled network, there are two options for implementing the switching.

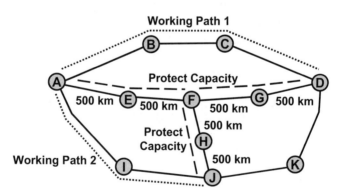

Fig. 7.2 The two working paths share the protection capacity along A-E-F. Switching is required at Node F to configure the proper protect path.

First, consider switching in the optical domain at Node F, where the OADM-MD or the all-optical switch at this node is configured to establish the desired protect path. With this option, no electronics are needed along the protection capacity, so that this solution would be similar in cost to dedicated protection (the actual cost difference would depend on how transponders are deployed at the connection endpoints for dedicated and shared protection, as is discussed in Section 7.2). However, the wavelength continuity constraint can make this challenging to design because the same wavelength would need to be assigned along each of the protection links shown in the figure. Furthermore, there are operational issues with utilizing switching in the optical domain for protection (e.g., optical transients), as is covered in Section 7.6.

To avoid these issues, it may be desirable to switch the protection capacity at Node F in the electrical domain. The wavelengths corresponding to the A-E-F, F-G-D, and F-H-J protection segments would be dropped from the optical network element at Node F to an electronic switch. To keep these three segments lit (to avoid optical transients), three transponders are needed at Node F. A corresponding number of electronic switch ports are utilized as well. Thus, because of this required electronics at Node F, this shared protection configuration would likely be *more* costly than dedicated protection (although, again, there could be fewer transponders required at the source and destination with shared protection, which would partially offset the switching cost). As this example illustrates, in an optical-bypass-enabled network, the switching required by shared protection may increase the cost of the network, especially if the optical reach is long enough that no (or just a few) regenerations are needed for dedicated protection.

Consider this same example with an O-E-O-based system. Sharing the protection capacity along A-E-F eliminates one regeneration at Node E. Furthermore, with shared protection, only three transponders are required at Node F, as opposed to four with dedicated protection. (There may be additional transponder savings at the endpoints depending on the scheme used.) While switching is needed at Node F to allow sharing of the A-E-F segment, the O-E-O network may already make use of electronic switches at each node. Thus, overall, using shared protection instead of dedicated protection should reduce the cost of an O-E-O network.

Another characteristic of shared protection, which holds for either O-E-O or optical-bypass-enabled networks, is that it is generally slower to recover due to the amount of required switching. In Fig. 7.2, restoration from a failure on either of the working paths requires that Node F be notified of the failure and the switch at that node be reconfigured (assuming it is not already in the desired configuration). This delays the restoration process and adds to the complexity of the restoration operation. Many shared protection schemes take hundreds of milliseconds (or even multiple seconds) to restore all failed demands. However, recent research on shared protection has focused on minimizing the amount of operations required at the time of failure, thereby conceivably reducing the restoration time to within 100 milliseconds for a continental-scale backbone network; this is discussed further in Sections 7.7 and 7.8.

To summarize, shared protection is more capacity-efficient than dedicated protection. In O-E-O networks, shared protection is typically less costly as well; with
optical bypass, the cost savings, if any, is typically small (or shared protection
may actually cost more). Shared protection is generally slower to recover from a
failure than dedicated protection and is more complex to implement. Finally,
shared protection leaves the network more vulnerable to a second failure. Clearly,
the major impetus behind shared protection is its capacity efficiency.

7.2 Client-Side vs. Network-Side Protection

Client-side and network-side protection refers to where the protection mechanism
is triggered; i.e., at the clients at the endpoints (client side) or in the optical layer
(network side). From a cost perspective, the main impact of the two schemes is
the amount of equipment required at the endpoints of protected demands. Either
type of protection can be operated in a dedicated or shared mode; however, for illustrative purposes, the discussion in this section will center around dedicated protection.

First, consider client-side 1+1 dedicated protection, where the client is responsible for generating two copies of the signal, both of which are transmitted over
the optical layer. A demand endpoint with this type of protection is illustrated in
Fig. 7.3(a), where an IP router is serving as the client. The router sends two copies of the signal to the optical layer, which routes them over disjoint paths. In the
reverse direction, the router receives two copies of the signal and is responsible for
selecting the better of the two copies.

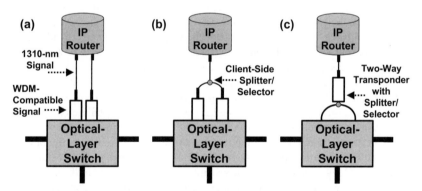

Fig. 7.3 (a) Client-side protection where the client (here an IP router) delivers the working and
protect signals to the optical layer in the transmit direction, and selects the better of the two signals in the receive direction. (b) Client-side protection where a splitter feeds the client signal
into two transponders, and a selector chooses the better of the two paths in the receive direction.
(b) Network-side protection where a two-way transponder generates the working and protect
paths in the transmit direction, and selects the better of the two paths in the reverse direction.

This configuration utilizes two transponders at either endpoint of the demand. The wavelengths on which the working and protect paths are carried do not have to be the same. In addition to protecting against link/node failures (assuming the two copies of the signal are routed over diverse paths), this architecture also provides protection against a transponder failure, a client-port failure, or a failure in the connection between the client and optical layers.

Another client-side protection configuration is shown in Fig. 7.3(b), where the client generates just one copy of the signal, which passes through a passive splitter that feeds two transponders. The optical layer routes the resulting two signals over disjoint paths. In the receive direction, there is decision circuitry to select the better of the two paths. This configuration is lower cost because it utilizes only one client port, but this also leaves it more vulnerable to failure.

A demand endpoint with network-side 1+1 dedicated protection is illustrated in Fig. 7.3(c). Here, a two-way (flexible) transponder is used at the demand endpoint to take a single copy of the client signal and generate two copies of the WDM-compatible signal, which are then routed over disjoint paths. (Two-way flexible transponders were covered in Section 2.8.2.) In the receive direction, the two-way transponder has decision circuitry to select the better of the two received signals and send it to the client. The advantage of this configuration is that only a single two-way transponder is required at either endpoint instead of two regular transponders. A two-way transponder is not much more expensive than a regular transponder (say 20% more expensive), resulting in a cost savings.

However, using a single two-way transponder leaves a connection vulnerable to a failure in the transponder itself. Thus, this type of protection is typically used in conjunction with 1:N shared protection of the transponders, where every group of N transponders is protected by one spare (or, more generally, M:N shared protection can be used). Three architectures for 1:N transponder protection are shown in Fig. 7.4 (these schemes can be used to protect transponders in general, not just two-way transponders; for simplicity, the figure shows regular transponders). Other 1:N transponder protection architectures are also possible as described in [GeRa00].

In Fig. 7.4(a), the client signal enters an edge switch that directs the signal to its primary transponder, or if that has failed, to the spare transponder. This architecture is well-suited for a node where an edge switch is deployed anyway for purposes of node flexibility. In Fig. 7.4(b), a passive splitter is used to direct the client signal to its primary transponder and to an Nx1 switch. If one of the primary transponders fails, the Nx1 switch is configured to select the client signal associated with the failed transponder and direct it to the spare transponder. A 1xN switch is used in the receive direction. The architectures of Fig. 7.4 (a) and (b) are still vulnerable to a client-port failure. Figure 7.4(c) is one method of addressing this vulnerability, where the client uses two ports to send two copies of the signal, one of which enters the Nx1 switch feeding the spare transponder. In any of these architectures, the spare transponder is ideally tunable so that it can tune to the wavelength of whichever transponder has failed, thereby allowing the affected connection to continue using the same wavelength.

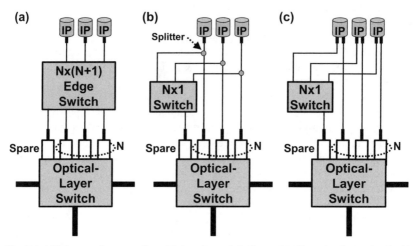

Fig. 7.4 1:N transponder protection. (a) An edge switch directs the client signal associated with the failed transponder to the spare transponder. (b) A passive splitter directs the client signal to its primary transponder and to an Nx1 switch that feeds the spare transponder. (c) Dual client signals are sent to the primary transponders and the Nx1 switch that feeds the spare transponder. In all three illustrations, the edge switch depicted is a photonic switch; an O-E-O switch could also be used.

The network-side 1+1 protection configuration of Fig. 7.3(c) poses interesting wavelength-assignment challenges in an optical-bypass-enabled network. Note that a two-way transponder has a single laser, such that the two generated paths are launched on the same wavelength. Similarly, it is assumed that the two received paths must be at the same wavelength (e.g., due a single filter in the receiver). The resulting wavelength constrictions depend on the amount of regeneration along the two paths, as illustrated in Fig. 7.5.

In Fig. 7.5(a), there is no regeneration in either the working or the protect path, thereby requiring that the same wavelength be assigned along both of the paths. In Fig. 7.5(b), there is one regeneration on the protect path and no regenerations on the working path. Normally, a regeneration affords the opportunity to change the wavelength; however, in this scenario, wavelength conversion on the protect path is not possible because it is not possible on the working path. Thus, again, the same wavelength must be assigned along both of the paths. In Fig. 7.5(c), there is one regeneration on both the working and protect paths. With this configuration, the same wavelength must be used on both A-B-C and A-E-F, but a different wavelength could be used on C-D-Z and F-G-Z. If there were additional regenerations along the paths, there would be further freedom in assigning the wavelengths (as long as the portions of the paths emerging from Node A are carried on the same wavelength, and the portions of the paths converging at Node Z are carried on the same wavelength).

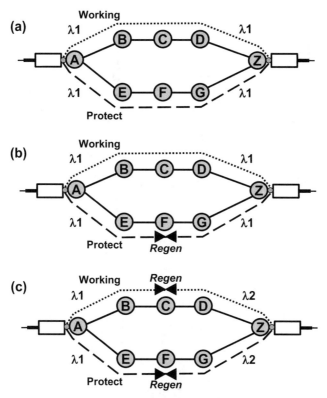

Fig. 7.5 Wavelength constraints with network-side 1+1 protection. (a) With no regenerations, the same wavelength must be assigned on all links of both the working and protect paths. (b) Even though the protect path has a regeneration, it cannot be used to change the wavelength, due to the constraints posed by the working path. (c) With a regeneration in both the working and protect paths, the wavelength used in one half of the working and protect paths can be different from the wavelength used in the other half.

Additionally, note that if the protected connection is bi-directional and the same wavelength is used in the A to Z direction as in the Z to A direction, then Nodes A and Z cannot be equipped with an OADM that does not support wavelength reuse. As described in Section 2.4.3, with a no-reuse OADM, any wavelength that drops from the OADM also continues on through the node. For example, in Fig. 7.5, assume that Node A is equipped with a no-reuse OADM. The Z to A working path enters Node A on λ_1. Thus, λ_1 would drop at Node A and continue on to Link AE; this would interfere with the protect path from A to Z, also carried on λ_1.

Protection based on two-way transponders can also operate in the 1:1 mode, where the protect side of the transponder is not turned on until after a failure occurs. If the transponders are not tunable, then the wavelength constraints depicted in Fig. 7.5 apply. If the transponders are tunable, then the working and protect paths may be carried on different wavelengths, requiring that the transponder be retuned at the time of failure.

Rather than using a two-way transponder for network-side protection, a regular transponder in conjunction with an all-optical switch can be used. For example, the 1:1 configuration is shown in Fig. 7.6. Under the no-failure condition, the transponder associated with the protected demand is directed to the working path. If the demand fails, the all-optical switch is reconfigured to direct the transponder to the protect path. Note that the lower cost OADM-MD is not capable of this type of protection because with this network element, transponders are tied to a particular network port. This scheme would likely be deployed with 1:N protection of the transponders, as discussed above. Additionally, if the transponder is not tunable, then the wavelength constraints depicted in Fig. 7.5 are applicable.

To accomplish network-side 1+1 protection, the all-optical switch needs to be capable of multicasting, where a single input connection is simultaneously directed to two output ports. Again, this configuration is not supported by an OADM-MD (or by non-multicasting all-optical switches).

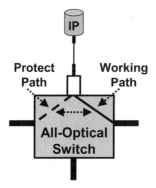

Fig. 7.6 An example of 1:1 network-side protection, where the all-optical switch directs the connection to the protect path after a failure occurs.

7.3 Ring Protection vs. Mesh Protection

Protection can be implemented using ring topologies or arbitrary mesh topologies. Ring protection tends to be simpler, although its additional constraints generally result in more required spare capacity as compared to mesh protection.

7.3.1 Ring Protection

In ring protection, the protection capacity is organized into ring structures. A ring is a simple survivable topology, such that if a single link or node failure occurs, all traffic that was routed over the failed link/node is routed in the reverse direction around the ring to avoid the failure. Note that a network with a mesh topology can use ring-based protection; a working path that traverses multiple rings is protected by spare capacity in each of the individual rings. There are many variations of ring-based optical protection, as described in [LSTN05]. For example, the implementations may differ on whether they use two-fiber rings or four-fiber rings, or whether they restore each path individually or several multiplexed wavelengths at once.

Dedicated ring protection is fairly straightforward: a bi-directional working path combined with its protection capacity occupies one wavelength around both directions of the ring. Shared ring protection allows greater capacity efficiency, although it may be more challenging to implement.

Network-side shared ring protection is described here in more detail to illustrate some of the interesting features, especially with regard to optical bypass. An example of this type of protection is shown in Fig. 7.7, where the ring is assumed to be a two-fiber ring. One fiber carries traffic in the clockwise direction, the other in the counterclockwise direction. In the example, there are two bi-directional working paths on the ring, between Nodes A and B and between Nodes D and F, as shown by the dotted lines. Both connections require shared protection. The protection capacity extends all the way around the ring, as indicated by the dashed lines. (Note that the working paths between A and B and between D and F could be portions of full end-to-end paths in a mesh network where multiple rings are used for protection.)

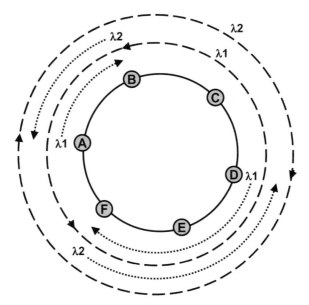

Fig. 7.7 Shared optical ring protection for two demands, AB and DF, under the no-failure condition. With this configuration, the wavelengths assigned to the two directions of a bi-directional connection are not the same.

The endpoints of the demands are equipped with two-way transponders as shown in more detail for Node A in Fig. 7.8. Assuming the transponder at Node A is not tunable, or assuming the tuning time is too slow for failure recovery, then the wavelength used to carry the working path along A-B must be the same wavelength assigned to protect the connection along the paths A-F-E-D-C-B. Assume that this wavelength is λ_1. Because the protection wavelength is shared by both demands, λ_1 must also be used by the two-way transponder at Node D, thereby requiring that the working path along D-E-F be carried on λ_1. Thus, λ_1 is used in the clockwise direction on one fiber to carry both working paths, and in the counterclockwise direction on the other fiber to carry the associated protection capacity. (If there are enough regenerations, this requirement can be relaxed to a degree; however, the most restrictive case is considered here.)

Now, consider the working paths in the reverse direction, e.g., B to A. Because λ_1 is used in the counterclockwise direction for protection, this same wavelength cannot be used to carry the working paths in this direction. Thus, the two directions of a connection (e.g., A to B vs. B to A) are forced to use different wavelengths. In the figure, it is assumed λ_2 carries the counterclockwise working paths and the corresponding clockwise protection.

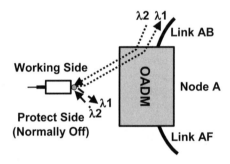

Fig. 7.8 Details of Node A from Fig. 7.7. If the connection between Nodes A and B has not failed, then the protect side of the two-way transmitter is off.

It is assumed that with no failures, the protect side of the two-way transmitters are off. If a failure affects a demand, the source and destination of the affected demand turn on the protect sides of their transponders. For example, Fig. 7.9 illustrates the configuration after Link EF has failed. The two-way transponders at Nodes D and F now transmit and receive on the protect path. All other transponders sharing the protection capacity remain in the working-path-only mode.

The configuration of the protection capacity under the no-failure condition is also of interest in an optical-bypass-enabled network. If all OADMs along the ring are configured to allow the protection wavelength to pass, and assuming no regenerations are deployed on the protection wavelength, then *lasing* may occur where noise present in the corresponding portion of the spectrum continues to loop around the ring, leading to system instabilities. One solution is to have at least one of the OADMs along the ring configured to block the protection wavelength. In Fig. 7.7, assume that the OADM at Node A is used for this purpose. If a failure occurs such that the protection wavelength is needed, but where Node A is not an endpoint of the failed connection, then the OADM at Node A must be reconfigured to allow the protection wavelength to pass. This slows down the restoration process.

A second approach is to install one regeneration at some node in the protection ring (even if it is not required based on optical reach), such that the protection wavelength enters the electronic domain at this node, breaking up the purely all-optical loop. Assume that the regeneration is placed at Node A in Fig. 7.7. If a failure occurs where Node A is not an endpoint of the failed connection, then no action is needed at this node. If, however, the demand between A and B fails, then Node A must simultaneously turn off the regeneration (e.g., the regenerator card) and turn on the protect side of its two-way transponder. This method incurs the cost of a regeneration; however, it has the advantage that the nodes not involved in the failure do not need to be reconfigured. (Note that it requires the OADM chassis to be able to accommodate both a regenerator and a two-way transponder at the same wavelength; thus, a fixed-slot system where each wavelength can be inserted in only one chassis slot is not compatible with this solution.)

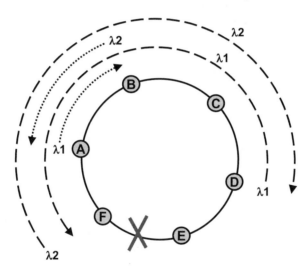

Fig. 7.9 Shared optical ring protection after Link EF fails. The protect sides of the transponders at Nodes D and F are turned on to restore the connection over the protect wavelength.

The type of optical ring protection described above is also called two-fiber Optical-Channel Shared Protection Ring (OChSPRing) protection [LSTN05]. The reference to an 'optical channel' emphasizes that each demand is restored individually. This is in contrast to two-fiber Optical Multiplex Section Shared Protection Ring (OMS-SPRing) protection [LSTN05], which restores all failed wavelengths on a fiber at once, as illustrated in Fig. 7.10. Assume that the odd-numbered wavelengths on the clockwise fiber carry working paths and the even-numbered wavelengths on that fiber carry the protect paths, with the opposite convention on the counterclockwise fiber. As shown in the figure, when a link fails, the two nodes adjacent to the failure, i.e., Nodes A and F, reconfigure a fiber-switch to interconnect the two fibers. Thus, a working path that had been routed clockwise along E-F-A-B is now routed E-F-E-D-C-B-A-B. The same wave-length can be used along the whole path, due to the assignment of odd and even wavelengths in the two fibers. The benefit of this type of protection is that many demands are restored at once; the downside is that the post-failure path may be excessively long. (Note that OMS-SPRing protection is analogous to SONET Bi-directional Line-Switched Ring (BLSR) protection.)

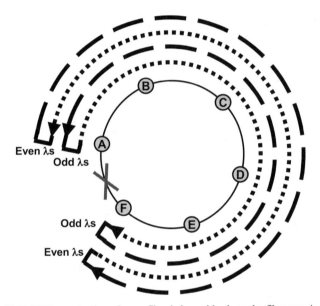

Fig. 7.10 OMS-SPRing protection where a fiber is looped back on the fiber carrying traffic in the opposite direction to avoid the failed link. The dotted lines indicate the working wavelengths and the dashed lines indicate the protect wavelengths.

7.3.2 Mesh Protection

Mesh protection allows the protect path to be routed in an arbitrary fashion rather than having to follow a precise topology such as a ring. The greater freedom gen-erally translates to greater capacity efficiency; i.e., mesh protection requires on the order of 20% to 60% less spare capacity as compared to rings [GeRa00]. How-ever, mesh protection requires more sophisticated tables to track the backup paths, and may require more communication among the nodes, especially with shared mesh protection.

An example of shared mesh protection was illustrated in Fig. 7.2. With mesh topologies, there are more opportunities for sharing capacity and transponders at a node. For example, Node A from Fig. 7.2 is shown in more detail in Fig. 7.11. In Fig. 7.11(a), the node is equipped with an OADM-MD and an edge switch. The two working paths share the protection capacity as well as the protect transponder. The edge switch at the node is used to direct the client signal to the shared transponder when a failure occurs. In Fig. 7.11(b), the node is equipped with an all-optical switch. There is no protect transponder associated with the protection wavelength. When a failure occurs, the transponder associated with the failed path is directed by the all-optical switch to the protection capacity. If the transponders are not tunable, then the two working paths, as well as the shared protection capacity, must use the same wavelength. If the transponders are tunable, then wavelengths can be assigned to the paths independently.

Shared mesh protection is covered in more detail, in the context of two specific schemes, in Sections 7.7 and 7.8.

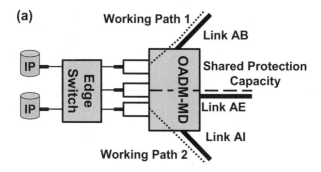

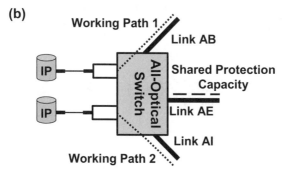

Fig. 7.11 Possible configurations of Node A from Fig. 7.2. (a) The two working paths share the protection capacity and its associated transponder, using the edge switch. (b) The all-optical switch is used to direct a failed connection to the shared protection capacity.

7.4 Fault-Dependent vs. Fault-Independent Protection

Another dichotomy among protection schemes is whether the protection mechanism for a connection depends on where the failure has occurred along its path. In *fault-dependent* schemes, the protection used depends on where the failure has occurred; in *fault-independent* schemes, the same protection mechanism is used regardless of the fault location. In O-E-O-based networks, where a signal can be electronically monitored at each node along a path (e.g., the SONET/SDH overhead bytes can be monitored for errors), isolating the location of a failure is relatively straightforward. In an optical-bypass-enabled network, where the signal is not converted to the electrical domain at each node, fault isolation is more challenging and may take longer. (Fault isolation is covered in Section 7.11.) Thus, fault-independent protection schemes are generally favored with optical bypass.

At one extreme of fault-dependent schemes is link-based protection, where the recovery mechanism depends on which specific link has failed. This is illustrated in Fig. 7.12 for a demand from Node A to Node Z. Under the no-failure condition, the demand is routed over the path shown by the dotted line in Fig. 7.12(a).

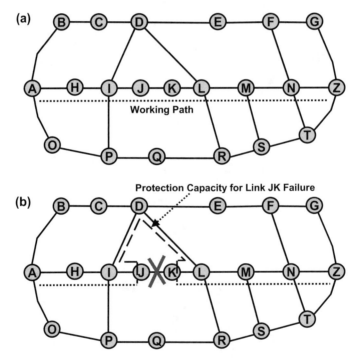

Fig. 7.12 Link-based protection. (a) The path between Nodes A and Z under the no-failure condition is shown by the dotted line. (b) After Link JK fails, the protection capacity shown by the dashed lines is used to avoid the failed link.

If a link fails, the recovery occurs between the two endpoints of the failed link. This is illustrated in Fig. 7.12(b), where it is assumed that Link JK has failed, and the protection resources along J-I-D-L-K are used to avoid the failure, as shown by the dashed line. Link-based protection is potentially very rapid, due to the proximity of the nodes involved with the recovery process.

At the other extreme is fault-independent path protection, where a diverse backup path running from the source to the destination is used whenever the working path fails, regardless of where the failure has occurred. Furthermore, the same wavelength (or wavelengths) is assigned along the backup path, independent of the fault location. Path-based protection is illustrated in Fig. 7.13. The demand from Node A to Node Z is routed over the same working path as in Fig. 7.12; the protect path is allocated along A-O-P-Q-R-S-T-Z. When a failure occurs anywhere along the working path, Node Z eventually detects errors in the received signal. It then communicates the failure condition to Node A to trigger the switchover to the protect path. The distance between the demand endpoints can be large, especially in a backbone network, so that this communication between nodes may take tens of milliseconds, delaying the onset of recovery. However, in a system where immediate fault isolation may not be possible, this communication delay is preferable to suffering an even longer delay to determine the exact location of the failure. Additionally, path protection is more capacity-efficient than link protection [IrMG98], further encouraging its use.

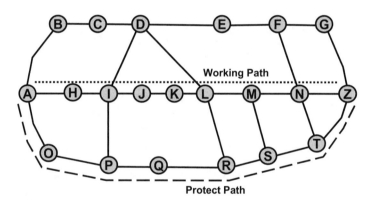

Fig. 7.13 Fault-independent path-based protection, where the same protect path is used regardless of the location of the failure on the working path.

Note that path protection does not necessarily imply fault-independent protection. For example, there can be scenarios where the same end-to-end protection path is used for recovery regardless of the failure, but where the wavelength assigned to the backup path depends on the location of the failure [EBRL03]. This is illustrated in the ring network of Fig. 7.14, where there are three working paths (AD, BE, and CF) that are pair-wise intersecting. Assume that there are two wavelengths for protection allocated around the ring using λ_1 and λ_2, and assume that wavelength conversion is not available along the protect path. This protection is sufficient to protect against any single failure, depending on the wavelength assignment. Assume that the AD working path is protected using λ_1 and that the BE working path is protected using λ_2. Then, the CF working path is protected using λ_1 if Link BC fails but is protected using λ_2 if Link AF fails (either wavelength can be used if Link AB fails). Because the wavelength assignment depends on the failure location, this is not considered failure-independent protection.

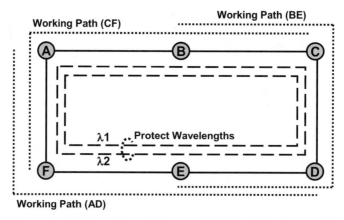

Fig. 7.14 Assume that the AD working path is protected with $\lambda 1$ and that the BE working path is protected with $\lambda 2$, and that wavelength conversion is not available on the protect paths. Then the wavelength used to protect the CF working path depends upon which link fails. Thus, the protection mechanism is not failure-independent.

A protection scheme that can be considered intermediary with respect to link and path protection is segment protection [HoMo02, GuPM03, XuXQ03, WLYK04]. After a working path is selected for a demand, it is divided up into multiple segments, where a backup path is independently provided for each segment. As discussed below, by dividing the path up into multiple shorter segments, the failure recovery time is reduced as compared to path protection.

Non-overlapping segment protection is illustrated in Fig. 7.15(a), where the working path is divided into three segments, A-H-I-J-K-L, L-M-N, and N-Z. The corresponding backup segments are A-O-P-Q-R-L, L-R-S-T-N, and N-T-Z. It is assumed that the endpoint of a segment is capable of detecting a failure that occurs in its associated segment. For example, if Link JK fails, Node L detects that a failure has occurred and signals Node A to switch to the backup segment; no switchover is required in the remaining segments. This should be faster than path protection, where Node Z would have to detect the failure and signal Node A to switch to the backup path. Furthermore, the backup segments typically have fewer links than an end-to-end backup path, such that fewer switch reconfigurations are likely required with shared segment protection.

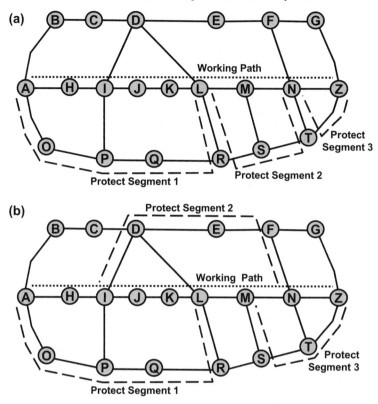

Fig. 7.15 Segment-based protection. (a) Non-overlapping segments (A to L, L to N, N to Z), where the nodes at the intersection of the segments are vulnerable to failure. (b) Overlapping segments (A to L, I to N, M to Z) where all of the intermediate nodes are protected.

Depending on the number of segments created, segment protection may be more capacity-efficient than path protection, especially when operated in a shared protection mode. With shared *path* protection, in order to share backup resources, the associated working *paths* must be link-and-node-disjoint. In shared *segment* protection, the associated working *segments* must be link-and-node-disjoint. Segments encompass fewer links than paths, leading to a greater opportunity for sharing. Furthermore, two working segments from the same demand path can share backup resources, assuming the working segments are completely disjoint. However, as the number of segments increases, the excess routing required to provide protection for each small segment begins to nullify the benefits of better sharing. (Note that in the extreme case where each link is a segment, the scheme is equivalent to link-based protection.)

Segment protection also provides protection against more failures than path protection. For example, it can recover from a multiple-failure scenario where no more than one failure occurs in any segment. Path protection fails if there is one failure anywhere on the working path and one failure anywhere on the protect path.

One drawback to the scheme of Fig. 7.15(a) is that it does not provide protection if a node at a segment boundary fails. This is rectified in Fig. 7.15(b), where the three segments (A-H-I-J-K-L, I-J-K-L-M-N, and M-N-Z) are overlapping such that all intermediate node failures are recoverable as well. (Failure on a link that falls in two segments is discussed below.)

Consider using segment protection in an optical-bypass-enabled network. Because the endpoints of the segments are responsible for detecting failures and triggering the protection mechanism, it is natural to regenerate the working path at these locations [ShGr04, KaAr04]; i.e., with O-E-O regeneration, the signal will be converted to the electrical domain to better enable error detection. Thus, in Fig. 7.15(a), the working path would be regenerated at Nodes L and N. (It may be possible to select the segment endpoints based on where the working path needs to be regenerated anyway due to the optical reach, so that no extra regenerations have to be added to the demand.) As noted above, however, the configuration of Fig. 7.15(a) does not provide protection against failures to Nodes L and N.

If the configuration of Fig. 7.15(b) is used instead, and the endpoints of the (bi-directional) segments are required to be O-E-O, then the working path would be regenerated at Nodes I, L, M, and N; with the number of regenerations doubled, much of the benefit of optical bypass is negated. Alternatively, regeneration could be implemented at different nodes in the two directions of the connection, where only the downstream segment endpoint is required to be O-E-O to detect failures. In the A to Z direction, regeneration is required at Nodes L and N, whereas in the Z to A direction, regeneration is required at Nodes M and I. The subconnections created by this regeneration pattern would be different in the two directions, which may ultimately lead to more wavelength assignment conflicts in the network.

Furthermore, with Fig. 7.15(b), consider a failure on Link JK, which belongs to both segments 1 and 2. Some convention can be adopted as to which segment recovers from such a failure. This is relatively simple to implement in an O-E-O network where it is assumed that the failed link can be readily determined. However, in an optical-bypass-enabled network, handling this scenario may be more challenging. If Link JK fails, segments 1 and 2 both detect a segment failure, but the exact location may not be readily determined. If both segments were to switch to their backups, the resulting backup path would be discontinuous. Thus, fault localization, at least to some degree, is needed. In essence, with overlapping segments, the protection mechanism may become more fault-dependent. Given this extra complexity with segment protection, path protection may remain the more desirable mechanism for optical-bypass-enabled networks (assuming protection against node failures is desired).

Another scheme, sub-path protection, is closely related to segment protection [OZSZ04]. Here, the network is partitioned into multiple areas or domains. For a path that crosses multiple domains, self-contained working and protect paths are found within each traversed domain. It differs from segment protection in that the working and protect paths within a domain can be searched for together, rather than first finding the working path, dividing it up into segments, and then looking for backup paths for each segment. Searching for the working and protect paths together within each domain is often more capacity-efficient. The recovery mechanism, however, is similar to segment protection.

7.5 Protection Against Multiple Concurrent Failures

As discussed in the chapter introduction, while the network may recover from a failure very rapidly, repairing a failure may take several hours. During the repair time, additional failures may occur. Thus, demands with very high availability requirements may need protection against multiple concurrent failures. The discussion in this section focuses on recovery from two concurrent failures.

Many of the protection methods described above can, at least in theory, be extended to protection against multiple failures. For example, 1+1 dedicated protection can be enhanced to 1+2 dedicated protection, where one working path is established along with two active backup paths. If all three paths are mutually link-and-node-disjoint, then 1+2 protection can recover from any combination of two failures, where failure recovery is almost immediate. With client-side 1+2 protection, the demand source and destination both require three transponders. With network-side 1+2 protection, the source and destination both require a three-way transponder, where the transponder feeds into three different network ports (or a multicasting all-optical switch can be used). One hindrance to 1+2 protection is that in most realistic networks, there are not three totally diverse paths between all source/destination combinations. Moreover, even if three diverse paths can be found, the amount of required protection capacity would be excessive. This protection method is likely only acceptable if the number of 1+2 protected demands is very small.

(Note that the routing algorithm code provided in Appendix B is capable of finding N link-and-node disjoint paths between two nodes, for arbitrary N. If such paths do not exist, it can be used to find N maximally disjoint paths between the nodes.)

Another option is to initially establish 1+1 dedicated protection to provide rapid protection against a first failure. Based on where the first failure occurs, a new protection path is set up, such that a new 1+1 dedicated protection scheme is established to allow rapid recovery from a second failure as well. Because the new protection path is established based on the location of the first failure, some sharing of the protection resources is possible, leading to less spare capacity requirements as compared to 1+2 protection. Note that the protection mechanism still operates in a failure-independent end-to-end mode for either the first or second failure. However, in between failures, a failure-dependent calculation is performed to determine the new backup path. The disadvantage of this scheme in relation to 1+2 protection is that it cannot protect against simultaneous dual failures along the initial two paths; however, this should be a rare occurrence.

A related option establishes 1+1 dedicated protection to protect against a first failure. However, after recovering from a first failure, it then relies on shared protection to recover from a second failure. Thus, recovery from the second failure is slower, but requires less spare capacity. This is especially well-suited to environments where two concurrent failures are rare.

A fourth option is to solely use shared protection to protect against either the first or second failure; this is the most capacity-efficient of the options, albeit the slowest to restore. With this scheme, it is common to make the assumption that there is sufficient time between restoration from a first failure and the onset of a second failure to allow re-computation of the protection paths [ScAF01, KiLu03, ZhZM06]. This potentially reduces the total amount of spare capacity required in the network, as illustrated in the example of Fig. 7.16. There are three demands, AD, EG, and IG; their respective working paths are routed over A-B-C-D, E-F-G, and I-J-H-G, as indicated by the dotted lines. It is assumed that the AD demand requires protection against any two network failures, whereas the other demands require protection against just one. There is one wavelength of shared protection capacity allocated on all links except EF and FG, as indicated by the dashed lines.

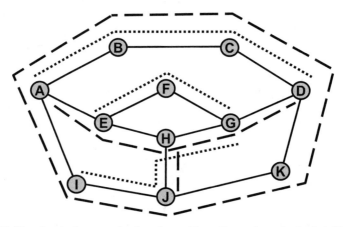

Fig. 7.16 Shared protection example where the working paths are shown by the dotted lines and the protection capacity by the dashed lines. The A-B-C-D connection requires protection against two failures. Depending on where the first failure occurs, the protect path for this path may change.

With no failures, assume that the protect path for the AD demand is planned as A-E-H-G-D. Assume that a first failure occurs that affects one of the three working paths. If the first failure occurs along A-B-C-D and the AD demand is recovered using A-E-H-G-D, then to prepare for a second failure, the AD protect path is recalculated to be A-I-J-K-D. Next, assume that the first failure occurs along E-F-G, and the EG demand is recovered along E-H-G. Again, the protect path of AD is recalculated to be A-I-J-K-D. Finally, consider the scenario where the first failure occurs along I-J-H-G, and the IG demand is recovered using I-A-B-C-D-G. A recalculation is performed to determine the AD protect path is now A-E-H-J-K-D. Thus, the location of the first failure determines how AD recovers if it is affected by a second network failure. Note that the protection mechanism still operates in a failure-independent end-to-end mode. However, in between failures, a failure-dependent calculation is performed to determine the proper backup path for AD, which allows AD to share its protection capacity with the other two demands.

Note that while the protect path may change after a first failure occurs, it is generally not acceptable to modify the working path of a demand that is otherwise unaffected by the failure, unless the demand has been contracted as preemptible traffic.

7.6 Effect of Optical Amplifier Transients on Protection

The previous sections addressed various classes of protection schemes that are relevant in both O-E-O and optical-bypass-enabled networks. This section discusses an operational issue related to protection that chiefly pertains to optical-bypass-enabled networks due to their susceptibility to optical amplifier transients. Such transients occur, for example, when there is a sudden change in the power level on a fiber, as conceptually illustrated in Fig. 7.17. In Fig. 7.17(a), three wavelengths are routed all-optically from Link AB to Link BC, via the OADM at Node B. A fourth wavelength, λ_1, is added at Node B onto Link BC. The power level of λ_1 is plotted as a function of time in Fig. 7.17(b). Assume that a failure occurs on Link AB, such that the three wavelengths on that link are suddenly brought down. Because these wavelengths had optically bypassed Node B, the failure brings down these wavelengths on Link BC as well, leaving λ_1 as the remaining wavelength on the fiber. This causes the power level of λ_1 to spike as the optical amplifiers attempt to maintain a constant total power level on the fiber, as shown in Fig. 7.17(b).

Transients arise with either EDFA [MoLS02] or Raman [KNSZ04] amplifiers. Transmission systems typically have dynamic controls to dampen such power variations and return the power level of the surviving wavelength(s) to the desired level. As shown in the plot, after a small amount of oscillation, the power level of λ_1 eventually returns to its pre-failure power level.

Excursions in the signal power level, even if brief, are undesirable as they lead to error bursts. Furthermore, in the presence of optical bypass, transients on one link may have a ripple effect, producing transients on other links, thereby causing errors to propagate. While Fig. 7.17 illustrates the effect of wavelengths suddenly being brought down, a similar, though inverse, effect occurs if wavelengths are suddenly added to a fiber.

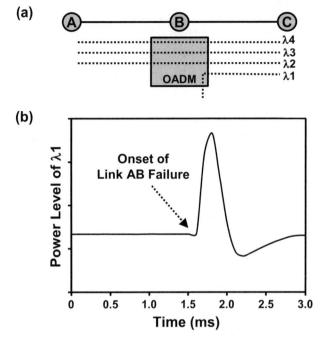

Fig. 7.17 (a) Initially, there are three wavelengths routed all-optically from Link AB to Link BC; λ1 is added at Node B. (b) The power level of λ1 is plotted as a function of time. After Link AB fails, bringing down the other three wavelengths, the power level of λ1 spikes.

Optical amplifier transients due to wavelengths being brought down by a failure are unavoidable (although the amplifier control mechanism should reduce their duration as much as possible). However, transients caused by a system operation, whether it be bringing up a new connection, tearing down an existing connection, or restoring service after a failure, are generally considered unacceptable by carriers. Thus, for optical-bypass-enabled systems, it is important that protection schemes be implemented such that transients are avoided, while still taking advantage of the economic benefits of optical bypass.

For 1+1 dedicated protection, transients should not be an issue in the recovery process; the backup path is active even under the no-failure condition, obviating the need for path turn-up at the time of failure. Transients are typically more problematic for shared protection (and 1:1 protection) schemes in which the backup path is 'lit' after a failure occurs. In order to avoid transients in these schemes, the power level of the protect path needs to be increased slowly, which slows down the restoration process.

To avoid contending with transients, the protection method should not encompass turning on or off the WDM-compatible lasers, tuning the lasers to different wavelengths, or switching signals on the WDM side (switching on the client side, however, is acceptable). A capacity-efficient shared mesh protection scheme that satisfies these constraints is described in the next section.

Note that O-E-O networks are more immune to such power-level transients as each link is essentially isolated by the O-E-O regeneration that occurs at each node. Thus, a failure on one link does not cause power-level variations on the adjacent links. For example, if Fig. 7.17 were O-E-O-based, λ_2, λ_3, and λ_4 would all be regenerated at Node B; if Link AB fails, the corresponding transponders (or regenerators) would still produce light on Link BC, leaving λ_1 unaffected.

7.7 Shared Protection Based on Predeployed Subconnections

The paradigm of predeployed subconnections can be used to avoid issues with power-level transients. The notion of a subconnection was introduced in Chapter 4, where regenerations along a path effectively break the end-to-end connection into smaller subconnections. Both ends of a subconnection are terminated in the electronic domain, with optical bypass at the intermediate nodes. In Chapter 4, the design process started with a connection and broke it into subconnections for regeneration and wavelength assignment purposes. With subconnection-based protection, the process is reversed; subconnections are predeployed in a network and concatenated as needed to form end-to-end backup paths.

A predeployed subconnection refers to a lit wavelength that is routed between two transponders, where the capacity is not currently being used to carry traffic. Thus, the transponders have been predeployed in the network and turned on for purposes of future traffic. Using predeployed subconnections as a building block for rapidly accommodating dynamic traffic or rapidly recovering from a failure was proposed in [SiSB01]. Shared mesh protection based on predeployed subconnections is described below; further details can be found in [Simm07].

Consider the network shown in Fig. 7.18(a), where it is assumed that the nodes are equipped with OADM-MDs and edge switches. Two working paths are established as indicated by the dotted lines; i.e., along A-B-C-D and A-J-K-I. There are three predeployed protection subconnections as indicated by the dashed lines: A-F-G-H, H-D, and H-I. The transponders at the endpoints of the working paths as well as the transponders at the endpoints of the protection subconnections are fed into the edge switch at the respective nodes. The details of Nodes A and H are shown in Figs. 7.18 (b) and (c), respectively.

The protection subconnection along A-F-G-H is shared by both working paths. If there is a failure along A-B-C-D, the edge switch at Node H concatenates the A-F-G-H subconnection with the H-D subconnection to form a backup path along A-F-G-H-D. In addition, the edge switches at Nodes A and D are reconfigured such that the client (in this case, an IP router) is connected to the backup path. Alternatively, if the failure occurs along A-J-K-I, then Node H concatenates A-F-G-H to H-I to form a protect path along A-F-G-H-I, and Nodes A and I reconfigure their edge switches to direct the client to the protect path.

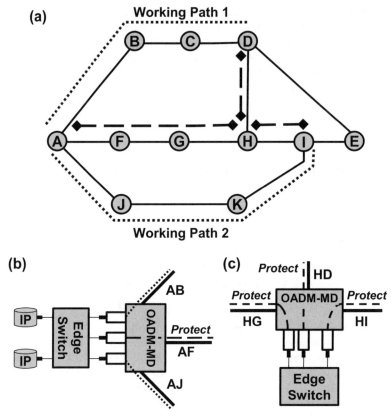

Fig. 7.18 (a) Shared protection based on predeployed subconnections. The three protection subconnections are indicated by the dashed lines. (b) Details of Node A. (c) Details of Node H. In both nodes, the edge switch as shown is photonic.

The most salient features of this scheme are that the transponders at either end of the protection subconnections are always on and at the desired wavelength, and that any switching occurs in the edge switch as opposed to in the OADM-MD (i.e., client-side signals are being switched, not network-side signals). Thus, the power levels on the fibers do not change as the protect path is formed, thereby avoiding issues with optical amplifier transients. After a destination detects that a connection has failed, the speed of the protection mechanism depends on the time it takes to notify the source and the intermediate switching locations (e.g., Node H) of the failure, and the time required to reconfigure the edge switches. No turning on, retuning, or switching of WDM-compatible signals is required. In a backbone network the size of the United States, recovery on the order of 100 milliseconds should be possible.

Furthermore, this protection scheme is compatible with the hierarchical protection paradigm [Simm99], such that it takes good advantage of optical bypass even for the protect wavelengths. In hierarchical protection, a subset of the nodes are chosen as 'high-level' nodes, where the bulk of the protection capacity extends between these nodes, optically bypassing the 'low-level' nodes. Nodes that generate a lot of protected traffic, nodes with a high degree, and nodes that are located strategically in the network (e.g., for regeneration) are generally favored as high-level nodes. Applying this paradigm to the subconnection scheme described above, the majority of the protection subconnections are predeployed with high-level nodes as endpoints (a small number of subconnections need to terminate on low-level nodes in order to provide protection for demands that originate at these nodes). This allows a significant amount of optical bypass to be realized as most of the protection capacity transits the low-level nodes.

The possible disadvantage of this scheme is the requirement for an edge switch at some, or all, of the nodes. However, as has been emphasized previously, an edge switch can improve the flexibility of a node, such that it may be desirable to deploy such switches anyway. Another predeployed-subconnection-based shared mesh protection scheme, which does not require an edge switch, is proposed in [LiCS05]. The scheme uses the combination of tunable regenerator cards and all-optical switches to concatenate the protection subconnections; the scheme is not compatible with an OADM-MD. Failure recovery requires retuning some transponders and regenerators, turning off some regenerators, and reconfiguring the all-optical switches at the connection endpoints. Thus, while the requirement for the edge switch is eliminated, the restoration process is somewhat slower and does not avoid the transient issue.

7.7.1 Cost vs. Spare Capacity Tradeoff

The predeployed-subconnection protection scheme inherently poses a tradeoff of cost versus capacity. To achieve better sharing of the protection capacity, shorter subconnections are predeployed (i.e., subconnections with fewer hops). This translates into a greater number of protection subconnections, where each subconnection incurs the cost of two transponders and two edge switch ports. Figure 7.19 illustrates this tradeoff. The dotted lines represent the working paths, and the dashed lines represent the protection subconnections. The same three working paths are shown in Fig. 7.19 (a) and (b): A-E, A-G-D, and C-G. In Fig. 7.19(a), there are three protection subconnections, whereas in Fig. 7.19(b) there are four. Either configuration is sufficient to provide protection from a single link or node failure. Figure 7.19(a) requires six protect transponders and seven wavelength-links of protection capacity, whereas Fig. 7.19(b) requires eight protect transponders but requires only six wavelength-links of protection capacity. Thus, Fig. 7.19(b) is more capacity-efficient, but more costly. Essentially, by dividing the A-B-C-D protection subconnection into A-B-C and C-D, as in Fig. 7.19(b), the C-D protection subconnection can be shared by all three working paths.

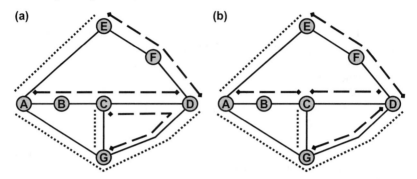

Fig. 7.19 The predeployed subconnections are indicated by the dashed lines. (a) Three predeployed subconnections, occupying seven links and requiring six transponders. (b) Four predeployed subconnections, occupying six links and requiring eight transponders.

This cost vs. capacity tradeoff is explored further here, using the 60-node backbone network that was shown in Fig. 5.8. The network was assumed to be optical-
bypass-enabled, with an optical reach of 2,500 km. Several shared mesh protection designs were performed, where in each design an increasing number of network nodes were selected as *protection hubs*. The protection hubs are akin to the
'high-level' nodes in the hierarchical protection scheme, where protection capacity
is 'chopped' into subconnections at the hubs. Thus, the greater the number of
hubs, the smaller the resulting protection subconnections, yielding more opportunities for sharing, but resulting in higher cost. (The working paths can optically
bypass the hubs, however.)

Demands requiring shared protection were added one-by-one to the network,
with no knowledge of future demands. Enough demands were added such that the
resulting capacity requirement on the most heavily loaded link was on the order of
100 wavelengths. All demands were at the line-rate (i.e., no traffic grooming was
needed). The paths of the demands were selected with an emphasis on sharing the
existing protection capacity.

Varying the number of protection hubs produces the "cost" vs. capacity curve
plotted in Fig. 7.20. (Each point in the curve represents an average of several
runs; the variance among the runs was very small.) The primary y-axis is the
normalized total number of transponders required for the working and protect
paths; this is used as a rough measure of network cost. (Any regeneration was tal-
lied as two transponders.) The x-axis is the total required capacity for the working
and protect paths, measured in wavelength-km. The percentages next to the data
points indicate the percentage of nodes that were selected as hub nodes. As expected, as the number of hubs increases, the required total capacity decreases but
the cost increases. From this graph, selecting roughly 15% to 20% of the nodes as
protection hubs represents a good tradeoff point, where the number of transponders and the required capacity are both within ~15% of their minimums. A study
was performed for several other networks in [Simm07], producing similar results.
(Note that, while not shown on the graph, selecting 100% of the nodes to be pro-

tection hubs reduces the total required capacity by less than 1% and increases the total number of transponders by almost 10%, as compared to the scenario where 55% of the nodes are protection hubs; thus, this is not an economically viable option.)

Given that the protection subconnections require transponders at the endpoints, and hence O-E-O conversion, it is interesting to investigate the amount of optical bypass attainable in the network as the number of hubs increases. The top curve in Fig. 7.20 plots the average optical bypass in the network (this is the percentage of working and protect wavelengths that enter a node that optically bypass the node). As the number of hubs increases, the average optical bypass decreases because the protection capacity is being electronically terminated more frequently. With 15% to 20% of the nodes as hubs, the average optical bypass is still significant, at about 65%, indicating that this shared protection scheme is able to take good advantage of optical bypass.

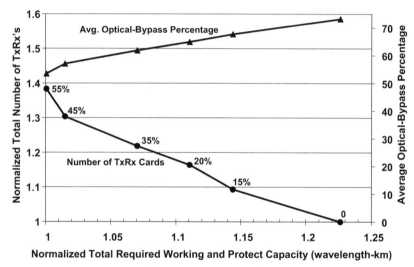

Fig. 7.20 The lower curve represents "cost" vs. capacity for shared mesh protection based on predeployed subconnections, for a 60-node backbone network. The total number of transponders (TxRx) required for the working and protect paths is used as a rough measure of cost. The percentages next to the data points indicate the percentage of nodes selected as protection hubs. The upper curve is the average optical-bypass percentage achieved in the network, assuming 2,500-km optical reach.

7.8 Shared Protection Based on Pre-cross-connected Bandwidth

The predeployed-subconnection protection scheme avoids problems with optical amplifier transients; however, it does require a small amount of switching to concatenate subconnections together to form the appropriate backup path. It is worthwhile to discuss a class of shared protection schemes based on pre-cross-connected bandwidth, where the need for *any* switching at the time of failure is eliminated except at the endpoints of the failed connection. Because of the minimal amount of switching required, these schemes are likely to be somewhat faster than the subconnection method, although the issue of transients does need to be addressed in these schemes.

The origin of this protection class is pre-connected protection cycles, or *p-cycles*, where the spare capacity is pre-connected to form cycles [GrSt98]. Each cycle protects against failures on the cycle itself, as well as failures on links that straddle the cycle. Restoration requires switching only at the nodes on either side of the failure. (The initial p-cycle proposal was for link-based protection; however, the idea was later extended to path protection [KoGr05].) P-cycle link-based protection is illustrated in Fig. 7.21, where there are two working paths as shown by the dotted lines: B-F-E-D and B-F-G. Only one cycle of protection capacity is shown, A-B-C-D-E-F-A, as indicated by the dashed line. This one cycle is not sufficient to protect against all possible working-path failures; however, for simplicity, the other cycles are not shown. (Note that because this is a closed protection ring, *lasing* could be an issue in an optical-bypass-enabled network; mitigating techniques such as adding a regeneration somewhere along the ring could be used, as described in Section 7.3.1.)

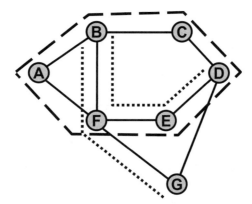

Fig. 7.21 P-cycle link-based shared protection, where just one of the required protection cycles is shown, as indicated by the dashed lines. It can protect against failures to links on the cycle (e.g., Link FE) and failures to links that are chords of the cycle (e.g., Link BF).

If Link FE fails, affecting the B-F-E-D connection, then the switches are configured at Nodes F and E to direct the connection around the protection cycle to avoid the failed link; i.e., the new path is B-F-A-B-C-D-E-D. This is similar to typical link-based ring recovery. However, the p-cycle can also be used to protect against failures on chords of the protection cycle. For example, if Link BF fails, affecting both working paths, the switches at Nodes B and F are reconfigured to redirect the two connections; e.g., one path is rerouted over B-A-F-G and the other over B-C-D-E-F-E-D. Taking advantage of chordal protection allows the spare capacity requirements of the scheme to be similar to that for mesh protection, while requiring switching only at the endpoints of the failure. As noted in [GrSt98], the scheme combines the speed of ring protection with the capacity efficiency of mesh protection.

As organizing the protection capacity into cycles may be too restrictive, the idea was extended to more general protection topologies in a path-based protection scheme called *pre-cross-connected trails* (PXT) [ChCF04]. (A trail can be considered an end-to-end path.) The protection capacity is laid out in arbitrary mesh formations similar to more general shared mesh path-based protection schemes, with the stipulation that all 'branch points' be eliminated. In the predeployed-subconnection scheme of Fig. 7.18, Node H represents a branch point; i.e., it is an intersection point of multiple protection subconnections, such that Node H is required to switch depending on which working path has failed. Thus, the arrangement of the protection capacity in Fig. 7.18 is not suitable for PXT protection. Figure 7.22 illustrates this same network, with the same two working paths, using PXT protection. Here, the protection capacity is routed over A-F-G-H-I-E-D; this path is considered a PXT. The key property is that under any failure scenario, only the endpoints of the failed connections need to participate in the recovery. For example, if the working path along A-B-C-D fails, the switches at Nodes A and D are reconfigured to direct the client traffic to the protection trail, with no other switching needed. Alternatively, if the working path along A-J-K-I fails, switch reconfigurations occur only at Nodes A and I. Various experiments in [ChCF04] showed PXT protection to be as efficient as more general shared mesh protection schemes.

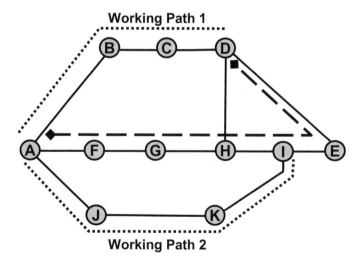

Fig. 7.22 PXT path-based shared protection. The PXT, shown by the dashed line, can protect either working path with switching required only at the connection endpoints.

A variety of architectures could be used at the connection endpoints. For example, there could be a transponder associated with the protection bandwidth; at the time of failure, the connection is directed to this transponder on the client side. Alternatively, the connection's transponder could be switched from the working path to the protect path on the network side at the time of failure, such that no transponders are dedicated to the protection bandwidth. The remainder of this section will assume the former architecture, specifically focusing on a network with optical bypass. However, transients are a concern with either architecture.

First, assume that under no failures, all protect transponders are off, and the optical network elements at the intermediate nodes of the PXT (i.e., F, G, H, I, and E) are configured to allow the corresponding wavelength to pass through. Assume that a failure occurs along A-J-K-I. The protect transponders at Nodes A and I must be turned on, where this process must occur gradually to avoid transients.

To avoid the transient issue, it is reasonable to consider pre-lighting the PXT similar to the predeployed-subconnection method, where the protect transponders at Nodes A and D are always on, all other protect transponders are off, and all intermediate optical network elements are initially in the through state. Node I, under no failures, is shown in Fig. 7.23(a). Assume that a failure occurs along A-J-K-I. Node I turns on its protect transponder and at the same time it must reconfigure its optical network element to block the PXT signal entering from the D to I direction (i.e., the transponder at Node D is on; its signal needs to be blocked at Node I; otherwise it would interfere with the protection signal being sent by Node I to Node A). Thus, again, transients are a concern; the process of simultaneously adding the power from the protect transponder at Node I while blocking the power from Node D's transponder must be done gradually. Node I, after recovery from the failure, is shown in Fig. 7.23(b). Note that in the A to D direction, the optical element at Node I ideally operates in a drop-and-continue mode, where it drops the protection signal at Node I and also allows it to continue to Node D so that the power level along I-E-D remains constant. As this discussion illustrates, pre-lighting the PXT still does not avoid having to be concerned with transients.

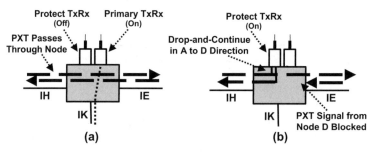

Fig. 7.23 Details of Node I from Fig. 7.22 in the mode where the PXT is lit under the no-failure condition. (a) With no failures, the network element at Node I passes the PXT wavelength in both directions. (b) When a failure occurs to the A-I connection, the protect transponder at Node I is turned on, and the network element at Node I must block the PXT signal from Node D.

The issue of transients with this scheme, as well as with most other shared protection schemes, does not imply that such schemes are inappropriate for an optical-bypass-enabled network; it just means that the adding/switching operations need to be executed gradually. This needs to be taken into account when designing the system and estimating the time for recovery.

7.9 Protection Planning Algorithms

As indicated by the discussion thus far in this chapter, there are numerous optical protection schemes that differ in aspects such as capacity efficiency, recovery speed, and coverage of various faults. Different network planning algorithms are needed to optimize the various schemes. This section addresses protection planning algorithms at a high level, to illustrate general techniques that may be used in the design process.

Section 3.6 of Chapter 3 covered routing algorithms that find link-and-node-disjoint paths between a source and destination; i.e., Shortest Pair of Disjoint Paths (SPDP) algorithms. If completely disjoint paths are not possible, such algorithms can find the maximally disjoint set of paths. SPDP algorithms are at the heart of most protection design strategies.

Section 3.6 also covered the notion of Shared Risk Link Groups (SRLGs), where a single failure may cause multiple links to fail due to shared conduit. More generally, Shared Risk Groups (SRGs) refers to any set of network resources that are part of the same failure group. Diversity with respect to SRGs is desirable as part of protection planning. As described in Section 3.6, there are various graph transformations that can be used to handle the most common SRG configurations.

7.9.1 Algorithms for Dedicated Protection

First, consider network planning with dedicated protection. Routing can be performed using the techniques of Chapter 3, where a SPDP algorithm is used to find a set of candidate working and protect paths that are link/node disjoint. As demand requests enter the network, one of the candidate path pairs is selected, typically based on the current network state. Regeneration sites are chosen for the working and protect paths, if needed, and the resulting subconnections are assigned wavelengths.

Depending on the protection scheme, the wavelength assignment process may need to be modified to account for additional constraints. For example, with the network-side 1+1 scheme described in Section 7.2, wavelengths cannot be assigned independently to the working and protect paths. One method of enforcing these constraints is to form 'subconnection groups' that are comprised of the subconnections that must be assigned the same wavelength. A subconnection group can then be treated as if it were one large subconnection composed of all the hops that are included in each of the individual subconnections in the group.

It is also possible to use one-step RWA methods with dedicated protection. However, as pointed out in Section 3.6.3, the graph transformations that accompany the one-step methods may produce SRLG-like situations that need to be handled when running a SPDP algorithm on the transformed graph. The transformation procedures also need to be modified to enforce situations where portions of the working and protect paths are required to be assigned the same wavelength.

7.9.2 Algorithms for Shared Protection

In contrast to dedicated protection, shared protection requires more advanced algorithms because sharing creates an interdependence among the protected demands. Two demands may share protection capacity only if their working paths are disjoint. Thus, the selection of the working path for a demand may affect the capacity required for protection. For example, a longer working path may be selected in order to make better use of the protection capacity already deployed. Consider the example of Fig. 7.24, where there is one protected demand between Nodes A and D, as shown by Working Path 1/Protect Path 1. The working path is routed over A-E-D and the protect path over A-B-C-D. Assume that a request for a second protected demand between Node A and Node D arrives. If the working path of this second demand is also routed over A-E-D, which is the shortest path, it will not be able to share the existing protection capacity along A-B-C-D. If, however, the new working path is routed over path A-F-G-D, which is 800 km longer, the protection capacity along A-B-C-D can be shared by both demands (because the working paths are disjoint).

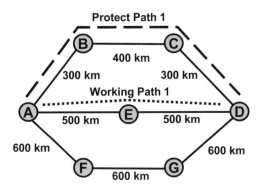

Fig. 7.24 The existing working and protect paths are shown. If another connection between A and D is added, routing it along A-F-G-D allows it to share the existing protection capacity.

While selecting the longer working path for the new demand may reduce the overall capacity requirements, it may ultimately result in a higher cost. Assume that the optical reach in Fig. 7.24 is 1,000 km. If the new working path is routed on A-F-G-D and the protect path shares the existing capacity along A-B-C-D, then two additional regenerations are required (on the working path). If the new working path is routed on A-E-D and the protect path is a new wavelength on A-B-C-D, then no additional regenerations are needed, although the additional protect wavelength may incur cost at its endpoints (e.g., a transponder and a switch port, depending on the protection mechanism). It may be less costly to add this extra protection capacity rather than select a working path with two extra regenerations. Thus, simply maximizing protection sharing may not always be the optimal strategy. These types of tradeoffs need to be evaluated as part of the design process.

The choice of the protect path also affects the amount of attainable sharing. Figure 7.25 shows a network with two existing protected demands, Working Path 1/Protect Path 1 and Working Path 2/Protect Path 2. If a new protected demand is added between Nodes A and G, and the working path is routed over A-J-K-G, there are several options for its protect path. Assume that the shortest possible protect path for the new demand is A-L-M-G. If the protect path is routed over this path, a new protection wavelength will need to be allocated along the path, with no sharing of existing protection resources. Alternatively, if the protect path is routed over A-B-C-D-E-F-G, then it can share the protection capacity that is already deployed, with no additional spare capacity needed. The second option is more capacity-efficient, although it produces a protect path that is twice as long.

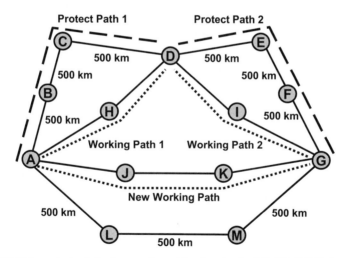

Fig. 7.25 If the protect path for the new demand is selected to be A-B-C-D-E-F-G, then the existing protection capacity can be shared. However, this path is twice as long as a protect path along A-L-M-G.

Routing the working path and/or the protect path over longer paths, or paths with more hops, generally increases the vulnerability to failure. As was discussed in Chapter 6 relative to grooming, the planning algorithm may need to enforce rules regarding how much excess routing can be tolerated in the working path and protect path in order to attain better sharing (the excess factors for the working and protect paths can be different). This can be based on the desired availability of the demands.

Next, three strategies for routing demands with shared protection are outlined.

7.9.2.1 Candidate-Path Strategy

Assume that a request arrives for a protected demand between a particular source and destination. One strategy is to consider each of the candidate path pairs that have been pre-calculated for the source/destination/protected class. (Generating candidate paths for protected demands using a SPDP algorithm was covered in Section 3.6. More candidate path pairs should be pre-calculated when the protection mode is shared as opposed to dedicated; e.g., at least five path pairs. Recall that while the paths comprising a pair should be maximally disjoint, the set of candidate path pairs may have some links/nodes in common.) For each candidate path pair, the amount of total capacity that would need to be added to the network to accommodate the working and protect paths is evaluated. Capacity needs to be added along the whole working path, whereas the protect path may be able to share protection capacity that is already deployed. The cost of the paths, due to added transponders and regenerators, along with the loading on the path links are also evaluated for each candidate path pair. Based on the relative weighting of factors such as capacity and cost, one of the candidate path pairs is selected.

If the selected candidate path pair yields little or no sharing of the protection capacity, then other paths can be considered. For example, the candidate paths can be examined again, where the working and protect paths are swapped. Using a candidate protect path as the working path, and vice versa, may allow better sharing, although it likely increases the distance of the working path.

7.9.2.2 Shareability-Metric Strategy

Another commonly used shared protection design strategy is to select a working path, and then perform a relatively simple graph modification that assigns link metrics based on the shareability of the protection capacity (e.g., [BLRC02]). To simplify the discussion, assume that the demands are at the wavelength level, and that protection capacity is allocated in units of wavelengths. The first step is to temporarily remove all links from the topology that are included in the working path. (To be more precise, any link that is part of an SRLG that contains a working-path link is removed.) Furthermore, any link that has no free capacity *and* no shared protection bandwidth is removed. Each remaining link is examined to determine if there is a shared protection wavelength already allocated on the link that could potentially be used to protect the new demand. For a protection wavelength to be shareable with the new demand, the working paths that already use this wavelength for protection must be disjoint from the working path of the new demand. In addition, if there is a limit specified on how many working paths can share a protection wavelength, the limit must not have been reached. Any link that is found to have at least one shareable protection wavelength is assigned a small cost metric. All other links are assigned their usual cost metric (e.g., link distance).

A shortest-path algorithm is then run on the modified graph to determine the protect path. Assigning a relatively low metric to the shareable links drives the protect path towards those links. Note that if the metric is too low for these links, then the path that is found may be excessively long (e.g., if the metric of the shareable links were zero, then there would be no penalty at all for traversing more links). As investigated in [BLRC02], the metric can be adjusted to strike the desired balance of sharing and path length; using a metric that is roughly 50% of the usual link metric was shown to achieve a good balance.

If this method is successful in finding a path, and the path distance and sharing are acceptable, then this path can be used as the protect path for the new demand. The path is guaranteed to be disjoint from the working path because the working-path links have been eliminated from the transformed graph. (As detailed in Section 3.6, a two-step process that searches for a protect path after eliminating the links of the working path may be suboptimal; thus, with this shareability-metric technique, it may be desirable to consider multiple candidates for the working paths.)

For shared protection based on subconnections, where the subconnections are terminated in a transponder at either endpoint, the above algorithm needs to be modified. Rather than looking at the shareability of links, the algorithm focuses on the shareability of the existing protection subconnections. If a protection subconnection is potentially shareable with the new demand, a link should be added to the graph that captures the subconnection, and the link should be assigned a relatively low metric. For example, if a shareable subconnection extends between Node X and Node Y, then a link between Nodes X and Y should be added to the graph. Any links in the true topology that do not have a shareable subconnection between them should be assigned their usual link metric. Running a shortest path algorithm on this modified graph then drives the protect path to the shareable subconnections.

7.9.2.3 Potential-Backup-Cost Strategy

Note that the previous method focuses on finding a good protect path given a particular working path. However, because of the constraint that two working paths that have a link or intermediate node in common cannot share a protection wavelength, the selection of the working path itself affects the amount of achievable sharing. There are various shared protection methodologies that take this into account, where the algorithm is proactive in searching for a working path that will likely yield a protect path that can take advantage of sharing.

For example, in [XuQX07], P_i is defined to be the maximum amount of protection capacity required on any link in the network in order to protect against a failure of link i. Let M be the maximum P_i over all links i in the network. Assume that the goal is to minimize the total utilized wavelength-links. Then each link i in the network is assigned a metric equal to $(1+\alpha)$, where α is proportional to P_i/M. The '1' term captures that the working path will occupy one wavelength on the link, whereas the α term represents the potential backup cost. A shortest path algorithm is run with this metric assignment, and the resulting path is taken as the working path. One can then use the methodology described in the previous section, where the working path links are removed and the remaining links are assigned a metric based on their shareability, to find a suitable protect path; this requires a second invocation of the shortest path algorithm.

7.9.2.4 Combining Strategies

A good overall design strategy is to use a combination of these techniques. The candidate-path strategy can be used initially to select the working and protect paths for a new demand. After the paths are selected, the amount of sharing can be evaluated. For a demand where there is little or no sharing of its protection capacity, the shareability-metric methodology described above (possibly in concert with the potential-backup-cost strategy) can be used to improve the sharing.

By using the candidate-path strategy first, demands are more likely to be routed over a preferred path. Additionally, this typically produces good sharing for the bulk of the demands, such that the other techniques, which are slower, are only needed for a relatively small number of demands.

(In the study of Section 7.7.1, the candidate-path strategy was used, combined with the shareability-based metric strategy for demands not achieving a high degree of sharing. Using the potential-backup-cost strategy, or using the shareability-based metric strategy for *all* demands, did not appreciably affect the results.)

7.9.2.5 Distributed Shared Protection with Partial Information

Thus far, it has been assumed that the shared protection algorithm has available to it knowledge of all working paths and their respective backup paths. This allows one to calculate whether certain protection bandwidth is shareable by a new working path. If shared protection is calculated in a distributed environment, updating each node with information on every working path and its respective backup path may be too onerous. Thus, schemes that operate on only partial information have been devised.

For example, [KoLa00] assumes that only aggregated link information is disseminated indicating how much bandwidth on link i has been allocated for working paths, how much has been allocated for backup paths, and how much bandwidth is unallocated; let these parameters be represented by W_i, B_i, and R_i, respectively. For any new working path that is chosen, W_i of the links on the new path increases. Let W_{max} be the maximum such W_i on the new working path. If the backup path is carried on link j, then W_{max} backup capacity is needed on link j to *guarantee* that there are sufficient protection resources for the new working path. Thus, B_j can be compared to W_{max} to determine how much more protection capacity would need to be allocated on link j if the backup path of the new demand is routed on this link. This is used to assign link metrics; i.e., link j is assigned a metric of $Max[(W_{max} - B_j), 0]$; if this value is greater than the residual bandwidth, R_j, it is instead assigned a metric of infinity. The links of the selected working path are also assigned a metric of infinity. By running a shortest path algorithm with these metrics, the likelihood of finding a backup path that minimizes the required amount of new capacity is enhanced. Multiple iterations can be run where different working paths are considered. Studies showed that while this scheme requires more capacity than when complete information is provided regarding working and backup paths, it performs significantly better than the case where protection capacity is not shared at all. (To accommodate node failures, each node is replaced by two dummy nodes interconnected by a link; the incoming nodal traffic terminates on one of the dummy nodes, and the outgoing nodal traffic is sourced by the other node. The link connecting the two dummy nodes is then included in the above algorithm.)

In [QiXu02], information that is somewhat more detailed is maintained, allowing the scheme to be more efficient in allocating protection bandwidth. Rather than simply tracking the total working capacity allocated on each link, the scheme tracks P_i^k, the amount of working capacity on link i that is protected by link k, for all links i and k in the network. When working paths are established, the signaling message specifies the protect path as well, so that P_i^k can be updated accordingly.

This can then be used to estimate the amount of additional protection bandwidth that would need to be allocated on a link to protect a particular new working path. This estimate is used as the link metric, similar to the scheme above, so that running a shortest path algorithm minimizes the estimated additional protection bandwidth. More details can be found in [QiXu02].

Related schemes that operate on partial information are proposed in [SrSS02]. The emphasis in these schemes is on finding a wavelength-continuous backup path. Thus, information regarding whether a given wavelength on a given link is used for a working path, for a protect path, or is unassigned is disseminated.

All of the methods described above avoid having to disseminate to all nodes the details of each working and protect path, and thus are more suitable for distributed computation of shared protection.

7.10 Protection of Subrate Demands

Chapter 6 covered subrate demands, where the bit-rate of the demand is less than that of a wavelength. As discussed there, the two most common ways of handling subrate demands are with end-to-end multiplexing or with grooming. With end-to-end multiplexing, subrate demands with the same source and destination are grouped together in a wavelength and carried as a unit from source to destination. With grooming, arbitrary subrate demands can be bundled together to form well-packed wavelengths, where repackaging of the wavelengths can occur at intermediate points along the demand paths. While Chapter 6 focused on the multiplexing and grooming aspects, this section specifically addresses protection for subrate demands.

There are generally two approaches to protecting subrate demands. First, there is wavelength-level protection, where the subrate demands are bundled into wavelengths, and then the wavelengths are routed with protection. Second, there is subrate-level protection, where the individual subrate demands are routed with protection, and then the working and the protect paths are bundled into wavelengths. Both methods are discussed below in the context of grooming (end-to-end multiplexing can be thought of as simplified grooming where bundling occurs at only the source and destination nodes). The examples given use SONET signal rates; however, the concepts are applicable to other grooming technologies.

7.10.1 Wavelength-Level Protection

The network shown in Fig. 7.26 is used to illustrate wavelength-level protection, where the wavelength rate is assumed to be OC-192. As shown in the figure, there are three protected OC-48 demands, A-E, A-F, and G-F, as well as one unprotected OC-48, A-F. Assume that Nodes A, B and E are equipped with grooming switches. The following grooming scheme is used (others are possible):

- all of the demands from Node A are grouped into a wavelength and routed to Node E

- the demand from Node G is routed on a wavelength to Node E

- all demands destined for Node F are carried on a wavelength from Node E to Node F

Using the terminology of Chapter 6, each of these three groupings is a *grooming connection* (GC). Each GC contains at least one protected subrate demand, thus each GC requires protection. With protection at the wavelength level, the GCs can be treated as three independent wavelength-level demands (A-E, G-E, and E-F) that require protection. Any of the protection schemes discussed in this chapter can be used to protect the GCs. For example, either dedicated or shared protection can be used.

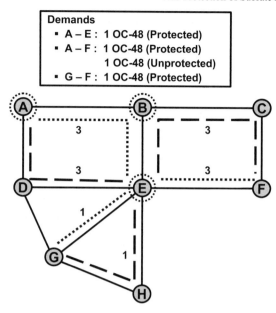

Demands
- A – E : 1 OC-48 (Protected)
- A – F : 1 OC-48 (Protected)
            1 OC-48 (Unprotected)
- G – F : 1 OC-48 (Protected)

Fig. 7.26 Protection of the specified subrate demands using wavelength-level protection. The circled nodes indicate those nodes equipped with grooming switches. The dotted lines indicate the working paths of the grooming connections, and the dashed lines indicate the corresponding protect paths. The numbers indicate how many subrate demands are carried in the grooming connections.

Figure 7.26 illustrates the working and protect paths for each of the GCs, where the working paths are shown by the dotted lines and the protect paths by the dashed lines. Thus, the protected A-F subrate demand follows A-B-E-F for the working path and A-D-E-B-C-F for the protect path, with grooming occurring at Node E. (Although the end-to-end working and protect paths for this connection are not link diverse, the scheme does provide protection against a single link failure, because the A-E and E-F GCs are protected independently.) The numbers next to the paths indicate how many OC-48s are carried in the GC. In the example of Fig. 7.26, only the grooming switch at Node E is being used.

Being able to treat the GCs like a wavelength service simplifies the planning and operation of the network; e.g., the system needs to track the protection path only for each GC as opposed to each subrate demand. One disadvantage of wavelength-level protection is that the subrate demands are vulnerable to failures at the grooming nodes. For example, if Node E fails, all of the traffic in the figure is brought down. Another disadvantage is that an unprotected subrate demand that has been groomed with a protected subrate demand will end up being protected. In the example, the demand between Nodes A and F that did not require protection is routed in protected GCs, essentially providing a higher level of service than the customer requested (and paid for). Another option would have been to route the unprotected subrate demand in a separate, unprotected GC, but this would have utilized more wavelengths in the network and required more transponders. Capacity-wise, another potential disadvantage is that the fill-rate of the GC applies to both its working and protect paths; i.e., the working and protect portions of the GC cannot be filled independently. Thus, in the example, unless additional subrate demands are added to the GC between Nodes G and E, the wavelength routed on G-E and the wavelength routed along G-H-E will remain only 25% full.

7.10.2 Subrate-Level Protection

With subrate-level protection, protection occurs on a per-subrate-demand basis. Each subrate demand requiring protection is initially routed along disjoint working and protect paths. The individual subrate paths are then groomed together into wavelengths. During the grooming process, the working and protect paths can, for the most part, be treated independently. However, as described in Chapter 6, efficient grooming may require that some paths be modified. When dealing with a protected subrate demand, and either the working or protect path is being modified in the grooming process, it is important to check that link/node diversity requirements will not be violated.

Either dedicated or shared protection can be used for the subrate demands. For example, with shared protection, two subrate demands can share the same subrate protection capacity, as long as the two working paths are link/node disjoint. The shared protection capacity must enter the grooming switch at the 'sharing points'. The grooming switch is actively involved in the recovery mechanism, to direct the failed subrate demand to the protection capacity.

Due to the large number of subrate demands that may be part of a network design (e.g., tens of thousands of protected subrate demands may be added at one time in a network design exercise), processing each demand individually may not be a scalable option. As discussed in Chapter 6, it is preferable to first group subrate demands that have the same source and destination and required protection, and select an initial working path and protect path for the group. The grooming operations described in Chapter 6 are then implemented to improve the packing of the wavelengths. In performing these operations, individual subrate demands may be shifted to different grooming connections. Thus, the bulk of the routing and grooming is performed on groups of subrate demands, whereas fine tuning occurs on a per-demand basis.

Because protection is at the subrate demand level, failure recovery requires more system memory and more signaling, as the system needs to track the protection path of each subrate demand and restore each one individually. One could use *protection groups* to reduce the complexity and restoration signaling overhead, where demands with the same working and protect paths are treated as a single unit for recovery [ADHN01].

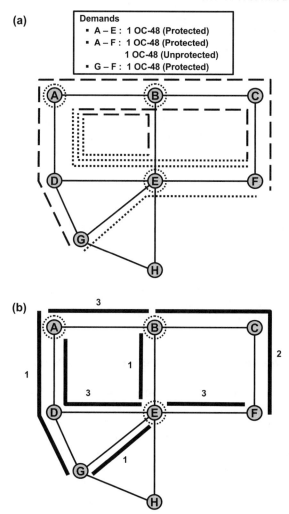

Fig. 7.27 Protection of the specified subrate demands using subrate-level protection. (a) The working paths of each subrate demand are shown by the dotted lines and the protect paths by the dashed lines. Grooming is performed at Nodes A, B, and E. (b) The thick lines indicate the grooming connections that result from bundling the subrate paths. The numbers indicate how many subrate demands are carried in the grooming connections.

Consider using subrate-level protection for the example of Fig. 7.26. Figure 7.27(a) illustrates the working and protect paths for each of the subrate demands (the working paths are shown with dotted lines, and protect paths with dashed lines). Note that the working and protect paths for each demand are link-and-node-disjoint, thereby providing greater protection against grooming node failures as compared to the wavelength-level protection scheme. The paths are groomed together, taking advantage of the grooming switches at Nodes A, B, and E, to form the GCs shown by the thick lines in Fig. 7.27(b) (other grooming arrangements are possible). The numbers indicate how many OC-48s are carried in a GC.

In addition to protecting against grooming node failures, subrate-level protection also provides finer control of the protection resources as compared to wavelength-level protection. This can be advantageous in that an unprotected demand does not end up unnecessarily protected simply because it has been groomed together with a protected demand. Furthermore, it allows the protect paths of some demands to be groomed together with the working paths of other demands. For example, in Fig. 7.27, if a new working path were added along A-B, it could be groomed together with the three protect paths that are in the GC along this link. Thus, a GC created with subrate-level protection cannot generally be designated as a 'working GC' or a 'protect GC'. However, with shared protection, note that mixing working paths with shared-protect paths in the same GC provides less sharing flexibility as the network evolves; e.g., if a wavelength carries both working and shared-protect paths, that wavelength cannot be manipulated (e.g., passed through more grooming points) to provide better sharing because it would disrupt a live connection. Ultimately, this may lead to lower overall capacity efficiency. Thus, in most scenarios, mixing working and shared-protect paths in the same GC is not recommended [ThSo02, OZZS03].

Further discussion on subrate-level protection can be found in [YaRa05b].

7.10.3 Wavelength-Level vs. Subrate-Level Protection

The designs of Fig. 7.26 and Fig. 7.27 offer insights into the relative performance of wavelength-level and subrate-level protection. These designs can be compared on a few measures; however, direct comparisons of the two schemes are somewhat skewed, as the subrate-level protection scheme protects against grooming-node failures and the wavelength-level scheme does not. In terms of capacity, as measured by wavelength-links, the subrate-level design requires 10 wavelength-links, whereas the wavelength-level design requires 11 wavelength-links. If measured by OC-48-links, the subrate-level design requires 20 OC-48-links, whereas the wavelength-level design requires 27 OC-48-links. In terms of number of GCs, the subrate-level design produces 7 GCs, each one terminating in a grooming port at either endpoint; the wavelength-level design generates 3 protected GCs. (The protected GCs utilize either one or two grooming ports at either endpoint, depending on whether protection of the grooming port is desired.) Thus, the subrate-level design utilizes more grooming switch ports and transponders. These trends tend to hold in more general networks, where subrate-level protection may be more capacity-efficient but wavelength-level protection requires less terminating equipment [OZZS03]. In addition, wavelength-level protection should be faster than subrate-level protection in restoring from a failure.

While the discussion thus far has focused on protection of SONET (or SDH) subrate demands, it applies equally well to an IP-over-optical architecture, where the IP layer is deployed directly over the optical layer. Providing protection for IP traffic solely at the wavelength level is fast; however, it is not able to recover from IP router failures. Implementing protection in the IP layer does protect against such failures and may be more capacity-efficient; however, with possibly thousands of IP flows brought down by a failure, the restoration process at the IP layer may be too slow and complex.

Given that each network layer has its own particular strengths and weaknesses, it is natural to consider implementing protection in multiple layers. This is discussed next, in the context of an IP-over-optical architecture.

7.10.4 Multilayer Protection

Consider the example of Fig. 7.28, where an IP connection is established between Nodes A and C, via the IP router at Node B, as shown by the dotted line. From the optical-layer viewpoint, there are two separate connections: one between A and B and one between B and C. If protection is desired in both the IP and optical layers, the question is how to operate the protection across the two layers.

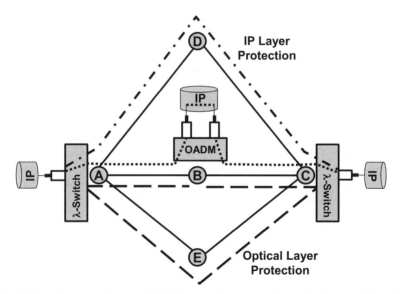

Fig. 7.28 An IP connection is established between Nodes A and C, via the IP router at Node B. With uncoordinated multilayer protection, the optical layer may provide protection capacity along Links AB, BC, AE, and EC, while the IP layer may protect the connection along A-D-C.

First, consider uncoordinated protection. In Fig. 7.28, assume that the IP connections between A and B and between B and C are established as (shared) protected connections in the optical layer. Assume that the optical layer provides protection for the two connections along A-E-C-B and B-A-E-C, respectively; this protection capacity is shown in the figure by the dashed lines. Additionally, assume that the IP layer plans protection for the A-C connection by requesting an unprotected connection from the optical layer between A and C. Assume that the optical layer routes this on the path A-D-C, as shown by the dotted/dashed line.

Assume that Link AB fails. Because there is no coordination among the layers, both protection mechanisms are triggered, such that the optical layer attempts to move the AB connection to A-E-C-B, while at the same time the IP layer attempts to move the AC connection to its alternate path, routed over A-D-C. This redundancy in protection is unnecessary. If both layers support preemptible traffic (where such traffic is carried on the protect path as long as the protection resources are not needed for failure recovery) then some of this traffic may be unnecessarily bumped. Furthermore, studies have shown that simultaneous recovery operations in the optical and IP layers can lead to slow convergence in the IP layer [PDCS06].

To avoid triggering simultaneous recovery mechanisms, the protection in the layers can be coordinated, e.g., via a bottom-up escalation strategy. The optical layer protection mechanism is given the opportunity to respond first to a failure. If this layer is not successful in recovering from the failure, then the IP layer takes action. Normally, such escalation strategies are controlled by a backoff timer, where the IP layer protection mechanism is not triggered until a certain time after the onset of the failure, to give time for the optical layer to respond. If the failure does require action at the IP layer, the backoff timer causes recovery delays. Proactive signaling schemes have also been considered to avoid the need for a backoff timer [EBRL02].

Further coordination can be implemented between the two layers. Note that in Fig. 7.28, there are two different paths allocated to protect the same connection. If the protection planning is coordinated between the IP and optical layers, then, for example, the backup IP path for the AC connection could be routed over A-E-C. Either layer can then use the same protection capacity along these links, where the escalation strategy is relied upon to ensure both layers do not attempt to grab the capacity at the same time. While more capacity-efficient, this scheme requires tighter interaction between the layers.

Another alternative is to allow the IP layer to dynamically control the optical layer in order to recover from a failure. For example, if the IP router at Node B fails, the IP layer could direct the optical layer to tear down the AB and BC connections and create an AC connection along this same path. The communication between the layers would occur via the control plane. Studies have shown this dynamic protection strategy to be more cost effective than a static protection scheme [PDCS06, ChAN03]. However, tearing down the old path and setting up the new path takes time, resulting in a longer restoration time as compared to having a pre-allocated restoration path.

For more details on dynamic IP-over-optical protection, and multilayer recovery in general, see the tutorial provided in [PDCS06]. Protection from two concurrent failures in a multilayer environment is discussed in [PrAS05].

7.11 Fault Isolation

The chapter up to this point has only dealt with failure recovery. Another equally important component of failure management is determining the actual cause and location of the failure. This operation is known as *fault isolation* or *fault localization*. As discussed in Section 7.4, with fault-dependent schemes, the recovery mechanism depends on what has failed, thereby requiring that fault isolation occur prior to failure recovery. Failure-independent protection allows fault isolation to occur after the connections have been recovered, thus providing more time for the fault isolation operation. However, in either case, the cause of the failure must ultimately be determined so that it can be fixed.

Fault isolation methods are often coupled to the system architecture. In O-E-O-based networks, signals are converted to the electrical domain at each node, allowing various error checks to be performed to determine the health of the signal. This link-by-link monitoring enhances the ability of the network management system to localize the root cause of a failure. In optical-bypass-enabled networks, an optical signal may traverse several links and nodes prior to it being converted back to the electrical domain. Thus, the same type of link-by-link electronic-based error checking that is performed in O-E-O networks is typically not possible. The remainder of this section specifically focuses on fault isolation in the presence of optical bypass.

Many failures (e.g., loss of light, or out-of-specification wavelength frequency) trigger a system alarm. Based on the alarms that are received, the network management system can determine what has failed. For example, decision trees can be established, where the particular combination of alarms is used to isolate the failure [MaTo03]. In principle, this appears straightforward; however, alarm correlation is somewhat more challenging in an optical-bypass-enabled network because an initial failure can cause a chain reaction of alarms in the network as the failure (e.g., the loss of light) propagates to other links. It is important for the system to be able to suppress alarms when appropriate, to avoid overwhelming the network management system. In [MaTT05], a methodology for selecting the network locations at which to deploy monitoring equipment is presented, with the goal of minimizing the possible failure events that could cause a given sequence of alarms.

Total link failures (e.g., fiber cuts) are generally easier to isolate in a network as compared to intermittent failures or failures that affect just a single wavelength. One method is to analyze all of the connections that have suddenly failed, and deduce on which link the failure has occurred. A more reliable method makes use of the optical supervisory channel (OSC) that is typically carried in-fiber, but out-of-band (e.g., 1510 nm), on each link of the network. The OSC is terminated in the electrical domain on each managed device in the network to provide a communications channel for remote management, monitoring, and control. For example, it can be terminated on each nodal network element and possibly on each optical amplifier. (There is usually an out-of-fiber means of communicating with the nodal network elements as well.) The OSC is typically carried at a rate that is much lower than the data line-rate so that the associated electronics are relatively low cost. The loss of the OSC channel between two endpoints, or error messages encoded on the OSC, can be used to isolate failures along the link [LiRa97].

Another proposed method of link-failure isolation is based on sending probes over specific paths in the network. In [WeCZ05], a sequence of probes is sent over a subset of the network links, where the path of a probe is selected based on which of the previous probes were not received successfully. More specifically, the probe path is chosen to maximize the network state information that is likely to be provided by the probe. By making such a choice, the number of probes that need to be sent to identify a single link failure is kept to a minimum. An alternative probe-based scheme, proposed in [HPWY07], relies on sending a predetermined set of probes at once, where the combined results of the probes (i.e., whether successfully received or not) uniquely determines which link has failed. This method is faster, although it typically requires more probes.

Finer granularity monitoring can be provided by various types of *optical performance monitors* (OPMs) [KBBE04, Will06]. For example, an OPM can tap off a small amount of the power from the WDM signal, and continually scan through each wavelength, checking signal parameters such as power level, wavelength accuracy, and OSNR. This provides at least some feedback on each individual wavelength. However, if an OPM is limited to checking the three aforementioned signal parameters, then some optical impairments, such as an increase in dispersion, will not be detected. More advanced OPMs, e.g., ones that can monitor the 'Q-factor' of the signal (the Q-factor is a measure of signal quality), are required to provide better fault detection capabilities [KBBE04]. From the point of view of enhancing the fault management capabilities of the network, one or more OPMs are ideally deployed on each link. However, due to the cost of OPMs, carriers may choose not to deploy them, or to deploy them only sparingly.

A topic related to fault isolation is detecting malicious attacks on a network. An introduction to this topic is provided in [MMBF97, MeCS98, ReLG06].

Chapter 8

Economic Studies

As a departure from previous chapters, which focused on the algorithmic aspects of optical networking, this chapter addresses the network economics. A range of network studies are presented that investigate various properties of optical networks, especially with regard to the economics of optical bypass. The studies are intended to provide guidance on how best to evolve a network as traffic levels continue to grow, and also to shed light on some of the desirable properties for a system vendor's portfolio.

The results of any economic study depend on the topology of the network, the traffic set and traffic growth model, and, of course, the cost assumptions used in the study. To simplify the presentation, the results of most of the studies are shown for a single reference network and reference traffic set that are representative of realistic carrier networks. However, the studies have been performed on a range of networks and traffic sets to probe dependencies on these factors. If warranted, results from other topologies are also presented. References to previously published network studies, based on different topologies and traffic, are also provided where appropriate. The baseline network topology and traffic set, along with the cost assumptions, are presented in Section 8.1.

The first two economic studies arise from the fact that extended-reach transmission and optical-bypass-enabling network elements come with a cost premium. For example, Raman-based amplifiers are generally needed for extended optical reach as opposed to lower-cost Erbium-based amplifiers. Dispersion compensation, equalization, and transient controls are also needed with optical-bypass systems, adding further to the cost of the infrastructure. The associated transponders are more costly as well due to the need for more complex modulation schemes, more precise lasers and filters, and more powerful error-correcting coding. The philosophy is that the extra cost of the optical-bypass-enabled infrastructure is more than compensated for by the elimination of a large percentage of the regenerations. However, this implies that whether optical-bypass technology is cost effective in a network depends on the traffic level (i.e., the greater the level of

J.M. Simmons, *Optical Network Design and Planning*,
DOI: 10.1007/978-0-387-76476-4_8, © Springer Science+Business Media, LLC 2008

traffic, the greater the number of regenerations that potentially can be removed). This relation is investigated in Section 8.2. The study is carried a step further in Section 8.3, where the benefits of optical bypass in an IP-over-optical architecture are explored for different levels of traffic.

The cost of technology also implies that there is a limit to the benefits that can be achieved from increasing the optical reach. After some point, the extra cost of extending the reach further is not offset by the additional reduction in the amount of regeneration. The optimal optical reach for a network, from a cost perspective, is explored in Section 8.4. The effect of optical reach on the amount of traffic that needs to add/drop at a node is also investigated in this section. This offers insights into the efficacy of optical network elements that limit the amount of add/drop.

The next two sections examine the benefits of certain elements that may be offered as part of a system portfolio. The economics of the OADM-MD are examined in Section 8.5, where this network element is compared to a nodal architecture composed of only OADMs and/or optical terminals. In Section 8.6, an analysis of providing two types of transponders, one with extended reach and one with more conventional reach, is presented.

One of the benefits of optical bypass is that it provides more flexibility in choosing the topology of the network for a given set of nodes. In Section 8.7, the effect of topology on network cost is compared for an optical-bypass-enabled system and an O-E-O-based system. The comparison is performed in the context of varying the number of links in a metro-core mesh network.

The final two sections are more forward looking. Section 8.8 looks at the optimal line-rate for a network as the aggregate network demand increases over a range from 1 Tb/s to 100 Tb/s. This provides guidance as to when higher-rate transponders may prove in economically. The impact of line-rate on the switching requirements is also considered. The final section examines an architecture where the bulk of the grooming occurs exterior to the network core, possibly in the optical domain. This is one possible evolution direction for future networks.

8.1 Assumptions

8.1.1 Reference Network Topology

The reference network used in most of the studies in this chapter is the same 60-node network that was shown in Fig. 5.8, and which is shown again in Fig. 8.1. This network is a realistic representation of a fairly large U.S. backbone network; the statistics are presented in Table 8.1. The emphasis of many of the studies is on backbone networks due to extended optical reach having more of an impact on such networks. However, where appropriate, the studies are related to regional and metro-core networks as well.

Fig. 8.1 Sixty-node backbone network used in many of the network studies. (While there are three diverse paths across the country in this topology, some carrier networks have only two.)

Table 8.1 Statistics of the Reference Network Topology

Number of Nodes	60
Number of Links	77
Average Nodal Degree	2.6
Number of Nodes With Degree 2	34
Largest Nodal Degree	5
Average Link Length	450 km
Longest Link Length	1,200 km
Optical Amplifier Spacing	80 km

8.1.2 Reference Traffic Set

The reference traffic set that accompanies the network of Fig. 8.1 represents realistic carrier traffic for a backbone network. Because the cost of grooming can be highly variable, e.g., due to large differences in cost between IP and SONET/SDH equipment, the baseline traffic set is restricted to purely line-rate traffic. This allows several of the studies to highlight the economics of transmission and core switching. However, when grooming costs are an integral part of a study, then appropriate subrate traffic is used (depending on the study, either a combination of line-rate and subrate traffic is used, or just subrate traffic is used).

The statistics of the baseline line-rate traffic set are shown in Table 8.2. When routed on the network of Fig. 8.1, several of the links are at or near a utilization of 80 wavelengths. This is assumed to represent a relatively full network. For many transmission systems, increases in traffic would require adding an extra fiber-pair on some links or upgrading the amplifiers to support more capacity.

The traffic set is far from uniform all-to-all traffic. If one were to designate the largest 20% of the nodes (based on traffic generated) as *Large*, with the remaining nodes designated as *Small*, then the traffic breakdown among node pairs is approximately: *Large/Large*: 30%; *Large/Small*: 50%; *Small/Small*: 20%.

Table 8.2 Statistics of the Reference Line-Rate Traffic Set

Number of Demands	400
Percentage of Demands Requiring Protection	50%
Average Length of a Primary Path	1,950 km
Average Length of a Secondary Path	3,400 km

8.1.3 Cost Assumptions

Network economics encompass two major classes of costs: the capital cost of the equipment and the operating cost to run the network. These are often referred to as *CapEx* (i.e., capital expenditures) and *OpEx* (i.e., operational expenditures). Capital cost is generally more straightforward to calculate for a given network design and is the focus of most of the network studies in this chapter. Furthermore, carriers tend to evaluate system proposals based on the capital costs, so it is reasonable to focus on this aspect.

For the studies in this chapter, the relative costs of the basic system components, for an 80-wavelength, 2,500-km optical-reach system with 10-Gb/s line-rate, were chosen as shown in Table 8.3. (All costs in the table are relative to the cost of a tunable 10-Gb/s transponder.) The following assumptions were made: the cost of the shelves for the transponders is amortized in the cost of the transponder; the cost of any equalization or fiber-based dispersion compensation is amortized in the cost of the in-line optical amplifier; the cost of the nodal network elements includes the cost of the pre- and post- nodal amplifiers. In general, OADM-MDs were used in the studies as opposed to more costly all-optical switches. The cost shown for an optical cross-connect (OXC) port assumes the OXC is a wavelength-level cross-connect. When used in an optical-bypass-enabled architecture, the OXC is assumed to serve as an adjunct edge switch.

For those studies that include subrate traffic, the cost of a grooming port is specified in the table as well. This cost is assumed to hold for a general grooming device. Historically, there has been a large cost difference between a SONET/SDH grooming switch port and an IP router port, with IP prices being as much as three times greater [SeKS03]. The cost specified in the table is closer to that of a SONET/SDH grooming switch port, with the expectation that IP port prices will gradually moderate.

Table 8.3 Relative Costs for an 80-Wavelength, 10-Gb/s, 2,500-km Optical-Reach System

Element	Relative Cost
Tunable Transponder	1X
Tunable Regenerator	1.4X
Bi-directional In-Line Amplifier	4X
Optical Terminal	5X
OADM	14X
Degree-3 OADM-MD	21X
Degree-4 OADM-MD	28X
Degree-5 OADM-MD	35X
Grooming Port	1.5X
Optical Cross-Connect (OXC) Port	0.2X

In many of the studies, several optical-reach distances are considered. To capture the fact that extended-reach transmission requires more advanced technology, it was assumed that the cost of the amplifiers, transponders, and regenerators increases by a factor of F for every doubling of the reach. Note that the network elements are equipped with nodal amplifiers, such that this portion of the nodal equipment was affected by this assumption as well. The parameter F is referred to here as the *cost increase factor*. Based on anecdotal evidence from vendors and carriers, F was assumed to be 25% in the studies in this chapter (sensitivity to this assumption was probed in several of the studies). Thus, relative to the costs in Table 8.3 for a 2,500-km reach system, the cost of amplification and transmission in a system with an optical reach of R km was scaled by a factor of:

$$1.25^{\wedge}(\log_2(R/2500)) \tag{8.1}$$

As an example, consider a conventional 600-km optical-reach system. Using the above formula, the amplification and transmission costs would be scaled down by approximately 35%.

Operating costs are notoriously more difficult to capture as they encompass a wide range of factors, such as the cost of electricity to run the equipment, leasing costs for office space, labor costs to install and maintain the equipment, as well as labor costs to sell the networking services. These aspects are enumerated in more detail in [Verb05], where it is reported that carriers estimate the ratio of total operating costs to capital costs to be on the order of 1.3 to 4.0. This wide range is in part indicative of the difficulty in tracking operational costs; it also reflects the dependence of these costs on the underlying networking technology. For example, one of the major selling points of optical-bypass technology is that by removing the bulk of the electronics from the network, there is a concomitant reduction in power costs, required office space, and installation and maintenance costs. This is further borne out by a study in [BaGe06] that included an analysis of the cost of operating a network based on different technologies. An optical-bypass-enabled network was shown to have lower operating costs as compared to various O-E-O architectures. As reinforced in [BaGe06], the absolute operational cost figures may be difficult to nail down; however, the relative trends should hold. Thus, to some extent, the number of transponders deployed in the network can be used as a rough indicator of relative operational costs.

8.2 Prove-In Point for Optical-Bypass Technology

As described in the introduction, the various components of an optical-bypass system are more costly than those in an O-E-O-based system. In order to attain extended reach, the amplifiers and transponders require more advanced technology. The elements to provide optical bypass in a node are also more costly; e.g., an OADM is more costly than two optical terminals. Thus, the 'first-deployed cost' of an optical-bypass-enabled network, prior to any traffic being added, is generally higher than that of an O-E-O network. However, the marginal cost of adding a demand to an O-E-O network is typically significantly higher due to the required amount of regeneration. Overall, as the traffic level increases, an optical-bypass-enabled network eventually becomes the lower-cost option, with the cost savings increasing as more traffic is added. The relative cost of optical-bypass and O-E-O technologies for different levels of traffic is investigated further here.

The 60-node network of Fig. 8.1 was used for the study. Two architectures were considered: an optical-bypass architecture with 2,500-km optical reach, and an O-E-O system with conventional 600-km optical reach. As indicated in Section 8.1.3, the cost of amplification and transmission in the 600-km reach system was assumed to be a factor of ~35% lower than that in the 2,500-km reach system (i.e., using the 25% cost increase factor as in Equation 8.1).

It was assumed that any regeneration was implemented with regenerator cards as opposed to back-to-back transponders. In the O-E-O architecture, there was a need for dedicated regeneration sites along the links that exceeded 600 km in length. Furthermore, in the O-E-O network, it was assumed that the nodes were equipped with patch-panels as opposed to automated switches (i.e., Fig. 2.3 was assumed, not Fig. 2.4). Thus, this was a non-configurable O-E-O architecture. Conversely, the optical-bypass architecture employed OADMs and OADM-MDs, which provide core configurability. This difference in agility is discussed further below.

The traffic level on the network was increased from 'empty' to 'full' to study the impact on the network cost. The reference traffic set discussed in Section 8.1.2 was used to represent a 'full' network; the traffic set was reduced to probe the costs at lower traffic levels. Note that all demands in this traffic set were at the line-rate so that grooming costs did not play a role in the analysis. (Section 8.3 considers subrate traffic.) The 50% of the demands that required protection were assumed to employ 1+1 dedicated client-side protection.

The resulting number of regenerations and capital costs for the two architectures are shown in Table 8.4, where the network costs are normalized to 1.0 for the optical-bypass-enabled network (the final column in the table is discussed below). The capital costs include the costs of the amplifiers, transponders, regenerators, and network elements. As shown in the table, for the assumptions that were made, the initial cost of the O-E-O network was roughly 33% lower than that of the optical-bypass-enabled network. However, as the traffic level increased, optical bypass gradually became the more cost-effective option. When the network was approximately full, with 80 wavelengths routed on the most heavily loaded link, optical bypass provided almost a 35% cost savings. (The results shown in the table are similar to results that were produced with different network topologies and traffic sets; more results can be found in [Simm04].)

This study indicates that whether optical-bypass technology is favored, at least from a cost perspective, depends on the level of traffic. For a network with little traffic, O-E-O may be the more cost-effective option, whereas optical bypass is more favored as the traffic level increases.

Table 8.4 O-E-O vs. Optical-Bypass Architectures (60-Node Network)

# of Demds.	~Max. # of λs on a Link	~Avg. # of λs on a Link	600 km, O-E-O, Non-Configurable		2,500 km, Optical Bypass, with Core Configurability		600 km, O-E-O, Non-Configurable (15% cost increase factor)
			# of Regens.	Normal. Cost	# of Regens.	Normal. Cost	Normal. Cost
0	0	0	0	0.67	0	1.00	0.75
100	20	10	754	0.88	54	1.00	1.00
200	40	20	1757	1.10	146	1.00	1.26
400	80	40	3373	1.34	272	1.00	1.55

The cost differences shown in Table 8.4 are of course dependent on the cost assumptions that were made. Thus, the costs in the table should be considered as indicative of trends rather than absolute cost numbers. For example, if the cost increase factor were 15% for every doubling of the optical reach, rather than 25%, the normalized capital costs for the O-E-O network would be as shown in the final column of Table 8.4. (This should be thought of as the 600-km O-E-O-system costs remaining the same, and the 2,500-km optical-bypass-system costs increasing by a smaller factor; the network capital costs are again normalized to 1.0 for the optical-bypass system.) Clearly, the benefits of optical bypass are realized sooner with this assumption.

The amount of regeneration is solely a function of the system reach and is not dependent on the cost assumptions. As shown in the table, optical bypass with 2,500-km optical reach consistently eliminated more than 90% of the regenerations as compared to an O-E-O system with 600-km reach. With O-E-O technology, the average unprotected demand required 4.4 regenerations and the average protected demand required 13.6 regenerations. With optical-bypass technology, these numbers were 0.3 and 1.2, respectively. Thus, the marginal cost of adding traffic in the optical-bypass-enabled network was clearly lower.

8.2.1 Comments on Comparing Costs

As emphasized above, the numbers in Table 8.4 should be focused on in terms of trends as opposed to the absolute cost differences between the two technologies. There are many factors that make direct cost comparisons difficult.

First, the level of configurability is very different in the two architectures that were considered. It was assumed that the O-E-O nodes were not equipped with switches, such that the nodes had no means of automated configurability. In the optical-bypass-enabled network, the OADMs and OADM-MDs provided core configurability; i.e., traffic could be configured as add/drop or through, or traffic could be routed in different directions through the node without requiring manual intervention. Thus, the optical-bypass solution was more configurable. Since static traffic was used in the study, these differences in configurability were not reflected in the costs of Table 8.4. (Note that, to obtain *edge* configurability with the optical-bypass solution, all-optical switches would need to be deployed instead of OADM-MDs, or adjunct edge switches would be needed in addition to the OADM-MDs.)

Second, the operating costs are likely to be significantly lower with optical-bypass technology. As discussed in Section 8.1.3, the amount of regeneration can be used as a rough measure of relative operational costs. With optical bypass, the amount of regeneration was reduced by over 90%, which would be accompanied by a large savings in deployment costs and space and power requirements. Furthermore, carriers should be able to provision demands more quickly using optical-bypass technology so that revenues can be generated sooner.

Another factor that should be considered is the 'effective capacities' of the two networks are slightly different. As was investigated in Section 5.8, optical bypass results in roughly 5% less network efficiency because of wavelength contention.

The time-value of money also needs to be considered. The up-front costs of an optical-bypass system are higher; the savings afforded by optical bypass are realized over time. Thus, the cost savings are somewhat mitigated depending on the rate of traffic growth in the network and the rate-of-return on investment capital.

8.2.2 O-E-O Technology with Extended Optical Reach

Another architectural option is to deploy technology that supports an extended optical reach, but continue to use O-E-O technology at the nodes to avoid any issues with wavelength contention. For example, consider the combination of 1,500-km optical reach and O-E-O nodes. For the 60-node network of Fig. 8.1, where the average link length is 450 km, this combination provided no cost benefit. The nodes were too densely packed to take much advantage of increasing the reach from 600 km to 1,500 km.

However, this architecture can provide some benefit in a less densely packed network. The study was repeated for the 30-node network shown in Fig. 8.2 (with a corresponding traffic set), where a 1,500-km O-E-O architecture was also considered. The average link length in this network is 700 km and the maximum link length is 1,450 km. It was assumed that the amplification and transmission costs increase by 25% for every doubling of the reach.

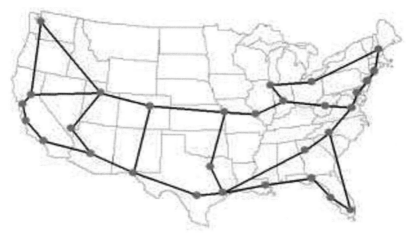

Fig. 8.2 Thirty-node backbone network. With a relatively low nodal density, increasing the reach from 600 km to 1,500 km can provide a cost benefit even in a network based on O-E-O technology.

The results are shown in Table 8.5. The comparison of the 600-km O-E-O architecture to that of the 2,500-km optical-bypass architecture is similar to that for the 60-node network. However, in the 30-node network, maintaining the O-E-O architecture but increasing the reach from 600 km to 1,500 km did provide a *small* cost benefit as the traffic level increased. For example, when the network was full, there was approximately a 5% cost savings afforded by the increased reach. Approximately 40% of the regenerations were removed by increasing the reach from 600 km to 1,500 km, which would result in operational-cost savings as well. (If the cost increase factor had been 15% rather than 25%, then the 1,500-km O-E-O architecture would have been 15% lower cost than the 600-km O-E-O architecture but 35% higher cost than the optical-bypass architecture, when the network was full.)

The caveats mentioned above with respect to comparing O-E-O and optical-bypass architectures are relevant here as well.

Table 8.5 O-E-O vs. Optical-Bypass Architectures (30-Node Network)

# of Demds.	~Max. # of λs on a Link	~Avg. # of λs on a Link	600 km, O-E-O, Non-Configurable		1,500 km, O-E-O, Non-Configurable		2,500 km, Optical Bypass, with Core Configurability	
			# of Regen.	Normal. Cost	# of Regen.	Normal. Cost	# of Regen.	Normal. Cost
0	0	0	0	0.72	0	0.79	0	1.00
60	20	12	653	0.98	392	1.00	78	1.00
120	40	25	1163	1.12	684	1.11	138	1.00
250	80	50	2277	1.35	1331	1.28	268	1.00

8.3 IP Transport Architectures

The previous section compared O-E-O and optical-bypass-enabled architectures for line-rate traffic. This section compares the architectures for subrate traffic; specifically, the focus is on an IP-based system. The three IP-transport architectures of interest are shown in Fig. 8.3 through Fig. 8.5. (There are several variations of these architectures that support varying degrees of reconfigurability; the 'purest' architectures are considered here, as shown in the figures.)

In Fig. 8.3, the IP router operates as a core switch, such that all wavelengths entering a node are terminated on the router. As shown, the system is clearly O-E-O-based. Because all traffic enters a router at every node, the wavelengths are maximally packed. This architecture is referred to here as the Pure IP architecture. (In a variation of this architecture, manual patch-cabling is used for traffic in a node that does not actually need to be processed by the IP router; however, this would not allow full automated nodal reconfigurability.)

An alternative O-E-O architecture is shown in Fig. 8.4. Here, a wavelength-level optical cross-connect (OXC) operates as the core switch; only the IP traffic that needs to be processed at the node is passed to the IP router. The OXC could be, for example, MEMS-based (the figure assumes such a photonic switch as there are no short-reach interfaces on the OXC ports). As specified in Table 8.3, it was assumed that the port of an OXC has a relative cost of 0.2X, as compared to 1.5X for an IP router port. The significant difference in port costs favors using the OXC in order to reduce the required size of the IP router. The disadvantage is that two 'boxes' are needed per node; moreover, any IP traffic that needs to be processed occupies ports on both the OXC and the IP router. This architecture is referred to here as the IP-over-OXC architecture.

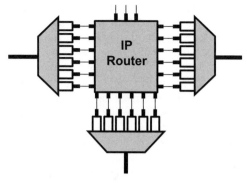

Fig. 8.3 In a pure IP architecture, the IP router serves as the core switch. All traffic entering the node is processed by the IP router.

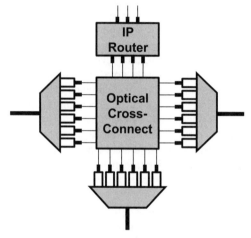

Fig. 8.4 In an IP-over-OXC architecture, transiting traffic is processed by the optical cross-connect (OXC) as opposed to the IP router. Here, the OXC is assumed to be a photonic switch (e.g., a MEMS-based switch).

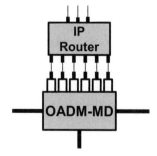

Fig. 8.5 In an IP-over-WDM architecture, transiting traffic remains in the optical domain.

The third architecture, which is optical-bypass-enabled, is shown in Fig. 8.5. This is referred to here as the IP-over-WDM architecture. Any traffic that is completely transiting the node remains in the optical domain, whereas the remaining traffic, including the regenerated traffic, is processed by the IP router. (The regenerated traffic is passed through the IP router to gain edge flexibility. If edge flexibility were not desired, this traffic could be handled with manually-patched-together transponders or with regenerator cards.) The benefit of the IP-over-WDM architecture is that it eliminates the electronics for the transiting traffic.

The three architectures were evaluated for aggregate network demand levels of 0.8 Tb/s, 1.7 Tb/s, and 3.4 Tb/s, on the network of Fig. 8.1. (The aggregate network demand is calculated by summing the total bi-directional traffic sourced in the network.) Assuming the network of Fig. 8.1 is equipped with a 10-Gb/s line-rate, the 3.4-Tb/s demand scenario approximately filled the network. All of the traffic was assumed to be subrate IP traffic. 50% of the traffic was protected, with protection implemented at the IP layer in all three architectures. The wavelengths were filled to a maximum of 75% capacity to allow for traffic bursts.

The optical reach for the two O-E-O architectures was assumed to be 600 km. There was a need for dedicated regeneration sites along the links that exceeded 600 km in length. At these locations, the regenerations were handled with regenerator cards; OXCs and IP routers were not deployed at these sites. The optical reach for the optical-bypass architecture was assumed to be 2,500 km.

In the first set of designs, all of the traffic-generating nodes were assumed to be equipped with IP routers, such that backhauling was not required. In the IP-over-OXC and the IP-over-WDM architectures, an IP service was allowed to be processed by at most five intermediate IP routers along its end-to-end path. As indicated above, with the Pure IP architecture, the traffic entered an IP router at every node along the end-to-end path to enable full automated reconfigurability at each node.

Table 8.6 indicates the total number of IP ports required for each scenario, where the IP-port count includes the ports used on the network-side and the client-side of the router. For the IP-over-OXC architecture, the OXC port counts are also shown. In this architecture, where an OXC is used as the core switch, the average required IP router size was reduced by approximately 45% with 0.8-Tb/s aggregate demand and approximately 60% with 3.4-Tb/s aggregate demand, as compared to the Pure IP architecture. The percentage savings increases with demand level due to relatively less grooming being needed with higher traffic levels. With the IP-over-WDM architecture, the average required IP router size was about 5% to 10% larger as compared with the IP-over-OXC architecture, due to regenerators being passed through the IP router. However, the average required router size was still significantly smaller than in the Pure IP architecture.

The maximum router size required at any node is also important to consider. These statistics are shown in Table 8.7, for the scenario with 3.4 Tb/s of aggregate demand. As expected, the Pure IP architecture required larger IP routers than the other architectures. (In the 3.4-Tb/s aggregate demand scenario, the maximum OXC size required at any node in the IP-over-OXC architecture was 246 ports.)

Table 8.6 Number of Ports and Normalized Costs for IP Transport Architectures

	600-km Reach, Pure IP		600-km Reach, IP-over-OXC			2,500-km Reach, IP-over-WDM	
Aggregate Demand	# of IP Ports	Normal. Cost	# of IP Ports	# of OXC Ports	Normal. Cost	# of IP Ports	Normal. Cost
0 Tb/s	0	0.73	0	0	0.75	0	1.00
0.8 Tb/s	1899	1.17	1007	2506	1.03	1053	1.00
1.7 Tb/s	3479	1.40	1614	4303	1.12	1681	1.00
3.4 Tb/s	6646	1.66	2751	7862	1.24	2995	1.00

Table 8.7 Maximum Number of IP Ports at a Node (3.4-Tb/s Aggregate Demand Scenario)

600-km Reach, Pure IP	600-km Reach, IP-over-OXC	2,500-km Reach, IP-over-WDM
198	150	154

The normalized capital costs for each of the scenarios is also shown in Table 8.6, where the costs are normalized to 1.0 for the IP-over-WDM architecture. (The costs of the transponders and the amplifiers in the 600-km-reach architectures were reduced by roughly 35% relative to the 2,500-km-reach architecture to account for the shorter reach. This corresponds to the assumed guideline of a 25% cost increase for every doubling of the reach.) Similar to the line-rate study of Section 8.2, optical-bypass technology was more cost effective as the traffic level increased. With an 'empty' network, the cost of the infrastructure was more expensive than either of the O-E-O designs. However, with a 'full' network, the Pure IP architecture was on the order of 65% more costly as compared to IP-over-WDM, whereas the IP-over-OXC architecture was roughly 25% more costly. With low demand, the two O-E-O architectures were fairly close in cost. As the traffic level increased, the IP-over-OXC architecture became significantly lower cost as the difference in cost between an OXC port and an IP port became a dominant factor.

Recall from Section 8.1.3 that the IP port cost used in the studies is significantly lower than the cost of commercial offerings, with the expectation that commercial costs will moderate with time. If higher costs had been used, the relative cost of the Pure IP architecture would have been appreciably higher than what is shown in the table.

A variation of Fig. 8.5 was also considered where an OXC is inserted between the OADM-MD and the IP router in order to process the regeneration. This allows the lower-cost OXC ports to be used for the regenerated traffic; however, it also burns two OXC ports for traffic that needs to be processed by the IP router. Overall, this architecture was approximately 5% more costly than the IP-over-WDM architecture shown in Fig. 8.5. If the optical reach were shorter such that the number of regenerations were significantly larger, this OXC-enhanced architecture might be more viable. However, it does require three 'boxes' in a node.

In a real deployment, a core IP router may be deployed at just a subset of the nodes, as discussed in Section 6.4. Thus, a second set of designs was performed where only 15 nodes were equipped with IP routers. Again, the benefits of optical bypass increased with the traffic level. (However, the cost of the Pure IP architecture was closer to that of the IP-over-OXC architecture in this environment.)

Other studies have been published regarding various IP transport architectures. For example, [SeKS03] compares a pure IP architecture to one where the IP router is deployed over an electronic grooming switch, where the grooming switch is a coarse SONET-based switch. (This is similar to the architecture of Fig. 8.4, except that the switch is a grooming switch as opposed to a wavelength-level OXC.) The latter architecture is shown to be the more cost-effective solution due to the lower cost of SONET ports relative to IP ports. A related study is performed in [TsMT05], with similar results. In [LWYD07], which assumes that only a subset of the nodes are equipped with core routers, it is shown that the relative cost of the Pure IP and IP-over-OXC architectures depends on the layer in which protection is implemented. Reference [BaGe06] looks at variations of the three IP transport architectures considered here. Similar to the results shown above, a significant cost savings was demonstrated in [BaGe06] for the IP-over-WDM architecture, where the savings increased with the level of traffic. A theoretical study that presents formulas for the number of ports and transponders that are potentially saved implementing the IP-over-WDM architecture as compared to the Pure IP architecture can be found in [SiSa99].

8.4 Optimal Optical Reach

Given that extended optical reach, combined with optical-bypass elements, provides a significant reduction in the number of required regenerations, it is tempting to assume that the longer the reach, the more cost savings are afforded by the system. However, increasing the reach typically requires amplifiers with more pumps or higher-powered pumps, transponders with more complex modulation formats and stronger error-correcting capabilities, and components with stricter tolerances. After some point, the cost of increasing the reach is not fully offset by the savings due to further reductions in regeneration, leading to a concave 'cost versus reach' curve, as investigated further in this section.

The 60-node network of Fig. 8.1 was used for this study, along with the traffic set of Section 8.1.2. Dedicated 1+1 protection was used for the demands that required protection. Optical-bypass-enabled designs were performed for the network where the optical reach was increased from 1,000 km to 6,000 km at 500-km intervals. The cost assumptions provided in Table 8.3 were used as the baseline for 2,500-km reach. As specified in Section 8.1.3, it was assumed that the cost of amplification and transmission increases by 25% for every doubling of the optical reach (i.e., Equation 8.1 was used as the cost adjustment for a reach of R km).

As expected, the number of regenerations decreased with increasing optical reach, as shown in Fig. 8.6. Initially, the decrease is fairly steep, with roughly

80% of the regenerations removed by increasing the reach from 1,000 km to 2,500 km. After 2,500 km, the curve begins to level off, indicating a diminishing 'rate of return' for increasing the reach yet further. Note that even with 6,000-km reach, not all regenerations were eliminated. For this particular network, a reach of 8,500 km was required to remove all regenerations from both the primary and protect paths.

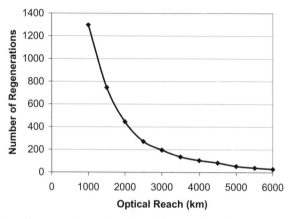

Fig. 8.6 Number of regenerations as a function of the optical reach. Most of the regenerations have been eliminated with 2,500-km optical reach.

The solid curve in Fig. 8.7 plots the normalized network capital cost as a function of the optical reach. The minimum cost was achieved with an optical reach in the range of 2,000 km to 2,500 km. As this graph illustrates, continuing to increase the reach beyond this point resulted in a more costly network. The minimum-cost point is clearly dependent on the assumption that the cost increase factor is 25% for every doubling of reach. As extended-reach technology matures, the cost premiums may decrease somewhat, which would shift the minimum-cost point to the right. For example, a cost increase factor of 15% produces the cost vs. reach curve shown by the dashed line in Fig. 8.7. With this assumption, the minimum cost was attained with a reach in the range of 2,500 km to 3,000 km.

Note that if connections need to be brought into the electrical domain for reasons other than regeneration, e.g., for grooming and/or shared protection, the optimal reach shifts to the left. To test this, another set of designs was performed where shared protection was used instead of dedicated protection (more specifically, subconnection-based shared protection, with 20% of the nodes selected as protection hubs, was used; see Section 7.7). The associated cost vs. reach curve, assuming a 25% cost increase factor, is shown by the dashed line in Fig. 8.8 (the solid-line curve in this figure is the same as that in Fig. 8.7). With shared protection, the minimum-cost optical reach was reduced to 1,500 km.

As investigated in Section 8.2, the economics of optical-bypass technology are dependent on the amount of traffic. While Fig. 8.7 shows the cost for a 'full' network, Fig. 8.9 includes additional cost curves corresponding to lower amounts of traffic (assuming a 25% cost increase factor and dedicated protection). As expected, with lower traffic levels, the optimal reach from a cost perspective decreased. For example, with only 50% of the demands added to the network, the lowest cost was achieved with a reach in the range of 1,000 km to 1,500 km.

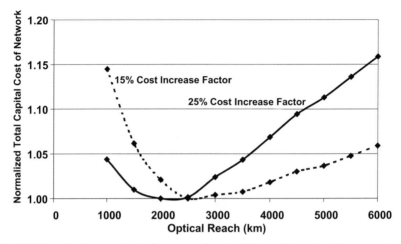

Fig. 8.7 Normalized network capital costs as a function of the optical reach. After some point, increasing the optical reach leads to a more costly network. The cost increase factor is the percentage increase in the cost of amplifiers and transponders for every doubling of the reach.

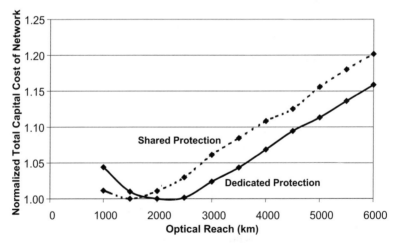

Fig. 8.8 Normalized network capital costs as a function of the optical reach, for dedicated protection and shared protection. The minimum-cost point shifts to the left with shared protection due to the protection capacity being 'chopped' at the protection hubs. (Both curves assume a cost increase factor of 25%.)

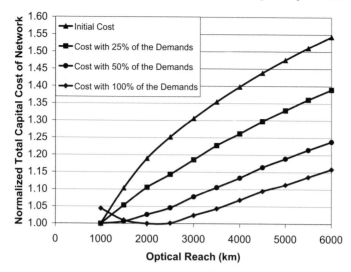

Fig. 8.9 Normalized network capital costs as a function of the optical reach, for different levels of traffic. With less traffic, shorter optical reach is more cost effective. Note that each curve is independently normalized to 1.0 at its minimum value.

In [Simm05], a similar study was performed for four other network topologies, ranging from a 16-node network to a 55-node network. The results for these other networks are similar to those shown here. Additionally, similar concave cost vs. reach results were presented in [SaSR05].

As has been emphasized repeatedly, the costs that are plotted are only the capital costs, not the operational costs. Operational costs are related to the number of regenerations, thus, the optimal reach may shift to the right when these costs are considered. However, given that the bulk of the regenerations have been removed with a reach of 2,500 km, including operational costs is likely to have only a small effect on the overall minimum-cost point.

8.4.1 Add/Drop Percentage as a Function of Optical Reach

As noted in Section 2.4.1, some commercially available OADMs, OADM-MDs, and all-optical switches limit the amount of wavelengths that can be added or dropped. With an OADM-MD, the limit is generally specified on a per-fiber basis; i.e., no more than P% of the fiber capacity can add/drop to/from each fiber entering a node. With an all-optical switch, the limit is generally specified on a per-node basis, which provides more flexibility; i.e., no more than P% of the total capacity of all fibers entering a node can be added/dropped at the node. An OADM could be of either type depending on its architecture (i.e., an OADM with edge flexibility would be similar to an all-optical switch, whereas an OADM without edge flexibility is similar to an OADM-MD). A typical value used commercially for P is 50%.

The network designs that were performed to test the effect of optical reach are used here to determine how the add/drop percentage varies with reach and whether 50% add/drop is sufficient to meet the needs of a typical network. Consider the case where 100% of the demands were added to the network, so that the network was essentially full, and where demands requiring protection employed 1+1 dedicated client-side protection. The fiber capacity was assumed to be 80 wavelengths. OADM-MDs and OADMs without edge flexibility were assumed in the design; thus, the important figure of merit is the add/drop percentage for each fiber. This percentage includes those wavelengths that add/drop for purposes of regeneration.

Figure 8.10 shows a histogram of the required fiber add/drop percentage when the optical reach was 2,500 km (there was one fiber-pair per link; thus, the 77 links correspond to 154 fibers). *The add/drop percentages are relative to 80 wavelengths per fiber.* For example, there were 81 fibers from which less than 10% add/drop was required (i.e., less than 8 add/drop wavelengths), whereas there were 4 fibers from which 50% to 60% add/drop was required. This indicates that a limit of 50% add/drop per-fiber would not have been sufficient for all nodes. If the system equipment had such a limit, then optical terminals would have needed to be deployed at the sites where the threshold was violated.

With 1,500-km optical reach, the 50% threshold was even more limiting, as shown by the histogram in Fig. 8.11. Here, 14 of the fibers required an add/drop percentage greater than 50% (i.e., more than 40 add/drop wavelengths).

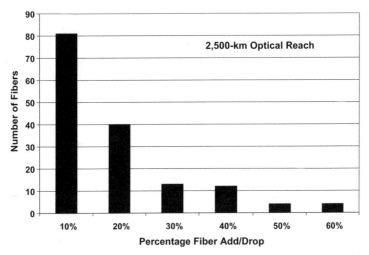

Fig. 8.10 A histogram of the required fiber add/drop percentage for the network design with dedicated protection and 2,500-km optical reach. For example, 81 of the fibers had less than 10% add/drop whereas 4 fibers had between 50% and 60% add/drop. The add/drop percentages are relative to 80 wavelengths per fiber.

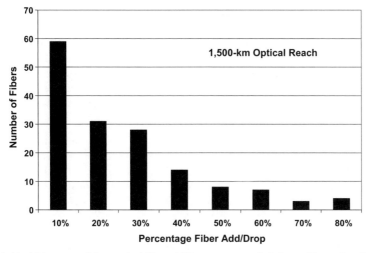

Fig. 8.11 A histogram of the required fiber add/drop percentage (relative to 80 wavelengths per fiber) for the network design with dedicated protection and 1,500-km optical reach.

If all-optical switches had been used, where the add/drop limit is on a per-node basis rather than on a per-fiber basis, then none of the nodes violated the 50% limit in the scenario with 2,500-km optical reach. However, with 1,500-km optical reach, two of the nodes required close to 60% add/drop and thus violated the 50% limit.

It is interesting to also consider the effect of shared protection on the per-fiber add/drop percentage. (With shared protection, the average utilization in the network decreased by 20%; however, the maximum link utilization decreased by less than 10%. Thus, a fiber capacity of 80 wavelengths is still assumed.) With 2,500-km reach, the average per-fiber add/drop percentage increased, due to breaking up the protection capacity to enable sharing. However, the *maximum* per-fiber add/drop percentage remained approximately the same so that there were still four fibers that violated the 50% limit. With 1,500-km reach and shared protection, the number of fibers violating the 50% add/drop limit was only 6, as opposed to 14 with dedicated protection. This decrease was due to sharing of the regenerations on the protection capacity, which resulted in fewer add/drop wavelengths.

8.5 Architecture of Higher-Degree Nodes

This section investigates the economics of various architectures for nodes of degree three and higher. As discussed in Section 2.5, there are three architectures that may be used at these higher-degree nodes: optical bypass in all directions through the node, where the node is equipped with an OADM-MD or an all-optical switch; optical bypass in a limited number of directions through the node, where the node is equipped with one or more OADMs (and possibly one optical terminal); and no optical bypass through the node, where the node is equipped only with optical terminals.

The network economics of these architectures were investigated using the 60-node network of Fig. 8.1 and the traffic set of Section 8.1.2. In all of the architectures considered, the degree-two nodes were equipped with OADMs; it is only the higher-degree nodes that were architected differently. In the 'pure optical bypass' design, a node of degree-N was equipped with a degree-N OADM-MD. In the 'OADM-only' design, a node of degree-N was equipped with $\lfloor N/2 \rfloor$ OADMs, and possibly one optical terminal; refer back to Fig. 2.10 for an illustration of this architecture at a degree-three node and a degree-four node. The OADMs were oriented to maximize the amount of optical bypass. In the 'O-E-O-at-the-hubs' design, a node of degree-N was equipped with N optical terminals.

The optical reach was assumed to be 2,500 km for all designs. In the first series of designs, 1+1 dedicated client-side protection was used for those demands requiring protection.

The resulting number of regenerations and normalized capital cost of each of these architectures is shown in the 'Dedicated Protection' columns of Table 8.8, where the costs are normalized to 1.0 for the OADM-MD (or pure bypass) design. The OADM-only design was 14% more costly than the OADM-MD design, whereas the O-E-O-at-the-hubs design was 38% more costly than the OADM-MD design. Furthermore, the OADM-MD architecture had 60% fewer regenerations than the OADM-only design, and 80% fewer regenerations than the O-E-O-at-the-hubs design. These results are consistent with those found in [JoFN04] for a European national network. (An earlier study, reported on in [ChYu03], showed less of a cost savings with pure bypass at the higher-degree nodes. However, this was due to the study using an all-optical switch with a small number of local access ports, as opposed to an OADM-MD. The all-optical switch is more costly; also, the small number of local access ports resulted in a significant amount of wavelength contention, which required that transponders be added to alleviate the contention.)

A second set of designs was performed where shared mesh protection was used for the demands that required protection. Subconnection-based shared protection was used, as described in Section 7.7, where 20% of the nodes were selected as protection hubs. All of the protection hubs were higher-degree nodes. Because of this additional regeneration at some of the higher-degree nodes, the pure bypass solution provided less of a cost benefit, as shown by the 'Shared Protection' columns of Table 8.8. The number of regenerations shown in the table includes those due to breaking the protection capacity at the protection hubs. (These added protection regenerations were handled with transponders as opposed to regenerator cards to allow sharing flexibility. It was assumed that an OXC was used to implement the switching for shared protection. The OXC costs are included in the normalized costs shown in the table.) Note that the total amount of regeneration increased with the OADM-MD architecture, as compared to dedicated protection, due to breaking the protection capacity at the protection hubs. For the other two architectures, the more dominant effect was that regenerations on the protection capacity were being shared, resulting in fewer total regenerations as compared to the dedicated protection scenario.

Table 8.8 Relative Network Capital Costs for Different Architectures at Higher-Degree Nodes

	Dedicated Protection		Shared Protection	
Architecture	# of Regen.	Normalized Cost	# of Regen.	Normalized Cost
OADM-MD	272	1.0	302	1.0
OADM-Only	712	1.14	575	1.09
O-E-O-at-the-Hubs	1485	1.38	1110	1.24

As emphasized in previous sections, Table 8.8 only compares the network capital costs. The fewer regenerations achieved with the OADM-MD architecture should translate into an associated reduction in operating costs as well. Furthermore, the OADM-MD architecture is more configurable, as it allows traffic to be redirected through a node in any direction. With either of the other architectures, any redirection of traffic through a node requires either manual intervention or the addition of an adjunct switch for automated reconfigurability. Thus, the network is more agile and lower cost with OADM-MDs at the higher-degree nodes.

Moreover, as stated above, in the OADM-only architecture, the OADMs were oriented to maximize the amount of optical bypass, where it was assumed that the traffic set was known ahead of time. Given that traffic forecasts are unlikely to be completely accurate, in a real deployment, some of the OADMs may end up being deployed in a less than optimal orientation, leading to additional regeneration. (Once the element is deployed, it is assumed that it cannot be reoriented later, as that would disrupt live traffic.)

Consider an OADM-only architecture where the OADMs are deployed in an *arbitrary* orientation at each higher-degree node. Using the same traffic as above, with dedicated protection, this architecture yielded a normalized cost that was 23% higher than the OADM-MD architecture. This compares with the 14% cost increase when the OADMs are oriented appropriately, as shown in Table 8.8. (Note that, to obtain the 23% figure, several trials were run with randomly chosen OADM orientations in each trial; the 90% confidence interval was ±0.5%.) While orienting all OADMs randomly is an extreme case, it demonstrates the greater potential uncertainty with an OADM-only design. The OADM-MD architecture is the most forecast-tolerant of all the architectures.

8.6 Reduced-Reach Transponders

While extended optical reach can lower the cost of the network, up to a point, not all connections make full use of the reach. For example, a 500-km connection does not take full advantage of a transponder that is designed to support a 2,500-

km all-optical path. Even a long connection may result in a subconnection that is of limited distance. For example, with 2,500-km reach, a 6,000-km connection would be nominally broken into two 2,500-km subconnections and one 1,000-km subconnection. The short subconnection does not take full advantage of the 2,500-km reach. A lower-cost transponder of reduced reach could be used for these shorter subconnections in order to reduce the overall network cost to some degree. This section investigates the magnitude of the cost savings that could be achieved using the reduced-reach transponder.

The 60-node network of Fig. 8.1 and the traffic set of Section 8.1.2 were used for the study. In the first series of tests, 1+1 dedicated client-side protection was used for those demands requiring protection. It was assumed that the optical reach of the system is 2,500 km; i.e., the optical amplifiers and the 'regular' transponder cards supported this reach. It was also assumed that a second transponder card was supported, where the reach of this card was varied between 500 km and 2,000 km to determine the optimal distance for this reduced-reach transponder.

The cost of the reduced-reach transponder was assumed to follow the same trends as assumed in previous sections: for every doubling of reach, the cost increases by 25%. With this cost assumption, it is always optimal from a cost perspective to break a connection into subconnections that are as close to 2,500 km as possible, leaving the shortest possible subconnection as the remainder. For example, with a 4,000-km connection, it is lower cost to break it into 2,500-km and 1,500-km subconnections as opposed to two 2,000-km subconnections. This is the strategy that was followed in the study, where possible. (Note, however, as discussed in Chapter 5, this may not always be the optimal regeneration strategy from the perspective of wavelength assignment.)

The normalized total *transponder* costs, as a function of the distance of the reduced-reach transponder, are shown by the solid line in Fig. 8.12. The point corresponding to a 'reduced-reach' of 2,500 km represents the scenario where *all* transponders had 2,500-km reach; i.e., a reduced-reach transponder was not supported in this scenario. In all of the scenarios, any regenerations were implemented with back-to-back transponders as opposed to regenerator cards. (If regenerator cards were used, there would need to be different types of cards. For example, one regenerator card would be needed for regeneration between subconnections that both require 2,500-km reach; one regenerator card would be needed for regeneration between subconnections where one subconnection requires 2,500-km reach but the other subconnection requires only the reduced reach; and one regenerator card would be needed for regeneration between two subconnections with reduced reach, which may be required due to the particular spacing of the nodes.)

With these assumptions, the lowest-cost total transponder cost was achieved with a reduced-reach transponder distance of 1,000 km. However, relative to a system with only 2,500-km-reach transponders, the cost savings was only about 6%. As a very rough rule-of-thumb, in a fully populated network, the transponder cost generally comprises no more than about 50% of the total optical-layer cost of the network, where amplification and network elements are the other remaining major costs. Thus, a 6% savings in transponder costs translates into no more than a 3% savings in the optical-layer cost.

A second set of tests were run where shared protection, based on the subconnection method of Section 7.7, was used for those demands requiring protection. Shared protection 'chops up' the protection capacity at the protection hubs to enable better sharing. In the study, 20% of the nodes were selected as protection hubs. The normalized total transponder costs for the network are shown by the dashed line in Fig. 8.12. Reduced-reach transponders were more effective in this scenario due to the smaller subconnections that resulted from chopping up the protection capacity. As with dedicated protection, the minimum cost was achieved with a reduced-reach transponder distance of 1,000 km. With shared protection, this resulted in a 12% savings in the total transponder cost, as compared to a system with just 2,500-km-reach transponders; this translates to no more than a 6% savings in the total optical-layer cost. (The presence of grooming would have a similar effect as shared protection, due to connections being converted to the electrical domain for purposes of entering a grooming switch.)

As this study demonstrates, the reduction in capital costs resulting from the reduced-reach transponder is fairly small. Furthermore, the operational costs may increase, as inventories need to be maintained for two types of cards. It may also be undesirable to use the limited-reach cards to handle dynamic traffic, as predeployment estimates would need to be performed for both types of cards. If the estimates are inaccurate, a connection may be forced to use reduced-reach cards, even if it results in additional required regenerations. All-in-all, this strategy is not effective in markedly reducing network costs.

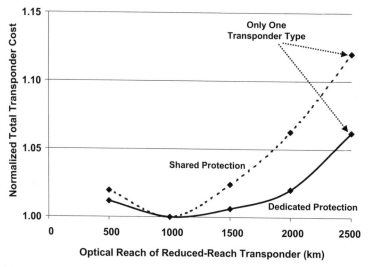

Fig. 8.12 Normalized total transponder cost as a function of the optical reach of the reduced-reach transponder. At 2,500 km, there is just one transponder type. With either dedicated or shared protection, the optimal reach for the reduced-reach transponder is 1,000 km.

8.7 Optimal Topology from a Cost Perspective

This section investigates the impact of the network topology on the overall network cost. The focus is on an 'overbuild' scenario, where a new system is being
deployed using existing fiber routes. The assumption is that various links are
available to interconnect the network nodes, but that from a capacity and protection perspective, not all of the links are necessary. The cost impact of removing
links from the topology depends on whether the system supports optical bypass or
whether it is O-E-O based. (In a greenfield scenario, where there is no existing fiber, an equivalent question is how should the links be laid out to meet the capacity
and protection requirements, while minimizing cost.)

To understand how the topology affects the network cost, consider the simple
network shown in Fig. 8.13, and assume that Link CF is being considered for removal. Removing this link eliminates any associated in-line amplifiers and reduces the amount of nodal equipment required at the endpoints. For example,
Node C and Node F would become degree-three nodes as opposed to degree-four
nodes. Any connection that was routed over Link CF must now be routed over
C-B-F or C-E-F. In an O-E-O network, because regeneration is required at every
intermediate node, the extra hop in the path translates to an extra regeneration.
Thus, whether removing Link CF reduces the network cost depends on whether
the additional regeneration cost is offset by the reduced amplifier and nodal
equipment costs. In an optical-bypass-enabled network, removing Link CF may
not result in any extra regeneration, depending on the length of the links and the
optical reach of the system. Thus, reducing the density of the topology is more
likely to produce a cost savings in a network with optical-bypass technology.

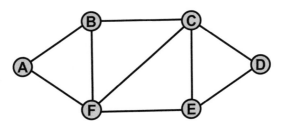

Fig. 8.13 Simple topology to illustrate the effects of removing a link. If Link CF is removed,
any amplifiers on the link are removed and the equipment at Nodes C and F can be simplified.
However, traffic routed on this link will now be routed through Node B or Node E. In an O-E-O
network, this will result in additional regeneration, whereas in an optical-bypass-enabled network
it may not.

To quantify this effect, the 15-node metro-core network of Fig. 8.14 was used.
This network was modeled after an actual 13-node metro-core network, which is
described in [WBSK03]. With all of links shown in Fig. 8.14, the average nodal
degree is 5.33. It was assumed that all links were less than 100 km in length, such
that no in-line amplifiers were required. A total of 800 Gb/s of subrate traffic was

added to the network, where the traffic pattern was similar to that specified in [WBSK03]. The line-rate was assumed to be 10 Gb/s. All nodes were assumed to be equipped with grooming switches so that backhauling was not required. All traffic was protected with 1+1 dedicated client-side protection.

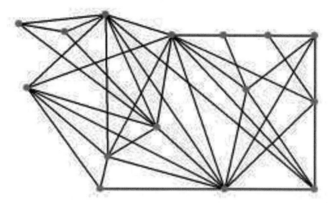

Fig. 8.14 Fifteen-node metro-core network used to study how the network capital costs change as links are removed.

Two system designs were performed, one based on O-E-O technology and one based on optical-bypass technology. Both systems were assumed to support conventional 600-km optical reach. In the O-E-O design, any regenerations were implemented with regenerator cards and patch-panels. In the optical-bypass design, the 600-km reach was long enough to eliminate all regenerations (however, some connections still entered the electrical domain at intermediate nodes for purposes of grooming). Links were systematically removed from the topology to determine the effect on required regeneration and network cost. The links with the least amount of traffic carried on them were selected for removal; however, a link was not removed if it was required to provide diverse paths for a connection.

Figure 8.15 plots the number of required regenerations in the O-E-O design as a function of the average nodal degree. As links were removed, the connections were forced to traverse more hops, leading to more regeneration. This was especially true once the average nodal degree was reduced below three. Similar results were presented in [Sale03] for a 12-node metro-core network. In the optical-bypass design, no regenerations were needed, even as links were removed.

Figure 8.16 plots the normalized capital cost of the network as a function of average nodal degree, for both the O-E-O and the optical-bypass designs. The relative costs of Table 8.3 were used. The cost of fiber was not included as it was assumed that the fiber routes already existed. (The required maximum fiber capacity is dependent on the topology; thus, the cost of the nodal amplifiers needed in the network elements may change as well. This effect was not included in the costs shown in Fig. 8.16, as the nodal amplifier costs would be the same for either system.)

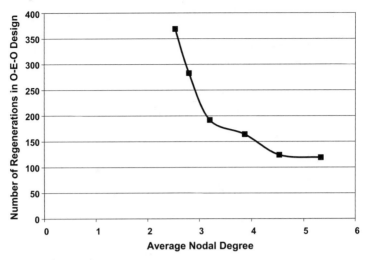

Fig. 8.15 Number of regenerations as a function of the average nodal degree for an O-E-O-based network, based on the original topology shown in Fig. 8.14.

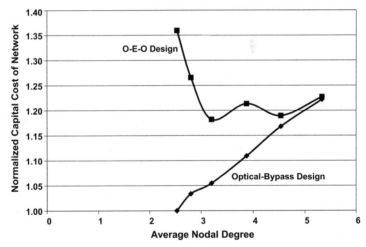

Fig. 8.16 Normalized capital cost of the network as a function of the average nodal degree for both O-E-O and optical-bypass designs, for the original topology shown in Fig. 8.14. With the O-E-O design, the cost wavers somewhat as links are initially removed, and then shoots up. With the all-optical design, the cost monotonically decreases.

With the O-E-O system, the initial removal of links had only a small effect on the network cost. The savings afforded by the reduction in optical terminals at the nodes was approximately offset by the addition of more regeneration. After the average nodal degree was reduced below three, however, the sharp increase in regeneration more than outweighed the benefits of fewer optical terminals, leading to a spike in network cost. With the optical-bypass system, the network cost monotonically decreased as links were removed due to lower-cost network elements being used. There was no concomitant increase in regeneration to offset this cost reduction.

Note that with the full topology, the difference in cost between the O-E-O system and the optical-bypass-enabled system was very small. In an O-E-O network, each link essentially provides 'fiber bypass', where a fiber deployed directly between two nodes can be used to avoid having to traverse intermediate nodes. With the full topology, the amount of fiber bypass achieved by the O-E-O network was significant, so that there was little cost disadvantage relative to a system with optical-bypass elements. However, with the topology pared down to an average nodal degree of 2.53, optical-bypass technology yielded a 35% cost savings.

This study demonstrates that the topology density has a large impact on network cost. Of course, cost is not the only factor that needs to be considered when laying out the network topology. It is important that enough links be present to meet the capacity requirements. For example, in the full topology, with an average nodal degree of 5.33, the average and maximum link load were 8 and 20 wavelengths, respectively; when the average nodal degree was reduced to 2.53, the average and maximum link load grew to 30 and 51 wavelengths, respectively. If the maximum supportable load on a link were 40 wavelengths, as is common in metro-core networks, this latter configuration would not be feasible. It is also important that the topology support the availability level required for each demand. Removing links results in longer paths for some connections, which increases their vulnerability to failures.

Another factor to consider is that of Shared Risk Link Groups (SRLGs). As described in Section 3.6.3, SRLGs arise when two or more links that appear to be disjoint partially lie in the same fiber conduit. If this shared portion of the conduit is damaged, all of the corresponding links are likely to fail. Thus, the underlying physical diversity of the fibers must be considered when selecting the topology; otherwise, true diverse paths may not be feasible for some of the demands.

While this study focused on a metro-core network, the same effect occurs in regional and backbone networks. With these geographically larger networks, removing a link not only simplifies the nodal equipment, but typically results in the removal of several in-line amplifiers as well. There tends to be less fiber routes from which to choose in a regional or backbone network; however, optical-bypass designs can generally benefit from the removal of some of the links, at least from a cost perspective.

8.8 Optimal Line-Rate

The line-rate of a system affects both the network cost and the equipment requirements. Historically, the line-rate speeds have followed the SONET/SDH hierarchy with each successive boost in technology yielding a four-fold increase in rate; e.g., 0.62 Gb/s to 2.5 Gb/s to 10 Gb/s to 40 Gb/s. The cost per bit/sec has generally decreased as the line-rate has increased and the technology has matured. For example, a 10-Gb/s transponder costs roughly twice as much as a 2.5-Gb/s transponder, resulting in a 50% reduction in the cost per bit/sec. While a corresponding cost benefit has not been achieved yet for 40-Gb/s technology, it is expected that once the technology matures, a 40-Gb/s transponder will cost 2.5 to 3 times more than a 10-Gb/s transponder. In addition to lower cost per bit/sec, higher line-rate cards require less power and space per bit/sec, resulting in a lower operational cost metric as well.

There is likely to be a divergence from the SONET/SDH hierarchy beyond 40 Gb/s, however. While 160 Gb/s was initially favored as the next line-rate, the industry has instead gravitated towards 100 Gb/s, to coalesce with the Ethernet rate hierarchy. (Ethernet technology has become almost ubiquitous in enterprise networks and offers a better price-point as compared to SONET/SDH technology.) Additionally, a 100-Gb/s system should be somewhat less technically challenging than a 160-Gb/s system.

With higher-speed line-rates, fewer wavelengths are required to carry the traffic (assuming traffic is efficiently packed), leading to fewer transponders and smaller nodal switching requirements. However, increasing the line-rate results in a greater need for grooming; e.g., it takes more 'bundling' effort to efficiently pack a 100-Gb/s wavelength as compared to a 40-Gb/s wavelength (assuming the typical service-rate has not increased accordingly). Furthermore, higher bit-rates are more susceptible to optical impairments such as dispersion, PMD, and nonlinear effects (see Section 4.1.1), leading to reduced optical reach, and hence more frequent regeneration. Thus, while the individual components may be relatively more cost effective as the bit-rate increases, the overall cost of the network may actually increase, depending on the traffic. The optimal line-rate to use for a particular network depends on the traffic level, where increased traffic favors higher rates. This is investigated further in this section.

8.8.1 Study Assumptions

The 60-node backbone network of Fig. 8.1 was used for the study, where the total aggregate network demand was gradually increased to determine which line-rate is most cost effective. Because the relationship of line-rate and grooming has an impact on the cost of the network, the traffic set consisted of a significant amount of traffic at relatively low bit-rates. This could represent, for example, SONET/SDH, IP, or Ethernet traffic. All nodes were assumed to be equipped with grooming/routing devices such that backhauling of the subrate traffic was not required. All of the traffic was assumed to undergo some type of grooming/routing at the edge of the network.

The aggregate network demand levels studied were: 1 Tb/s, 5 Tb/s, 10 Tb/s, 20 Tb/s, 50 Tb/s, and 100 Tb/s. Each successive demand set was not simply a multiple of the previous one. While some demands were assumed to scale up in traffic rate, e.g., from a 2.5-Gb/s connection to a 10-Gb/s connection, new demands were also added where the endpoints for the new traffic were selected from a distribution based on the existing traffic. Additionally, a small percentage of the demands were modified to have arbitrarily chosen endpoints.

At each traffic level, approximately 50% of the traffic required protection, which was implemented with 1+1 dedicated client-side protection. In tallying the aggregate network demand, protected traffic counted twice; e.g., a protected 1-Gb/s connection contributed 2 Gb/s to the aggregate demand. In the time-frame of 2007, backbone networks typically carried on the order of 5 Tb/s to 10 Tb/s of traffic. Thus, this study covered at least a ten-fold growth in aggregate demand over this level.

The line-rates considered were: 2.5 Gb/s, 10 Gb/s, 40 Gb/s, and 100 Gb/s. The networks were all designed with optical-bypass technology, although the optical reach depended upon the line-rate. The optical reach assumed for each line-rate is shown in Table 8.9, reflecting the expected shorter reach with higher bit-rates. The values in this table were used over the whole range of traffic levels. This implicitly assumes that the technology at each line-rate will continue to improve, as extended reach will be more challenging to accomplish as the fiber capacity increases. (In all cases, it was assumed that a single fiber-pair was deployed on each link.)

Table 8.9 Optical Reach for Each Line-Rate Used in the Study

Line-Rate	Optical Reach
2.5 Gb/s	4,000 km
10 Gb/s	3,000 km
40 Gb/s	2,000 km
100 Gb/s	1,500 km

The relative costs listed in Section 8.1.3 were assumed to hold for a 10-Gb/s line-rate system with a link capacity of 800 Gb/s. (In previous sections, these costs were used for a 2,500-km optical reach; in this study, they were used for a 10-Gb/s line-rate with a reach of 3,000 km.) Various adjustments were made for other line-rates and capacities. First, the cost ratios of the transponders (and regenerator cards) assumed for the different line-rates are shown in Table 8.10. The bottom two rows of the table are purposely aggressive in order to probe the best-case prove-in point for these technologies. Sensitivity to these cost ratios is considered in Section 8.8.3.

Table 8.10 Transponder Cost Ratios for Each Increase in Line-Rate Speed

Line-Rate Increase	Cost Ratio
2.5 Gb/s to 10 Gb/s	2.0
10 Gb/s to 40 Gb/s	2.5
40 Gb/s to 100 Gb/s	1.5

Second, for each scenario, the costs of the in-line amplifiers and the nodal amplifiers were a function of the maximum required fiber capacity of any link in the network, in total bits/sec; for every doubling of the maximum fiber capacity, the amplifier cost was increased by 25%. (This reflects that total signal power is, to a first order, proportional to the total bit-rate per fiber. Pump laser costs scale sublinearly with power output; thus, amplifier costs typically scale sub-linearly with power output as well.) This cost assumption penalizes a system that inefficiently grooms the traffic, as that will result in greater required fiber capacity.

In previous sections, there has been an assumption that every doubling of reach incurs a 25% increase in the cost of the amplifiers and the transponders. However, that assumption was made with respect to a constant line-rate. Here, it is assumed that the optical reaches specified in Table 8.9 represent 'equivalent' technology. Thus, there was no further adjustment to the costs based on reach.

8.8.2 Study Results

The resulting normalized network costs for each line-rate and traffic scenario are shown in Fig. 8.17. Some line-rate/traffic combinations were not considered because they would have resulted in a tremendous number of wavelengths and/or an extremely high network cost. Based on these results, the optimal line-rate (from a cost perspective) at each aggregate demand level is shown in the middle column of Table 8.11. (In some scenarios, two line-rates resulted in approximately the same lowest cost.) As discussed previously, higher-speed technology is cheaper on a per bit/sec basis. With low aggregate demand, this benefit was offset by inefficiencies in packing the wavelengths. As the aggregate demand increased, the wavelengths were packed more efficiently, and the lower relative transponder costs dominated. Thus, the line-rate that produced the lowest-cost network increased as the level of traffic increased. Not surprisingly, due to the aggressive cost assumptions made for 100-Gb/s technology, this line-rate proved in relatively quickly. It was the lowest-cost line-rate for 20-Tb/s aggregate demand and higher. Results with less aggressive cost assumptions are discussed in the next section.

Note that the network was relatively more cost effective as the traffic level increased. The 100-Tb/s network cost on the order of 15 times more than the 1-Tb/s network even though the aggregate demand increased by a factor of 100.

The results of this study are in-line with what was shown in [SaSi06] for a 55-node backbone network. However, the study of [SaSi06] considered 160 Gb/s rather than 100 Gb/s as the next line-rate after 40 Gb/s; thus, 40 Gb/s was the optimal line-rate for a wider range of traffic levels.

As the network evolved, the relative cost contribution of the individual components changed significantly. Table 8.12 indicates the breakdown of the costs for the major network components for the 1-Tb/s aggregate demand scenario with 2.5-Gb/s line-rate, and the 100-Tb/s aggregate demand scenario with 100-Gb/s line-rate. As shown in the table, the cost of the electronics (grooming/routing, transponders, and regenerators) dominated as the network grew in size.

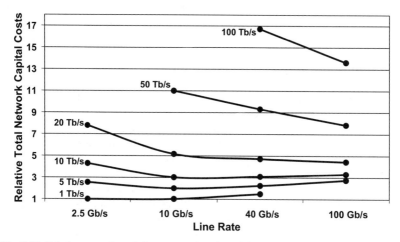

Fig. 8.17 Relative network capital costs as a function of the line-rate. The labels next to each plot indicate the aggregate network demand. As the demand level increased, the line-rate that produced the lowest cost network also increased.

Table 8.11 Optimal Line-Rate from a Cost Perspective

Aggregate Demand	Aggressive Costs	Less Aggressive Costs (Section 8.8.3)
1 Tb/s	2.5 Gb/s, 10 Gb/s	2.5 Gb/s, 10 Gb/s
5 Tb/s	10 Gb/s	10 Gb/s
10 Tb/s	10 Gb/s, 40 Gb/s	10 Gb/s
20 Tb/s	100 Gb/s	10 Gb/s, 40 Gb/s
50 Tb/s	100 Gb/s	40 Gb/s, 100 Gb/s
100 Tb/s	100 Gb/s	40 Gb/s, 100 Gb/s

Table 8.12 Cost Breakdown by Equipment Type

	1-Tb/s Aggregate Demand (2.5-Gb/s Line-Rate)	100-Tb/s Aggregate Demand (100-Gb/s Line-Rate)
Optical Amplifiers	30%	10%
Nodal Optical Network Elements	20%	1%
Grooming/Routing	31%	53%
Transponders	18%	20%
Regenerator Cards	1%	16%

The switching requirements are another factor to consider when selecting the line-rate. Table 8.13 indicates the number of bi-directional wavelength ports required on the largest core optical *switch fabric* at any of the nodes, for each scenario in the study. (As emphasized, this is the size of the switch fabric, not the number of external fiber ports.) The counts include ports for all wavelengths on the network fibers entering the node, as well as ports for the actual number of add/drop wavelengths at the node, including regeneration. In all scenarios, the largest required switch occurred at a degree-five node.

The feasibility of the switch size may dictate the line-rate to be used. Consider the 10-Tb/s aggregate demand scenario, and assume that switch fabrics no larger than 512x512 are possible. With this level of traffic, both the 10-Gb/s line-rate and the 40-Gb/s line-rate yielded approximately the same lowest cost. However, with 10-Gb/s line-rate, the required switch-fabric size was on the order of 768x768; with 40-Gb/s line-rate, it was a little more than 256x256. Thus, the 40-Gb/s line-rate is preferable from a switching perspective.

If the switch size is problematic, another possible solution is to switch at a waveband granularity rather than a wavelength granularity. As described in Sections 2.4.2 and 2.7, a waveband is a group of wavelengths that are treated as a single unit. By switching wavebands instead of wavelengths, the required switch fabric is correspondingly smaller. The tradeoff is that there may be some inefficiencies due to the coarser switching granularity; however, as mentioned in Section 2.4.2, studies have shown that wavebands have a relatively small effect on the network efficiency (assuming intelligent algorithms are used).

Table 8.13 Largest Optical Switch Fabric Required in Each Scenario (# of Wavelength Ports)

Line-Rate	Aggregate Network Demand					
	1 Tb/s	5 Tb/s	10 Tb/s	20 Tb/s	50 Tb/s	100 Tb/s
2.5 Gb/s	347	1538	3195	5969	-	-
10 Gb/s	148	424	762	1564	3996	-
40 Gb/s	99	173	273	447	1113	2068
100 Gb/s	-	146	173	248	508	822

8.8.3 Less Aggressive Cost Assumptions

As noted above, the cost ratios assumed in Table 8.10 for the 40-Gb/s and the 100-Gb/s line-rates were aggressive. Figure 8.18 shows the relative network costs for more moderate cost assumptions. This figure assumes that the cost ratio of 40-Gb/s to 10-Gb/s technology is 3 (instead of 2.5), and the cost ratio of 100-Gb/s to 40-Gb/s technology is 2 (instead of 1.5). The optimal line-rates from a cost perspective are shown in the right-most column of Table 8.11. With these cost assumptions, 40 Gb/s did not prove in until the aggregate network demand was 20 Tb/s (although, at this level, it yielded approximately the same cost as the 10-Gb/s line-rate). For 50-Tb/s to 100-Tb/s aggregate demand, the 40-Gb/s and 100-Gb/s line-rates produced approximately the same network cost. The switch fabric sizes of Table 8.13 still hold, however; thus, 100-Gb/s line-rate may be more desirable from a switching perspective.

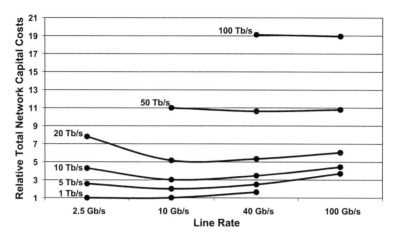

Fig. 8.18 Relative network capital costs as a function of the line-rate, where less aggressive cost estimates were used for 40-Gb/s and 100-Gb/s technology as compared to Fig. 8.17.

8.9 Optical Grooming in the Edge Networks

As the level of network traffic grows while the number of network nodes remains approximately fixed, the average amount of traffic between node pairs increases. Thus, an increasing amount of traffic can be efficiently packed into wavelengths at the edge of the network without requiring any intermediate grooming in the core (i.e., backbone network). This implies that efficient grooming in the edge net-works (i.e., regional and metro-core networks) can be used to offload much of the grooming burden from the core network. Furthermore, edge grooming may be able to take advantage of optical-grooming techniques. As discussed in Section 6.9, some of the optical-grooming strategies that are being developed are more suitable for edge networks than they are for large core networks, due to these techniques requiring scheduling and/or collision management.

To investigate the efficacy of grooming at the edge, the 100-Tb/s aggregate demand scenario of Section 8.8 was considered, with a line-rate of 40 Gb/s. Roughly 95% of the traffic was limited to being groomed solely at the network edge; i.e., intermediate grooming in the backbone core was not permitted for this traffic. The remaining 5% of the traffic was allowed to undergo intermediate grooming as usual. This limited grooming was still quite effective in packing the wavelengths; the overall network capacity requirements increased by only 1% as compared to a design where all traffic is eligible for intermediate grooming. Furthermore, assuming that all of the traffic was IP and that the traffic that was solely groomed at the edge completely bypassed the IP routers (e.g., via the use of optical grooming in the edge networks), then the average required IP router size was reduced by 85%. (An explicit cost analysis was not performed for the overall design because the cost of optical grooming and the cost of the edge/core interface would be too speculative.)

The study was repeated for the 100-Tb/s aggregate demand scenario with 100-Gb/s line-rate. Due to the higher line-rate, 20% of the traffic was allowed to undergo intermediate grooming in the core network. With this configuration, the network capacity requirements increased by roughly 5%; the average required IP router size was reduced by about 65%, again assuming that the traffic groomed at the edge completely bypassed the IP routers. As expected, with a higher line-rate, the scheme is not as effective in reducing the required amount of grooming; however, a significant benefit can still be realized.

These results show that grooming at the edge is potentially a viable scheme for the bulk of the traffic in a heavily loaded network. In the remainder of this section, it is assumed that the grooming in the edge networks is performed in the optical domain; e.g., schemes based on fast optical switching or on passive optical broadcasting may be used to aggregate the traffic. (The OBS scheme of Section 6.9.3 or the TWIN scheme of Section 6.9.4 are possible aggregation architectures that may be suitable for this application.) With optical aggregation at the edge, consideration must be given to the architecture of the interface between the edge and core networks. The following discussion follows that of [SaSi06].

First, assume that traffic is routed in the optical domain between the edge and core networks without any O-E-O conversion, as shown in Fig. 8.19. The advantage of this approach is that transponders are not needed at the edge/core interface. However, the traffic sources in the edge network would need to be equipped with transponders that are compatible with the transmission system of the core network. Due to the stringent performance requirements in the core, such equipment may be too expensive for customer premises equipment. Another possible disadvantage of an all-optical interface is that any voids between the multiplexed data bursts will remain as signal voids in the core network, which could possibly cause amplifier transients.

A second option is to isolate the transmission systems of the edge and core networks via an O-E-O interface, as shown in Fig. 8.20. Edge networks generally have shorter optical reach and lower capacity requirements than the core; thus, lower-cost transponders can be used for transmission solely within the edge network. Transponders at the edge/core interface would be used to convert the aggregated traffic to optical signals that meet the stringent requirements of the core. Furthermore, by converting the aggregated traffic to the electrical domain, bit-stuffing can be used to eliminate any voids in the optically multiplexed signal. The tradeoff is the cost of the transponders at the edge/core interface. However, these transponders are required for the aggregated traffic, not for the individual traffic streams; thus, this scheme may be ultimately lower cost than requiring high-quality transponders at all of the traffic sources.

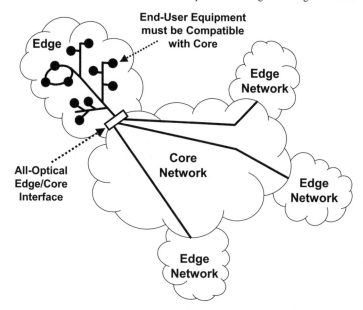

Fig. 8.19 If the edge/core interface is all-optical, then the end-users must be equipped with trans-ponders that are compatible with the stringent requirements of the core network. (Adapted from [SaSi06]. © 2006 IEEE)

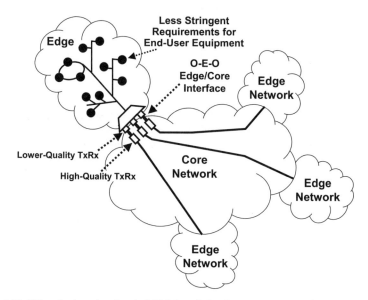

Fig. 8.20 If the edge/core interface is O-E-O-based, then lower-cost transponders (TxRx) can be used in the edge network. (Adapted from [SaSi06]. © 2006 IEEE)

As discussed in Section 6.9, the various optical grooming schemes being studied need to manage resource contention issues. Due to the complexities of scheduling across large networks, it is assumed that scheduling would be performed independently within each edge network attached to the core. This can lead to conflicts however; e.g., two different regions could send traffic on the same wavelength that arrives at the same destination regional network at the same time. The architecture of the edge/core interface affects how such conflicts are managed. With an all-optical interface, rapidly tunable all-optical wavelength converters could be used to shift one of the conflicting streams to another wavelength. However, this could potentially produce two simultaneous streams that are destined for the same end-user. Thus, each customer premise would need to be capable of receiving multiple wavelengths at one time, e.g., with an array receiver. With an O-E-O edge/core interface, the wavelengths in the edge networks can be chosen independently from those in the core network such that wavelength contention issues can be minimized (i.e., wavelength conversion can be obtained by tuning the transmitter on the edge-network side to a different wavelength). However, an array receiver would still be needed at the customer premises. Alternatively, once the signal is in the electronic domain at the edge/core interface, electronic buffering could be used to resolve conflicts, eliminating the need for array receivers.

Clearly, there are still issues that need to be resolved with such optical grooming architectures. However, these schemes provide another example of how operating primarily in the optical domain, with O-E-O conversion at key junctures, can provide an opportunity for reduced electronics, along with the attendant savings in capital and operational costs. It is likely that such applications of optical networking will be the key enablers of continued scalable network growth.

8.10 General Conclusions

As the studies of this chapter have shown, there is no single technology that is optimal in all scenarios. The transmission and switching technologies for a network need to be chosen based on factors such as the topology, the amount of traffic, and the desired amount of network agility.

Electronic-based and optical-based technologies clearly have their associated strengths. Thus far, there has been a trend for optics to scale better than electronics as the line rates increase; however, there are still functions that are best performed in the electrical domain. Thus, it is likely that networks will remain a mix of technologies.

Appendix A

Suggestions for RFI/RFP Network Design Exercises

Carriers often issue a network design exercise as part of a Request for Information (RFI) or a Request for Proposal (RFP). System vendors perform the network designs using their respective equipment to provide carriers with information regarding architecture, technology, and pricing. To streamline the design process and to enable carriers to glean relevant information when comparing results, some suggestions to assist carriers in preparing a design exercise are provided here.

1. Provide all data (e.g., nodes, links, demands) in text files or in spreadsheets. Do not require any manual entry of data by the vendors, as this is likely to lead to errors.

2. Provide the latitude/longitudes of the network nodes. Most design tools can use this information to position nodes on the screen, to help visualize the design.

3. If demand sets for multiple time periods are provided, specify whether the demand sets are incremental or cumulative (e.g., are the demands in set #2 added to the demands that already exist from set #1, or does set #2 represent all of the demands). Additionally, when adding demands to the network in subsequent time periods, specify whether the demands already in the network need to be kept fixed, or whether an optimization can be performed over all of the demands.

4. If the design exercise is for extended-reach technology, then fiber span information should be provided, including fiber type, span distances, and span losses. (Information such as PMD can be helpful as well, if available.)

5. Ideally, there would be no more than three design exercises:

    a. A baseline demand set for which the design should be optimized

    b. A modified demand set, where some percentage of the baseline demands have different sources/destinations. The design is run using the equipment configuration that was chosen for the baseline design. For example, if an OADM-only architecture is being used, the orientation of the OADMs selected for the baseline design must be used in the modified design. This tests the forecast tolerance of the architecture.

    c. A projected growth demand set, to test the scalability of the design

6. Be specific about what type of protection is desired (e.g., is shared protection suitable, or must dedicated protection be used).

7. If the exercise includes subrate demands, specify whether grooming/routing devices are required at all nodes, or whether traffic backhauling is acceptable. Additionally, be specific about what types of demands can be muxed or groomed together in a wavelength (e.g., what services, what protection types).

8. Provide some guidelines regarding the routing to ensure that comparisons across designs are valid. However, forcing all connections to always use the shortest path or explicitly specifying a path for each connection is too constrictive. It is preferable to specify a guideline such as the routed path for each connection should be no longer than P% longer than the shortest possible path.

9. Request design output in a specified format so that comparisons across system vendors can be readily performed.

Appendix B
C-Code For Routing Routines

The C-code for several useful routing functions is provided in this appendix. The first set of routines uses the Breadth-First-Search method to find a shortest path between a source and a destination. The code for these routines follows the algorithm outlined in [Bhan99]. The second set of routines finds the K-shortest paths between a source and a destination. This code follows the equivalence method of [HeMS03]. The final set of routines finds N mutually disjoint paths between a source and a destination. This follows the algorithm outlined in [Bhan99]. Various parameters can be set to indicate whether the paths should be link-disjoint or link-and-node-disjoint. In scenarios where N mutually disjoint paths do not exist, the function can be used to find the N maximally disjoint paths.

For the most part, memory is pre-allocated for the routines, as this tends to run faster. The maximum size of the network can be adjusted if necessary by redefining the appropriate parameters, which appear at the start of the code.

A small *main* function is provided to demonstrate how to specify the network topology and how to call the three major routines.

Minimal amount of error checking has been added.

DISCLAIMER: NO WARRANTY OF ANY KIND IS EXPRESSED OR IMPLIED. YOU USE THE CODE AT YOUR OWN RISK. IN NO EVENT SHALL THE AUTHOR, THE AGENTS OF THE AUTHOR, OR THE PUBLISHER BE LIABLE FOR DATA LOSS, LOSS OF PROFITS, LOSS OF BENEFITS, OR OTHER INCIDENTAL OR CONSEQUENTIAL DAMAGES WHILE USING THIS CODE.

```c
#include <stdio.h>
#include <stdlib.h>
#include <string.h>
#include <limits.h>

/* Adjust these parameters as needed */
#define MaxNodes 250
#define MaxLinks 1000
#define MaxNodeName 30
#define MaxNodeDegree 20
#define MaxPathHops 150

#define FALSE 0
#define TRUE 1

/* when performing graph transformations, need to add dummy nodes and links */
#define MaxNodesWithDummies MaxNodes+MaxPathHops
#define MaxLinksWithDummies MaxLinks+2*MaxPathHops

const double CommonNodePenalty = 1500000.0;      /* make greater than sum of link distances */
const double CommonLinkPenalty = 160000000.0;   /* make much greater than
                                                         CommonNodePenalty */
const double INFINITY = 17000000000.0; /* make much greater than CommonLinkPenalty */
const double SMALL = .0001; /* small number relative to link distances */

typedef unsigned short USHORT ;
typedef char BOOL ;

typedef struct tagNodeT {
    char Name[MaxNodeName+1];
    USHORT     OutgoingLinks[MaxNodeDegree];
    USHORT     IncomingLinks[MaxNodeDegree];
    USHORT     NumOutgoingLinks;
    USHORT     NumIncomingLinks;
    } NodeT, *NodeTP;

typedef struct tagLinkT {
    USHORT          LinkNode1; /* node at one end of link */
    USHORT          LinkNode2; /* node at other end of link */
    USHORT          Status; /* active link or not (true or false) */
    double          Length; /* any metric could be used here */
    } LinkT, *LinkTP;
```

```
typedef struct tagNetworkT {
    NodeT           Nodes[MaxNodesWithDummies];
    USHORT          NumNodes;
    LinkT           Links[MaxLinksWithDummies];
    USHORT          NumLinks;
    } NetworkT, *NetworkTP;

typedef struct tagPathT{
    USHORT          PathHops[MaxPathHops];
    USHORT          NumPathHops;
    double          PathDistance;
    } PathT, *PathTP;

USHORT PredecessorNode[MaxNodesWithDummies];
USHORT PredecessorLink[MaxNodesWithDummies];
double NodeDistance[MaxNodesWithDummies];
NetworkT TempNetwork;   /* use for graph transformations */

BOOL BreadthFirstSearchShortestPath (NetworkTP NetworkP, USHORT SourceNode,
    USHORT DestinationNode, PathTP ShortestPath);
BOOL BreadthFirstSearchRelax (NetworkTP NetworkP, USHORT NodeA, USHORT Link,
    USHORT FinalDestination);
USHORT KShortestPaths (NetworkTP NetworkP, USHORT Source, USHORT Dest, USHORT K,
    PathTP KPaths);
USHORT NShortestDiversePaths (NetworkTP NetworkP, USHORT Source, USHORT Destination,
    USHORT NumDisjointPaths, BOOL LinkDisjointOnly, BOOL CommonLinksAllowed,
    BOOL CommonNodesAllowed, PathTP NPaths);
USHORT AddLinkToTopology (NetworkTP NetworkP, USHORT Node1, USHORT Node2,
    double Length, BOOL ReverseLinkStatus);


main ()
{
    USHORT n;
    NetworkT network;
    PathT Paths[10];

    /* create a simple network to demonstrate the functions */
    strcpy(network.Nodes[0].Name, "A");
    strcpy(network.Nodes[1].Name, "B");
    strcpy(network.Nodes[2].Name, "C");
    strcpy(network.Nodes[3].Name, "D");
    strcpy(network.Nodes[4].Name, "Z");
    network.NumNodes = 5;
    network.NumLinks = 0;
```

```
for (n=0; n<network.NumNodes; n++)
    network.Nodes[n].NumIncomingLinks = network.Nodes[n].NumOutgoingLinks = 0;

AddLinkToTopology(&network, 0, 1, 10.0, TRUE);
AddLinkToTopology(&network, 1, 2, 10.0, TRUE);
AddLinkToTopology(&network, 2, 3, 10.0, TRUE);
AddLinkToTopology(&network, 3, 4, 10.0, TRUE);
AddLinkToTopology(&network, 0, 2, 10.0, TRUE);
AddLinkToTopology(&network, 2, 4, 10.0, TRUE);

BreadthFirstSearchShortestPath(&network, 0, 4, &Paths[0]);

KShortestPaths(&network, 0, 4, 10, Paths);

NShortestDiversePaths(&network, 0, 4, 3, FALSE, TRUE, TRUE, Paths);
}




/****************************************************************/
/*      Code for Breadth First Search Shortest Paths           */
/*      Follows algorithm of [Bhan99]                          */
/*      Finds the shortest path from source to destination     */
/*      Returns the shortest path in ShortestPathP             */
/*      Returns TRUE/FALSE depending on whether a path exists  */
/****************************************************************/

BOOL BreadthFirstSearchShortestPath (NetworkTP NetworkP, USHORT SourceNode,
    USHORT DestinationNode, PathTP ShortestPathP)
{
    USHORT n, m, numNodesOnListCurrent, numNodesOnListNew, node, link;
    USHORT nodesOnList1[MaxNodesWithDummies], nodesOnList2[MaxNodesWithDummies];
    USHORT* currentNodeList;
    USHORT* newNodeList;
    USHORT* temp;
    BOOL addedNodeToList1[MaxNodesWithDummies];
    BOOL addedNodeToList2[MaxNodesWithDummies];
    BOOL* currentAdded;
    BOOL* newAdded;
    BOOL* temp2;
    PathT tempPath;
```

```
for (n = 0; n < NetworkP->NumNodes; n++) {
    NodeDistance[n] = INFINITY;
    PredecessorNode[n] = PredecessorLink[n] = USHRT_MAX;
    addedNodeToList1[n] = addedNodeToList2[n] = FALSE;
    }

NodeDistance[SourceNode] = 0.0;
numNodesOnListCurrent = 1;
nodesOnList1[0] = SourceNode;

currentNodeList = nodesOnList1;
newNodeList = nodesOnList2;
currentAdded = addedNodeToList1;
newAdded = addedNodeToList2;

while (numNodesOnListCurrent > 0) {
    numNodesOnListNew = 0;
    for (n = 0; n < numNodesOnListCurrent; n++) {
        node = currentNodeList[n];
        currentAdded[node] = FALSE;
        for (m = 0; m < NetworkP->Nodes[node].NumOutgoingLinks; m++) {
            link = NetworkP->Nodes[node].OutgoingLinks[m];
            if (!NetworkP->Links[link].Status)
                continue;
            if (!BreadthFirstSearchRelax(NetworkP, node, link, DestinationNode))
                continue;
            if ((!newAdded[NetworkP->Links[link].LinkNode2]) &&
                (NetworkP->Links[link].LinkNode2 != DestinationNode)) {
                    newNodeList[numNodesOnListNew] = NetworkP->Links[link].LinkNode2;
                    numNodesOnListNew++;
                    newAdded[NetworkP->Links[link].LinkNode2] = TRUE;
                    }
            }
        }
    numNodesOnListCurrent = numNodesOnListNew;
    temp = currentNodeList; currentNodeList = newNodeList; newNodeList = temp;
    temp2 = currentAdded; currentAdded = newAdded; newAdded = temp2;
    }

ShortestPathP->NumPathHops = 0;
ShortestPathP->PathDistance = NodeDistance[DestinationNode];

if (NodeDistance[DestinationNode] > (INFINITY-SMALL))
    return(FALSE);
```

```
    /* the predecessor path traces from the destination back to the root */
    /*    reverse it to have it start at root */
    node = DestinationNode;
    while (node != SourceNode) {
        if (ShortestPathP->NumPathHops >= MaxPathHops) {
            printf("Need to increase MaxPathHops\n");
            exit(-1);
            }
        link = PredecessorLink[node];
        tempPath.PathHops[ShortestPathP->NumPathHops] = link;
        node = PredecessorNode[node];
        ShortestPathP->NumPathHops++;
        }

    for (m=0; m<ShortestPathP->NumPathHops; m++)  /* reverse the order */
        ShortestPathP->PathHops[m] = tempPath.PathHops[(ShortestPathP->NumPathHops)-m-1];

    return(TRUE);
}


BOOL BreadthFirstSearchRelax (NetworkTP NetworkP, USHORT NodeA, USHORT Link,
    USHORT FinalDestination)
{
    USHORT nodeB;
    double newDistanceAB;

    nodeB = NetworkP->Links[Link].LinkNode2;
    newDistanceAB = NodeDistance[NodeA] + NetworkP->Links[Link].Length;

    if ((NodeDistance[nodeB] > (newDistanceAB + SMALL)) &&
        (NodeDistance[FinalDestination] > (newDistanceAB + SMALL))) {
            NodeDistance[nodeB] = newDistanceAB;
            PredecessorNode[nodeB] = NodeA;
            PredecessorLink[nodeB] = Link;
            return(TRUE);
            }

    return(FALSE);
}
```

```
/******************************************************************/
/*       Code for K Shortest Paths between a source and destination      */
/*       Follows equivalence method of [HeMS03]                          */
/*       The K paths are returned in KPaths                              */
/*       The function returns the number of paths actually found         */
/******************************************************************/

#define MaxEquivalences MaxNodes
#define MaxSplit MaxNodes
#define NodeType 0
#define LinkType 1

typedef struct tagEquivalenceClassT{
    char EquivalenceType;
    PathT    PrefixPath;   /* source to most recent split */
    PathT    SuffixPath;   /* split to next split (or to end); Don't track distance for suffix */
    PathT    ShortestPath;
    USHORT   FirstLink[MaxSplit];
    USHORT   NumFirstLinks;
    USHORT   SplitNode;
    } EquivalenceClassT, *EquivalenceClassTP;

EquivalenceClassT Equivalences[MaxEquivalences];

void CreatePathFromEquivalence (EquivalenceClassTP EquivalenceP, double Distance,
    PathTP PathP);
void FindBestPathInEquivalence (NetworkTP NetworkP, EquivalenceClassTP EquivalenceP,
    USHORT Dest);
void UpdateEquivalences (NetworkTP NetworkP, EquivalenceClassTP Equivalences,
    USHORT* NumEquivalences, USHORT BestPath, USHORT Dest);


USHORT KShortestPaths (NetworkTP NetworkP, USHORT Source, USHORT Dest, USHORT K,
    PathTP KPaths)
{
    USHORT j, n, bestPath, numEquivalences, firstLink;
    double minDist;

    for (j=0; j<K; j++) {
        KPaths[j].NumPathHops = 0;
        KPaths[j].PathDistance = 0.0;
        }

    if (Source == Dest)
        return(K);

    /* first find shortest path */
```

```c
if (!BreadthFirstSearchShortestPath(NetworkP, Source, Dest, &KPaths[0]))
    return (0);  /* didn't find any paths from source to dest */

if (K == 1)
    return(1);

firstLink = KPaths[0].PathHops[0];


Equivalences[0].EquivalenceType = NodeType;
Equivalences[0].FirstLink[0] = firstLink;
Equivalences[0].NumFirstLinks = 1;
Equivalences[0].PrefixPath.NumPathHops = 0;
Equivalences[0].PrefixPath.PathDistance = 0.0;
Equivalences[0].SuffixPath.NumPathHops = 0;
Equivalences[0].SuffixPath.PathDistance = 0.0;
Equivalences[0].SplitNode = Source;


Equivalences[1].EquivalenceType = LinkType;
Equivalences[1].FirstLink[0] = firstLink;
Equivalences[1].NumFirstLinks = 1;
Equivalences[1].PrefixPath.NumPathHops = 0;
Equivalences[1].PrefixPath.PathDistance = 0.0;
Equivalences[1].SuffixPath = KPaths[0];
Equivalences[1].SplitNode = Source;
numEquivalences = 2;
FindBestPathInEquivalence(NetworkP, &Equivalences[0], Dest);
FindBestPathInEquivalence(NetworkP, &Equivalences[1], Dest);


for (j=1; j<K; j++) {
    minDist = INFINITY; bestPath = USHRT_MAX;
    for (n=0; n<numEquivalences; n++) {
        /* could use a heap to make this faster */
        if (Equivalences[n].ShortestPath.PathDistance < (minDist-SMALL)) {
            bestPath = n;
            minDist = Equivalences[n].ShortestPath.PathDistance;
            }
        }
    if (bestPath == USHRT_MAX)
        break;

    CreatePathFromEquivalence(&Equivalences[bestPath], minDist, &KPaths[j]);
    if (j < (K-1))
        UpdateEquivalences(NetworkP, Equivalences, &numEquivalences, bestPath, Dest);
    }
return(j);

}
```

```
void CreatePathFromEquivalence (EquivalenceClassTP EquivalenceP, double Distance,
    PathTP PathP)
{
    USHORT h, numHops;

    *PathP = EquivalenceP->PrefixPath;
    numHops = PathP->NumPathHops;

    if (EquivalenceP->EquivalenceType == LinkType) {
        PathP->PathHops[numHops] = EquivalenceP->FirstLink[0];
        numHops++;
        }

    for (h=0; h<EquivalenceP->ShortestPath.NumPathHops; h++) {
        PathP->PathHops[numHops] = EquivalenceP->ShortestPath.PathHops[h];
        numHops++;
        }

    PathP->NumPathHops = numHops;
    PathP->PathDistance = Distance;
}

void FindBestPathInEquivalence (NetworkTP NetworkP, EquivalenceClassTP EquivalenceP,
    USHORT Dest)
{
    USHORT h, n, linkID, nodeID, splitNode;
    double minDist, distToAdd;
    PathT tempPath;

    /* make a copy of the network because will perform graph transformations */
    TempNetwork = *NetworkP;

    EquivalenceP->ShortestPath.PathDistance = INFINITY;

    /* eliminate nodes along prefix path so don't get loops */
    for (h=0; h<EquivalenceP->PrefixPath.NumPathHops; h++) {
        linkID = EquivalenceP->PrefixPath.PathHops[h];
        nodeID = TempNetwork.Links[linkID].LinkNode1;
        for (n=0; n<TempNetwork.Nodes[nodeID].NumIncomingLinks; n++) {
            linkID = TempNetwork.Nodes[nodeID].IncomingLinks[n];
            TempNetwork.Links[linkID].Status = FALSE;
            }
        }

    distToAdd = EquivalenceP->PrefixPath.PathDistance;
```

```c
    if (EquivalenceP->EquivalenceType == NodeType) {
        for (n=0; n<EquivalenceP->NumFirstLinks; n++) {
            linkID = EquivalenceP->FirstLink[n];
            TempNetwork.Links[linkID].Status = FALSE;
            }
        BreadthFirstSearchShortestPath(&TempNetwork, EquivalenceP->SplitNode, Dest,
            &(EquivalenceP->ShortestPath));
        }
    else { /* LinkType */
        /* kick out the vertex node also */
        nodeID = EquivalenceP->SplitNode;
        for (n=0; n<TempNetwork.Nodes[nodeID].NumIncomingLinks; n++) {
            linkID = TempNetwork.Nodes[nodeID].IncomingLinks[n];
            TempNetwork.Links[linkID].Status = FALSE;
            }

        linkID = EquivalenceP->FirstLink[0];
        distToAdd += NetworkP->Links[linkID].Length;

        splitNode = NetworkP->Links[linkID].LinkNode2;
        minDist = INFINITY;
        for (h=1; h<EquivalenceP->SuffixPath.NumPathHops; h++) {
            /* if there are many hops, [HeMS03] has a method to speed it up */
            linkID = EquivalenceP->SuffixPath.PathHops[h];
            if (!TempNetwork.Links[linkID].Status)
                continue;
            TempNetwork.Links[linkID].Status = FALSE;
            BreadthFirstSearchShortestPath(&TempNetwork, splitNode, Dest, &tempPath);
            TempNetwork.Links[linkID].Status = TRUE; /* put back the link */
            if (tempPath.PathDistance < (minDist-SMALL)) {
                EquivalenceP->ShortestPath = tempPath;
                minDist = tempPath.PathDistance;
                }
            }
        }

    if (EquivalenceP->ShortestPath.PathDistance < INFINITY-SMALL)
        EquivalenceP->ShortestPath.PathDistance += distToAdd;
}


void UpdateEquivalences (NetworkTP NetworkP, EquivalenceClassTP EquivalencesP,
    USHORT* NumEquivalences, USHORT BestPath, USHORT Dest)
{
    USHORT linkID, splitNode, h, h2, numHops;
```

```
if (EquivalencesP[BestPath].EquivalenceType == NodeType) {
    if (*NumEquivalences >= MaxEquivalences) {
        printf("Need to increase number of equivalence classes\n");
        exit(-2);
    }

    linkID = EquivalencesP[BestPath].ShortestPath.PathHops[0];
    EquivalencesP[BestPath].FirstLink[EquivalencesP[BestPath].NumFirstLinks] = linkID;
    EquivalencesP[BestPath].NumFirstLinks++;

    EquivalencesP[*NumEquivalences].EquivalenceType = LinkType;
    EquivalencesP[*NumEquivalences].FirstLink[0] = linkID;
    EquivalencesP[*NumEquivalences].NumFirstLinks = 1;
    EquivalencesP[*NumEquivalences].PrefixPath = EquivalencesP[BestPath].PrefixPath;
    EquivalencesP[*NumEquivalences].SuffixPath = EquivalencesP[BestPath].ShortestPath;
    EquivalencesP[*NumEquivalences].SuffixPath.PathDistance = 0.0;
    EquivalencesP[*NumEquivalences].SplitNode = EquivalencesP[BestPath].SplitNode;
    FindBestPathInEquivalence(NetworkP, &EquivalencesP[*NumEquivalences], Dest);
    (*NumEquivalences)++;

    /* can't set new shortest of BestPath until done with previous shortest above */
    FindBestPathInEquivalence(NetworkP, &EquivalencesP[BestPath], Dest);
}
else { /* link type */
    if (*NumEquivalences >= (MaxEquivalences-2)) { /* will add three more */
        printf("Need to increase number of equivalence classes\n");
        exit(-2);
    }

    /* find where shortest path diverges from suffix */
    /* suffix starts at branch node; shortest starts at 2nd node */
    for (h=1; h<EquivalencesP[BestPath].SuffixPath.NumPathHops; h++) {
        if (EquivalencesP[BestPath].ShortestPath.PathHops[h-1] !=
            EquivalencesP[BestPath].SuffixPath.PathHops[h])
                break;
    }

    /* add one for new split node;  position h is where the paths diverge */
    EquivalencesP[*NumEquivalences].EquivalenceType = NodeType;
    EquivalencesP[*NumEquivalences].FirstLink[0] =
        EquivalencesP[BestPath].ShortestPath.PathHops[h-1];
    EquivalencesP[*NumEquivalences].FirstLink[1] =
        EquivalencesP[BestPath].SuffixPath.PathHops[h];
    EquivalencesP[*NumEquivalences].NumFirstLinks = 2;
```

```
EquivalencesP[*NumEquivalences].PrefixPath = EquivalencesP[BestPath].PrefixPath;
for (h2=0; h2<h; h2++) {  /* add in 1st link plus the part in common */
    linkID = EquivalencesP[BestPath].SuffixPath.PathHops[h2];
    EquivalencesP[*NumEquivalences].PrefixPath.
        PathHops[Equivalences[BestPath].PrefixPath.NumPathHops+h2] = linkID;
    EquivalencesP[*NumEquivalences].PrefixPath.PathDistance +=
        NetworkP->Links[linkID].Length;
    }
EquivalencesP[*NumEquivalences].PrefixPath.NumPathHops += h;
EquivalencesP[*NumEquivalences].SuffixPath.NumPathHops = 0;
EquivalencesP[*NumEquivalences].SuffixPath.PathDistance = 0.0;
linkID = EquivalencesP[BestPath].SuffixPath.PathHops[h];
splitNode = NetworkP->Links[linkID].LinkNode1;
EquivalencesP[*NumEquivalences].SplitNode = splitNode;
FindBestPathInEquivalence(NetworkP, &EquivalencesP[*NumEquivalences], Dest);
(*NumEquivalences)++;

/* add one for one split path */
EquivalencesP[*NumEquivalences].EquivalenceType = LinkType;
EquivalencesP[*NumEquivalences].FirstLink[0] =
    Equivalences[BestPath].ShortestPath.PathHops[h-1];
EquivalencesP[*NumEquivalences].NumFirstLinks = 1;
EquivalencesP[*NumEquivalences].PrefixPath =
    Equivalences[(*NumEquivalences)-1].PrefixPath;
EquivalencesP[*NumEquivalences].SuffixPath.PathDistance = 0.0;
numHops = 0;
for (h2=h-1; h2<Equivalences[BestPath].ShortestPath.NumPathHops; h2++) {
    linkID = Equivalences[BestPath].ShortestPath.PathHops[h2];
    Equivalences[*NumEquivalences].SuffixPath.PathHops[numHops] = linkID;
    numHops++;
    }
EquivalencesP[*NumEquivalences].SuffixPath.NumPathHops = numHops;
EquivalencesP[*NumEquivalences].SplitNode = splitNode;
FindBestPathInEquivalence(NetworkP, &EquivalencesP[*NumEquivalences], Dest);
(*NumEquivalences)++;

/* add one for other split path */
EquivalencesP[*NumEquivalences].EquivalenceType = LinkType;
EquivalencesP[*NumEquivalences].FirstLink[0] =
    Equivalences[BestPath].SuffixPath.PathHops[h];
EquivalencesP[*NumEquivalences].NumFirstLinks = 1;
EquivalencesP[*NumEquivalences].PrefixPath =
    Equivalences[(*NumEquivalences)-1].PrefixPath;
EquivalencesP[*NumEquivalences].SuffixPath.PathDistance = 0.0;
numHops = 0;
```

```
for (h2=h; h2<EquivalencesP[BestPath].SuffixPath.NumPathHops; h2++) {
    linkID = EquivalencesP[BestPath].SuffixPath.PathHops[h2];
    EquivalencesP[*NumEquivalences].SuffixPath.PathHops[numHops] = linkID;
    numHops++;
    }
EquivalencesP[*NumEquivalences].SuffixPath.NumPathHops = numHops;
EquivalencesP[*NumEquivalences].SplitNode = splitNode;
FindBestPathInEquivalence(NetworkP, &EquivalencesP[*NumEquivalences], Dest);
(*NumEquivalences)++;

EquivalencesP[BestPath].SuffixPath.NumPathHops = h;
FindBestPathInEquivalence(NetworkP, &EquivalencesP[BestPath], Dest);
    }
}


/*******************************************************************/
/*      Code for N Shortest Diverse Paths                        */
/*      Follows method of [Bhan99]                               */
/*      The N paths are returned in NPaths                       */
/*      The function returns number of paths found               */
/*      The paths are not necessarily in order from shortest to longest  */
/*******************************************************************/

void KDualPathGraphTranformation (NetworkTP NetworkP, PathTP PathP, USHORT Source,
    USHORT Destination, BOOL LinkDisjointOnly, BOOL CommonLinksAllowed,
    BOOL CommonNodesAllowed, USHORT DummyNodeThreshold);
void GenerateTwoRealPaths (NetworkTP NetworkP, PathTP TempPath1P, PathTP TempPath2P,
    PathTP RealPath1P, PathTP RealPath2P);
void CleanPath (NetworkTP NetworkP, USHORT DummyLinkThreshold, USHORT Source,
    USHORT Destination, PathTP NewPathP);
void AdjustNodeInfoForNewLink (NetworkTP NetworkP, USHORT LinkID);
void ChangeLinkSource (NetworkTP NetworkP, USHORT LinkID, USHORT OldSource,
    USHORT NewSource);
void ChangeLinkDestination (NetworkTP NetworkP, USHORT LinkID, USHORT OldDest,
    USHORT NewDest);
USHORT GetReverseLink (NetworkTP NetworkP, USHORT LinkID);

USHORT NShortestDiversePaths (NetworkTP NetworkP, USHORT Source, USHORT Destination,
    USHORT NumDisjointPaths, BOOL LinkDisjointOnly, BOOL CommonLinksAllowed,
    BOOL CommonNodesAllowed, PathTP NPaths)
    /* looks for N paths that are mutually disjoint, or maximally disjoint depending on the settings */
    /* if LinkDisjointOnly is TRUE, then it doesn't try to avoid common nodes among the paths */
    /* if CommonLinksAllowed is TRUE, then if fully disjoint paths can't be found, it will accept */
    /*    paths with common links (although it still minimizes the number of common links) */
    /* if CommonNodesAllowed is TRUE, then if fully disjoint paths can't be found, it will accept */
    /*    paths with common nodes (althouth it still minimizes the number of common nodes) */
```

```
{
    USHORT h, n, numFoundPaths;
    int j, k;
    PathTP tempPaths;

    for (n=0; n<NumDisjointPaths; n++) {
        NPaths[n].NumPathHops = 0;
        NPaths[n].PathDistance = 0.0;
        }

    if (Source == Destination)
        return(NumDisjointPaths);

    /* first find shortest path */
    if (!BreadthFirstSearchShortestPath(NetworkP, Source, Destination, &NPaths[0]))
        return (0);  /* didn't find any paths */
    if (NumDisjointPaths == 1)
        return(1);

    tempPaths = (PathTP)malloc(NumDisjointPaths*sizeof(PathT));
    if (tempPaths == NULL) {
        printf("Out of Memory\n");
        exit(-20);
        }

    tempPaths[0] = NPaths[0];
    for (n=1; n<NumDisjointPaths; n++) {
        /* make a copy of the network because will perform graph transformations */
        TempNetwork = *NetworkP;
        for (j=0; j<n; j++)
            KDualPathGraphTranformation(&TempNetwork, &NPaths[j], Source, Destination,
                LinkDisjointOnly, CommonLinksAllowed, CommonNodesAllowed,
                NetworkP->NumNodes);

        /* run shortest path on the transformed graph */
        if (!BreadthFirstSearchShortestPath(&TempNetwork, Source, Destination, &tempPaths[n]))
            break;

        /* clean path up; may have dummy nodes in it */
        CleanPath(&TempNetwork, NetworkP->NumLinks, Source, Destination, &tempPaths[n]);

        /* paths found above may not be the true paths; need to unravel any interleaving */
        for (j=n; j>0; j--) {
            for (k=j-1; k>=0; k--) {
```

```
                GenerateTwoRealPaths(&TempNetwork, &tempPaths[j],
                    &tempPaths[k], &NPaths[j], &NPaths[k]);
                tempPaths[j] = NPaths[j];
                tempPaths[k] = NPaths[k];
                }
            }
        }

    numFoundPaths = n;
    for (n=0; n<numFoundPaths; n++) {
        NPaths[n].PathDistance = 0.0;
        for (h=0; h<NPaths[n].NumPathHops; h++)
            NPaths[n].PathDistance += NetworkP->Links[NPaths[n].PathHops[h]].Length;
        }
    free(tempPaths);

    return(numFoundPaths);
}


void KDualPathGraphTranformation (NetworkTP NetworkP, PathTP PathP, USHORT Source,
    USHORT Destination, BOOL LinkDisjointOnly, BOOL CommonLinksAllowed,
    BOOL CommonNodesAllowed, USHORT DummyNodeThreshold)
{
    USHORT n, prevNode, link, newLink, dummyID, forwardLink, reverseLink, reverseLink2;
    int h;
    double tempLength;

    for (h=PathP->NumPathHops-1; h>=0; h--) {
        forwardLink = PathP->PathHops[h];
        if (h == 0) { /* don't split source node, but handle the link */
            if (NetworkP->Links[forwardLink].Length > (CommonLinkPenalty - SMALL)) {
                /* must be a common link in the previous paths that have been found */
                /* increase penalty if use it again */
                NetworkP->Links[forwardLink].Length += CommonLinkPenalty;
                continue;
                }

            reverseLink = GetReverseLink(NetworkP, forwardLink);
            tempLength = NetworkP->Links[forwardLink].Length;

            NetworkP->Links[reverseLink].Length = -1.0*tempLength;
            NetworkP->Links[reverseLink].Status = TRUE;

            NetworkP->Links[forwardLink].Status = FALSE;
            /* add big amount to length to discourage its use */
```

```
        NetworkP->Links[forwardLink].Length = tempLength + CommonLinkPenalty;
        if (CommonLinksAllowed) { /* common links OK if can't be avoided */
            NetworkP->Links[forwardLink].Status = TRUE;
            }
        continue;
        }


    prevNode = NetworkP->Links[forwardLink].LinkNode1;

    if (prevNode >= DummyNodeThreshold) {
        /* must be a common node in the previous paths that have been found */
        /* don't add another dummy node */

        if (NetworkP->Links[forwardLink].Length > (CommonLinkPenalty - SMALL)) {
            /* must also be a common link in the previous paths that have been found */
            /* increase penalty if use it again */
            NetworkP->Links[forwardLink].Length += CommonLinkPenalty;
            continue;
            }

        reverseLink = GetReverseLink(NetworkP, forwardLink);
        tempLength = NetworkP->Links[forwardLink].Length;

        NetworkP->Links[reverseLink].Length = -1.0*tempLength;
        NetworkP->Links[reverseLink].Status = TRUE;

        NetworkP->Links[forwardLink].Status = FALSE;
        /* add big amount to length to discourage its use */
        NetworkP->Links[forwardLink].Length = tempLength + CommonLinkPenalty;
        if (CommonLinksAllowed) { /* common links OK if can't be avoided */
            NetworkP->Links[forwardLink].Status = TRUE;
            }

        if ((!LinkDisjointOnly) && (CommonNodesAllowed)) {
            /* increase penalty if use the node again */
            for (n=0; n<NetworkP->Nodes[prevNode].NumIncomingLinks; n++) {
                link = NetworkP->Nodes[prevNode].IncomingLinks[n];
                if (NetworkP->Links[link].Length > (CommonNodePenalty-SMALL)) {
                    NetworkP->Links[link].Length += CommonNodePenalty;
                    break;
                    }
                }
            }

        continue;
```

```
    }

/* split prevNode */
if (NetworkP->NumNodes >= MaxNodesWithDummies) {
    printf("Need to increase MaxNodes\n");
    exit(-4);
    }
if (NetworkP->NumLinks >= (MaxLinksWithDummies-1)) { /* will add 2 links */
    printf("Need to increase MaxLinks\n");
    exit(-5);
    }

dummyID = NetworkP->NumNodes;
NetworkP->NumNodes++;
strcpy(NetworkP->Nodes[dummyID].Name, "DummyAdd");
NetworkP->Nodes[dummyID].NumOutgoingLinks = 0;
NetworkP->Nodes[dummyID].NumIncomingLinks = 0;

reverseLink = GetReverseLink(NetworkP, forwardLink);
tempLength = NetworkP->Links[forwardLink].Length;

NetworkP->Links[reverseLink].Length = -1.0*tempLength;
NetworkP->Links[reverseLink].Status = TRUE;
ChangeLinkDestination(NetworkP, reverseLink, prevNode, dummyID);

NetworkP->Links[forwardLink].Status = FALSE;
/* add big amount to length to discourage its use */
NetworkP->Links[forwardLink].Length = tempLength + CommonLinkPenalty;
if (CommonLinksAllowed) { /* common links OK if can't be avoided */
    NetworkP->Links[forwardLink].Status = TRUE;
    ChangeLinkSource(NetworkP, forwardLink, prevNode, dummyID);
    }

reverseLink = GetReverseLink(NetworkP, PathP->PathHops[h-1]);
for (n=0; n<NetworkP->Nodes[prevNode].NumOutgoingLinks; n++) {
    link = NetworkP->Nodes[prevNode].OutgoingLinks[n];
    if (link == reverseLink) {
        continue;
        }
    else { /* for all other links emanating from prevNode, change the source to dummy */
        ChangeLinkSource(NetworkP, link, prevNode, dummyID);
        n--;
        }
    }
```

```
                /* add dummy link to point from dummy to prevNode - give it distance 0 */
                newLink = AddLinkToTopology(NetworkP, dummyID, prevNode, 0.0, FALSE);

                if (LinkDisjointOnly) {
                    reverseLink2 = GetReverseLink(NetworkP, newLink); /* prevNode to dummy */
                    NetworkP->Links[reverseLink2].Status = TRUE;
                    NetworkP->Links[reverseLink2].Length = SMALL;
                    }
                else if (CommonNodesAllowed) { /* common nodes OK if can't be avoided */
                    reverseLink2 = GetReverseLink(NetworkP, newLink);/* prevNode to dummy */
                    NetworkP->Links[reverseLink2].Status = TRUE;
                    /* add big amount to length to discourage its use */
                    NetworkP->Links[reverseLink2].Length = CommonNodePenalty;
                    }
                }
        }
}


void GenerateTwoRealPaths (NetworkTP NetworkP, PathTP TempPath1P, PathTP TempPath2P,
    PathTP RealPath1P, PathTP RealPath2P)
{
    int i, j, k, lastpos, lastj;
    USHORT link;
    BOOL exchange;

    /* to generate the shortest paths, look for overlapping sections in the two paths */
    RealPath1P->NumPathHops = RealPath2P->NumPathHops = 0;
    exchange = FALSE;
    lastpos = 0;
    for (i=0; i<TempPath1P->NumPathHops; i++) {
        for (j=0; j<TempPath2P->NumPathHops; j++) {
            if (GetReverseLink(NetworkP, TempPath2P->PathHops[j]) ==
            TempPath1P->PathHops[i]) {
                /* found an overlapping section; remove it and then exchange links */
                lastj = j;
                for (i++,j--; i<TempPath1P->NumPathHops, j>=0; i++, j--) {
                    if (GetReverseLink(NetworkP, TempPath2P->PathHops[j]) !=
                        TempPath1P->PathHops[i])
                            break;
                }
                j++; i--; /* go back to last position of overlap */
                for (k=lastpos; k<j; k++) {
                    link = TempPath2P->PathHops[k];
                    if (!exchange) {
                        RealPath2P->PathHops[RealPath2P->NumPathHops] = link;
                        RealPath2P->NumPathHops++;
```

```
                                  }
                          else {
                                  RealPath1P->PathHops[RealPath1P->NumPathHops] = link;
                                  RealPath1P->NumPathHops++;
                                  }
                          }
                  exchange = !exchange;
                  lastpos = lastj + 1;
                  break;
                  }
              }
      if (j == TempPath2P->NumPathHops) { /* not an overlapping link */
          if (!exchange) {
                  RealPath1P->PathHops[RealPath1P->NumPathHops] = TempPath1P->PathHops[i];
                  RealPath1P->NumPathHops++;
                  }
          else {
                  RealPath2P->PathHops[RealPath2P->NumPathHops] = TempPath1P->PathHops[i];
                  RealPath2P->NumPathHops++;
                  }
              }
          }
      for (k=lastpos; k<TempPath2P->NumPathHops; k++) {
          link = TempPath2P->PathHops[k];
          if (!exchange) {
              RealPath2P->PathHops[RealPath2P->NumPathHops] = link;
              RealPath2P->NumPathHops++;
              }
          else {
              RealPath1P->PathHops[RealPath1P->NumPathHops] = link;
              RealPath1P->NumPathHops++;
              }
          }
}


void CleanPath (NetworkTP NetworkP, USHORT DummyLinkThreshold, USHORT Source,
      USHORT Destination, PathTP NewPathP)
{
      USHORT h, linkID, numHops;

      numHops = 0;
      for (h=0; h<NewPathP->NumPathHops; h++) {
          linkID = NewPathP->PathHops[h];
          if (linkID >= DummyLinkThreshold)
              continue;
```

```
            NewPathP->PathHops[numHops] = linkID;
            numHops++;
            }
        NewPathP->NumPathHops = numHops;
}


USHORT AddLinkToTopology (NetworkTP NetworkP, USHORT Node1, USHORT Node2,
    double Length, BOOL ReverseLinkStatus)
{
    USHORT newLinkID, reverseLinkID;

    /* assumes links always added in pairs */
    /* if reverse link direction not needed, pass in its status as FALSE */

    if (NetworkP->NumLinks >= (MaxLinksWithDummies-1)) {
        printf("Need to increase MaxLinks\n");
        exit(-5);
        }

    newLinkID = NetworkP->NumLinks;
    NetworkP->NumLinks += 2;

    NetworkP->Links[newLinkID].LinkNode1 = Node1;
    NetworkP->Links[newLinkID].LinkNode2 = Node2;
    NetworkP->Links[newLinkID].Length = Length;
    NetworkP->Links[newLinkID].Status = TRUE;
    AdjustNodeInfoForNewLink(NetworkP, newLinkID);

    reverseLinkID = GetReverseLink(NetworkP, newLinkID);
    NetworkP->Links[reverseLinkID].LinkNode1 = Node2;
    NetworkP->Links[reverseLinkID].LinkNode2 = Node1;
    NetworkP->Links[reverseLinkID].Length = Length;
    NetworkP->Links[reverseLinkID].Status = ReverseLinkStatus;
    AdjustNodeInfoForNewLink(NetworkP, reverseLinkID);

    return(newLinkID);
}


void AdjustNodeInfoForNewLink (NetworkTP NetworkP, USHORT LinkID)
{
    USHORT node;

    node = NetworkP->Links[LinkID].LinkNode1;
    if (NetworkP->Nodes[node].NumOutgoingLinks >= MaxNodeDegree) {
        printf("Need to increase MaxNodeDegree\n");
        exit(-6);
        }
```

```
        NetworkP->Nodes[node].OutgoingLinks[NetworkP->Nodes[node].NumOutgoingLinks] =
            LinkID;
        NetworkP->Nodes[node].NumOutgoingLinks++;

        node = NetworkP->Links[LinkID].LinkNode2;
        if (NetworkP->Nodes[node].NumIncomingLinks >= MaxNodeDegree) {
            printf("Need to increase MaxNodeDegree\n");
            exit(-6);
            }
        NetworkP->Nodes[node].IncomingLinks[NetworkP->Nodes[node].NumIncomingLinks] =
            LinkID;
        NetworkP->Nodes[node].NumIncomingLinks++;
}


void ChangeLinkSource (NetworkTP NetworkP, USHORT LinkID, USHORT OldSource,
        USHORT NewSource)
{
        USHORT n;

        if (NetworkP->Links[LinkID].LinkNode1 != OldSource) {
            printf("Inconsistent\n");
            exit(-7);
            }
        if (NetworkP->Nodes[NewSource].NumOutgoingLinks >= MaxNodeDegree) {
            printf("Need to increase MaxNodeDegree\n");
            exit(-6);
            }

        NetworkP->Links[LinkID].LinkNode1 = NewSource;
        NetworkP->Nodes[NewSource].OutgoingLinks[NetworkP->Nodes[NewSource].
            NumOutgoingLinks] = LinkID;
        NetworkP->Nodes[NewSource].NumOutgoingLinks++;

        /* remove it from OldSource list*/
        for (n=0; n<NetworkP->Nodes[OldSource].NumOutgoingLinks; n++)
            if (NetworkP->Nodes[OldSource].OutgoingLinks[n] == LinkID) break;
        for (; n<NetworkP->Nodes[OldSource].NumOutgoingLinks-1; n++) {
            NetworkP->Nodes[OldSource].OutgoingLinks[n] =
                NetworkP->Nodes[OldSource].OutgoingLinks[n+1];
            }
        NetworkP->Nodes[OldSource].NumOutgoingLinks--;
}
```

```c
void ChangeLinkDestination (NetworkTP NetworkP, USHORT LinkID, USHORT OldDest,
    USHORT NewDest)
{
    USHORT n;

    if (NetworkP->Links[LinkID].LinkNode2 != OldDest) {
        printf("Inconsistent\n");
        exit(-8);
        }
    if (NetworkP->Nodes[NewDest].NumIncomingLinks >= MaxNodeDegree) {
        printf("Need to increase MaxNodeDegree\n");
        exit(-6);
        }

    NetworkP->Links[LinkID].LinkNode2 = NewDest;
    NetworkP->Nodes[NewDest].IncomingLinks[NetworkP->Nodes[NewDest].
        NumIncomingLinks] = LinkID;
    NetworkP->Nodes[NewDest].NumIncomingLinks++;

    /* remove it from OldDest list*/
    for (n=0; n<NetworkP->Nodes[OldDest].NumIncomingLinks; n++)
        if (NetworkP->Nodes[OldDest].IncomingLinks[n] == LinkID) break;

    for (; n<NetworkP->Nodes[OldDest].NumIncomingLinks-1; n++) {
        NetworkP->Nodes[OldDest].IncomingLinks[n] =
            NetworkP->Nodes[OldDest].IncomingLinks[n+1];
        }
    NetworkP->Nodes[OldDest].NumIncomingLinks--;
}

USHORT GetReverseLink (NetworkTP NetworkP, USHORT LinkID)
{
    /* Assumes links are always added in pairs */
    if (LinkID % 2 == 0)
        return(LinkID + 1);
    else return(LinkID - 1);
}
```

Abbreviations

ADM - Add/Drop Multiplexer
ANSI - American National Standards Institute
ASE - Amplified Spontaneous Emission
ASON - Automatically Switched Optical Network
ATM - Asynchronous Transfer Mode
AWG - Arrayed Waveguide Grating
BFS – Breadth First Search
BLSR - Bi-directional Line-Switched Ring
CapEx – Capital Expenditure
CDS – Connected Dominating Set
CSP - Constrained Shortest Path
dB – Decibel
DCE - Dynamic Channel Equalizer
DCF - Dispersion Compensating Fiber
DSE - Dynamic Spectral Equalizer
EDC - Electronic Dispersion Compensation
EDFA - Erbium Doped Fiber Amplifier
E-NNI - External Network-Network Interface
FEC - Forward Error Correction
FWM - Four-Wave Mixing
Gb/s – Gigabit per second (10^9 bits per second)
GC – Grooming Connection
GHz – Gigahertz
GMPLS - Generalized Multi-Protocol Label Switching
IETF - Internet Engineering Task Force
I-NNI – Internal Network-Network Interface
IP - Internet Protocol
ITU - International Telecommunication Union
JET - Just Enough Time
JIT - Just In Time
km – kilometer
MAC - Media Access Control

Mb/s – Megabit per second (10^6 bits per second)
MCP - Multi-Constrained Path
MEMS – Micro-electro-mechanical System
MLSE - Maximum Likelihood Sequence Estimation
ms - millisecond
NDSF - Non Dispersion-Shifted Fiber
NF - Noise Figure
nm – nanometer
NNI – Network-Network Interface
NRZ - Non-Return-to-Zero
NZ-DSF - Non-Zero Dispersion-Shifted Fiber
OADM - Optical Add/Drop Multiplexer
OADM-MD - Multi-Degree Optical Add/Drop Multiplexer
OAM - Operations, Administration, and Maintenance
OBS – Optical Burst Switching
OC - Optical Carrier
OChSPRing - Optical-Channel Shared Protection Ring
O-E-O – Optical-Electrical-Optical
OFS - Optical Flow Switching
OIF - Optical Internetworking Forum
OMS-SPRing - Optical Multiplex Section Shared Protection Ring
OOK - On-Off Keying
OpEx – Operational Expenditure
OPM - Optical Performance Monitor
OPS – Optical Packet Switching
OSC - Optical Supervisory Channel
OSNR - Optical-Signal-to-Noise-Ratio
OTN - Optical Transport Network
OTU - Optical channel Transport Unit
OXC – Optical Cross-Connect
PCE - Path Computation Element
PDL - Polarization Dependent Loss
PIC - Photonic Integrated Circuit
PMD - Polarization-Mode Dispersion
PON - Passive Optical Network
ps - picosecond (10^{-12} second)
RFC - Request for Comments
RFI – Request for Information
RFP – Request for Proposal
ROADM – Reconfigurable Optical Add/Drop Multiplexer
RSP - Restricted Shortest Path
RWA - Routing and Wavelength Assignment
RZ - Return-to-Zero
SDH - Synchronous Digital Hierarchy
SLA - Service Level Agreement

SONET - Synchronous Optical Network
SPDP - Shortest Pair of Disjoint Paths
SPM - Self-Phase Modulation
SPRing – Shared Protection Ring
SRG - Shared Risk Group
SRLB - Selective Randomized Load Balancing
SRLG - Shared Risk Link Group
STM - Synchronous Transport Module
STS - Synchronous Transport Signal
Tb/s – Terabit per second (10^{12} bits per second)
THz – Terahertz
TWIN - Time-Domain Wavelength Interleaved Networking
TxRx – Transmitter/Receiver Card (Transponder)
UNI - User-Network Interface
VCAT - Virtual Concatenation
WDM - Wavelength Division Multiplexing
WGR – Wavelength Grating Router
WSS - Wavelength-Selective Switch
WSXC - Wavelength-Selective Cross-connect
XPM - Cross-phase Modulation

Bibliography

[ADHN01] G. P. Austin, B. T. Doshi, C. J. Hunt, R. Nagarajan, and M. A. Qureshi, "Fast, scalable, and distributed restoration in general mesh optical networks," *Bell Labs Technical Journal*, pp. 67-81, Jan.-Jun. 2001.

[ABGL01] D. Awduche, L. Berger, D. Gan, T. Li, V. Srinivasan, and G. Swallow, "RSVP-TE: Extensions to RSVP for LSP Tunnels," Internet Engineering Task Force, Request for Comments (RFC) 3209, Dec. 2001.

[BaGe06] R. Batchellor and O. Gerstel, "Cost effective architectures for core transport networks," *Proceedings, Optical Fiber Communication/National Fiber Optic Engineers Conference (OFC/NFOEC'06)*, Anaheim, CA, Mar. 5-10, 2006, Paper PDP42.

[BaHu96] R. A. Barry and P. A. Humblet, "Models of blocking probability in all-optical networks with and without wavelength changers," *IEEE Journal of Selected Areas in Communications*, vol. 14, no. 5, pp. 858-867, Jun. 1996.

[BaKi02] P. Bayvel and R. Killey, "Nonlinear optical effects in WDM transmission," in *Optical Fiber Telecommunications IV B*, I. Kaminow and T. Li, Editors, San Diego: Academic Press, 2002, pp. 611-641.

[BaLe02] N. Barakat and A. Leon-Garcia, "An analytic model for predicting the locations and frequencies of 3R regenerations in all-optical wavelength-routed WDM networks," *Proceedings, IEEE International Conference on Communications (ICC'02)*, New York, NY, Apr. 28-May 2, 2002, vol. 5, pp. 2812-2816.

[BCRV06] G. Bernstein, D. Caviglia, R. Rabbat, and H. Van Helvoort, "VCAT-LCAS in a clamshell," *IEEE Communications Magazine*, vol. 44, no. 5, pp. 34-36, May 2006.

[Berg03] L. Berger, Editor, "Generalized Multi-Protocol Label Switching (GMPLS) Signaling Functional Description," Internet Engineering Task Force, Request for Comments (RFC) 3471, Jan. 2003.

[BeRS03] G. Bernstein, B. Rajagopalan, and D. Saha, *Optical Network Control: Architecture, Protocols, and Standards*, Reading, MA: Addison-Wesley Professional, 2003.

[Bhan99] R. Bhandari, *Survivable Networks: Algorithms for Diverse Routing*, Boston, MA: Kluwer Academic Publishers, 1999.

[BhSF01] N. M. Bhide, K. M. Sivalingam, and T. Fabry-Asztalos, "Routing mechanisms employing adaptive weight functions for shortest path routing in optical WDM networks," *Photonic Network Communications*, vol. 3, no. 3, pp. 227-236, Jul. 2001.

[BLRC02] E. Bouillet, J.-F. Labourdette, R. Ramamurthy, and S. Chaudhuri, "Enhanced algorithm cost model to control tradeoffs in provisioning shared mesh restored lightpaths," *Proceedings, Optical Fiber Communication (OFC'02)*, Anaheim, CA, Mar. 17-22, 2002, Paper ThW2.

[Blum04] D. J. Blumenthal, "Optical packet switching," *Proceedings, 17th Annual Meeting of the IEEE LEOS*, Puerto Rico, Nov. 7-11, 2004, Paper ThU1.

[BoUh98] A. Boroujerdi and J. Uhlmann, "An efficient algorithm for computing least cost paths with turn constraints," *Information Processing Letters*, vol. 67, pp. 317-321, 1998.

[BSAL02] A. Boskovic, M. Sharma, N. Antoniades, and M. Lee, "Broadcast and select OADM nodes application and performance trade-offs," *Proceedings, Optical Fiber Communication (OFC'02)*, Anaheim, CA, Mar. 17-22, 2002, Paper TuX2.

[BuWW03] P. Bullock, C. Ward, and Q. Wang, "Optimizing wavelength grouping granularity for optical add-drop network architectures," *Proceedings, Optical Fiber Communication (OFC'03)*, Atlanta, GA, Mar. 23-28, 2003, Paper WH2.

[CaAQ04] X. Cao, V. Anand, and C. Qiao, "Multi-layer versus single-layer optical cross-connect architectures for waveband switching," *Proceedings, IEEE INFOCOM 2004*, Hong Kong, Mar. 7-11, 2004, vol. 3, pp. 1830-1840.

[CaCH04] H. S. Carrer, D. E. Crivelli, and M. R. Hueda, "Maximum likelihood sequence estimation receivers for DWDM lightwave systems," *Proceedings, IEEE Global Telecommunications Conference (GLOBECOM'04)*, Dallas, TX, Nov. 29-Dec. 3, 2004, vol. 2, pp. 1005–1010.

[ChAN03] C. Chigan, G. W. Atkinson, and R. Nagarajan, "Cost effectiveness of joint multilayer protection in packet-over-optical networks," *Journal of Lightwave Technology*, vol. 21, no. 11, pp. 2694-2704, Nov. 2003.

[ChCF04] T. Y. Chow, F. Chudak, and A. M. Ffrench "Fast optical layer mesh protection using pre-cross-connected trails," *IEEE/ACM Transactions on Networking*, vol. 12, no. 3, pp. 539-548, Jun. 2004.

[ChGn06] S. Chandrasekhar and A. H. Gnauck, "Performance of MLSE receiver in a dispersion-managed multispan experiment at 10.7 Gb/s under nonlinear transmission," *IEEE Photonics Technology Letters*, vol. 18, no. 23, pp. 2448–2450, Dec. 1, 2006.

[ChLH06] A. L. Chiu, G. Li, and D.-M. Hwang, "New problems on wavelength assignment in ULH networks," *Proceedings, Optical Fiber Communication/National Fiber Optic Engineers Conference (OFC/NFOEC'06)*, Anaheim, CA, Mar. 5-10, 2006, Paper NThH2.

[ChQY04] Y. Chen, C. Qiao, and X. Yu, "Optical burst switching: A new area in optical networking research," *IEEE Network*, vol. 18, no. 3, pp. 16-23, May/Jun. 2004.

[ChSc07] M. W. Chbat and H.-J. Schmidtke, "Falling boundaries from metro to ULH optical transport equipment," *Proceedings, Optical Fiber Communication/National Fiber Optic Engineers Conference (OFC/NFOEC'07)*, Anaheim, CA, Mar. 25-29, 2007, Paper NTuA3.

[ChWM06] V. W. S. Chan, G. Weichenberg, and M. Médard, "Optical flow switching," *Workshop on Optical Burst Switching (WOBS)*, San Jose, CA, Oct. 2006.

[ChYu94] K. Chan and T. P. Yum, "Analysis of least congested path routing in WDM lightwave networks," *Proceedings, IEEE INFOCOM 1994*, Toronto, Ontario, Jun. 12-16, 1994, vol. 2, pp. 962-969.

[ChYu03] A. Chiu and C. Yu, "Economic benefits of transparent OXC networks as compared to long systems with OADMs," *Proceedings, Optical Fiber Communication (OFC'03)*, Atlanta, GA, Mar. 23-28, 2003, Paper WQ2.

[CMSG04] T. J. Carpenter, R. C. Menendez, D. F. Shallcross, J. W. Gannett, J. Jackel, and A. C. Von Lehmen, "Cost-conscious impairment-aware routing," *Proceedings, Optical Fiber Communication (OFC'04)*, Los Angeles, CA, Feb. 22-27, 2004, Paper MF88.

[CoLR90] T. H. Cormen, C. E. Leiserson, and R. L. Rivest, *Introduction to Algorithms*, Cambridge, MA: MIT Press, 1990.

[Conr02] J. Conradi, "Bandwidth-efficient modulation formats for digital fiber transmission systems," in *Optical Fiber Telecommunications IV B*, I. Kaminow and T. Li, Editors, San Diego: Academic Press, 2002, pp. 862-901.

[CSGJ03] T. Carpenter, D. Shallcross, J. Gannett, J. Jackel, and A. Von Lehmen, "Maximizing the transparency advantage in optical networks," *Proceedings, Optical Fiber Communication (OFC'03)*, Atlanta, GA, Mar. 23-28, 2003, Paper FA2.

[CXBT02] N. Chi, L. Xu, K. S. Berg, T. Tokle, and P. Jeppesen, "All-optical wavelength conversion and multichannel 2R regeneration based on highly nonlinear dispersion-imbalanced loop mirror," *IEEE Photonics Technology Letters*, vol. 14, no. 11, pp. 1581-1583, Nov. 2002.

[DGGM02] N. G. Duffield, P. Goyal, A. Greenberg, P. Mishra, K. K. Ramakrishnan, and J. E. van der Merwe, "Resource management with hoses: Point-to-cloud services for virtual private networks," *IEEE/ACM Transactions on Networking*, vol. 10, no. 5, pp. 679-692, Oct. 2002.

[DoOk06] C. R. Doerr and K. Okamoto, "Advances in silica planar lightwave circuits," *Journal of Lightwave Technology*, vol. 24, no. 12, pp. 4763-4789, Dec. 2006.

[DuRo02] R. Dutta and G. N. Rouskas, "Traffic grooming in WDM networks: Past and future," *IEEE Network,* vol. 16, no. 6, pp. 46-56, Nov./Dec. 2002.

[EBRL02] G. Ellinas, E. Bouillet, R. Ramamurthy, J.-F. Labourdette, S. Chaudhuri, and K. Bala, "Restoration in layered architectures with a WDM mesh optical layer," *International Engineering Consortium (IEC) Annual Review of Communications*, vol. 55, Jun. 2002.

[EBRL03] G. Ellinas, E. Bouillet, R. Ramamurthy, J.-F. Labourdette, S. Chaudhuri, and K. Bala, "Routing and restoration architectures in mesh optical networks," *Optical Networks Magazine*, vol. 4, no. 1, pp. 91-106, Jan./Feb. 2003.

[ElBa06] T. S. El-Bawab, Editor, *Optical Switching*, New York, NY: Springer, 2006.

[ElMW06] A. Elwalid, D. Mitra, and Q. Wang, "Distributed nonlinear integer optimization for data-optical internetworking," *IEEE Journal on Selected Areas in Communications*, vol. 24, no. 8, pp. 1502-1513, Aug. 2006.

[EMSW03] A. Elwalid, D. Mitra, I. Saniee, and I. Widjaja, "Routing and protection in GMPLS networks: From shortest paths to optimized designs," *Journal of Lightwave Technology,* vol. 21, no. 11, pp. 2828-2838, Nov. 2003.

[Epps94] D. Eppstein, "Finding the *k* shortest paths," *Proceedings, 35th Annual Symposium on Foundations of Computer Science*, Santa Fe, NM, Nov. 20-22, 1994, pp. 154-165.

[FaVA06] J. Farrel, J.-P. Vasseur, and J. Ash, "A Path Computation Element (PCE)-Based Architecture," Internet Engineering Task Force, Request for Comments (RFC) 4655, Aug. 2006.

[FoTC97] F. Forghieri, R. W. Tkach, and A. R. Chraplyvy, "Fiber nonlinearities and their impact on transmission systems," in *Optical Fiber Telecommunications III A*, I. Kaminow and T. Koch, Editors, San Diego: Academic Press, 1997, pp. 196-254.

[GaJo79] M. R. Garey and D. S. Johnson, *Computers and Intractability: A Guide to the Theory of NP-Completeness*, New York, NY: W.H. Freeman and Co., 1979.

[GeRa00] O. Gerstel and R. Ramaswami, "Optical layer survivability – an implementation perspective," *IEEE Journal on Selected Areas in Communications*, vol. 18, no. 10, pp. 1885-1899, Oct. 2000.

[GeRa04] O. Gerstel and H. Raza, "Predeployment of resources in agile photonic networks," *Journal of Lightwave Technology*, vol. 22, no. 10, pp. 2236-2244, Oct. 2004.

[GeRS98] O. Gerstel, R. Ramaswami, and G. Sasaki, "Cost effective traffic grooming in WDM rings," *Proceedings, IEEE INFOCOM 1998,* San Francisco, CA, Mar. 29-Apr. 2, 1998, pp. 69-77.

[GoLY02] M. Goyal, G. Li, and J. Yates, "Shared mesh restoration: a simulation study," *Proceedings, Optical Fiber Communication (OFC'02)*, Anaheim, CA, Mar. 17-22, 2002, Paper ThO2.

[Gora02] W. J. Goralski, *SONET/SDH*, Third Edition, New York, NY: McGraw-Hill, 2002.

[Grov03] W. Grover, *Mesh-based Survivable Transport Networks: Options and Strategies for Optical, MPLS, SONET and ATM Networking*, Upper Saddle River, NJ: Prentice-Hall, 2003.

[GrSt98] W. Grover and D. Stamatelakis, "Cycle-oriented distributed preconfiguration: Ring-like speed with mesh-like capacity for self-planning network restoration," *Proceedings, IEEE International Conference on Communications (ICC'98)*, Atlanta, GA, Jun. 7-11, 1998, vol. 1, pp. 537–543.

[GuCh03] A. Gumaste and I. Chlamtac, "Mesh implementation of light-trails: A solution to IP centric communication," *Proceedings, International Conference on Computer Communication and Networks (ICCCN'03),* Dallas, TX, Oct. 20-22, 2003, pp. 178-183.

[GuKh98] S. Guha and S. Khuller, "Approximation algorithms for connected dominating sets," *Algorithmica*, vol. 20, no. 4, 1998, pp. 374-387.

[GuOr99] R. A. Guerin and A. Orda, "QoS routing in networks with inaccurate information: Theory and algorithms," *IEEE/ACM Transactions on Networking,* vol. 7, no. 3, pp. 350-364, Jun. 1999.

[GuOr02] R. Guerin and A. Orda, "Computing shortest paths for any number of hops," *IEEE/ACM Transactions on Networking,* vol. 10, no. 5, pp. 613-620, Oct. 2002.

[GuPM03] K. P. Gummadi, M. J. Pradeep, and C. S. R. Murthy, "An efficient primary-segmented backup scheme for dependable real-time communication in multihop networks," *IEEE/ACM Transactions on Networking,* vol. 11, no. 1, pp. 81-94, Feb. 2003.

[HeBr06] J. He and M. Brandt-Pearce, "RWA using wavelength ordering for crosstalk limited networks," *Proceedings, Optical Fiber Communication/National Fiber Optic Engineers Conference (OFC/NFOEC'06),* Anaheim, CA, Mar. 5-10, 2006, Paper OFG4.

[HeMS03] J. Hershberger, M. Maxel, and S. Suri, "Finding the k shortest simple paths: A new algorithm and its implementation," *Proceedings, Fifth Workshop on Algorithm Engineering and Experiments,* Baltimore, MD, Jan. 11, 2003, pp. 26-36.

[HoMo02] P. H. Ho and H. T. Mouftah, "A framework for service-guaranteed shared protection in WDM mesh networks," *IEEE Communications Magazine,* vol. 40, no. 2, pp. 97-103, Feb. 2002.

[HPWY07] N. J. A. Harvey, M. Patrascu, Y. Wen, S. Yekhanin, and V. W. S. Chan "Non-adaptive fault diagnosis for all-optical networks via combinatorial group testing on graphs," *Proceedings, IEEE INFOCOM'07,* Anchorage, AK, May 6-12, 2007, pp. 697-705.

[HSKO99] K. Harada, K. Shimizu, T. Kudou, and T. Ozeki, "Hierarchical optical path cross-connect systems for large scale WDM networks," *Proceedings, Optical Fiber Communication (OFC'99),* San Diego, CA, Feb. 21-26, 1999, Paper WM55.

[HuDu07] S. Huang and R. Dutta, "Dynamic traffic grooming: The changing role of traffic grooming," *IEEE Communications Surveys and Tutorials,* 1st Quarter 2007, vol. 9, no. 1, pp. 32-49.

[HXGB05] Y.-K. Huang, L. Xu, I. Glesk, V. Baby, B. Li, and P. R. Prucnal, "Simultaneous all-optical 3R regeneration of multiple WDM channels," *Proceedings, 18th Annual Meeting of the IEEE LEOS,* Sydney, Australia, Oct. 23-27, 2005, Paper MM2.

[IGKV03] R. Izmailov, S. Ganguly, V. Kleptsyn, and A. C. Varsou, "Non-uniform waveband hierarchy in hybrid optical networks," *Proceedings, IEEE INFOCOM 2003,* San Francisco, CA, Mar. 30-Apr. 3, 2003, vol. 2, pp.1344-1354.

[IrMG98] R. R. Iraschko, M. H. MacGregor, and W. D. Grover, "Optimal capacity placement for path restoration in STM or ATM mesh-survivable networks," *IEEE/ACM Transactions on Networking,* vol. 6, no. 3, pp. 325-336, Jun. 1998.

[ITU01] International Telecommunication Union, Architecture of Optical Transport Networks, ITU-T Rec. G.872, Nov. 2001.

[ITU03] International Telecommunication Union, Interfaces for the Optical Transport Network (OTN), ITU-T Rec. G.709, Mar. 2003.

[ITU06] International Telecommunication Union, Architecture for the Automatically Switched Optical Network (ASON), ITU-T Rec. G.8080/Y.1304, Jun. 2006.

[JoFN04] G. L. Jones, W. Forysiak, and J. H. B. Nijhof, "Economic benefits of all-optical cross connects and multi-haul DWDM systems for European national networks," *Proceedings, Optical Fiber Communication (OFC'04),* Los Angeles, CA, Feb. 22-27, 2004, Paper WH2.

[KaAr04] E. Karasan and M. Arisoylu, "Design of translucent optical networks: Partitioning and restoration," *Photonic Network Communications,* vol. 8, no. 2, pp. 209-221, Mar. 2004.

[KaAy98] E. Karasan and E. Ayanoglu, "Effects of wavelength routing and selection algorithms on wavelength conversion gain in WDM optical networks," *IEEE/ACM Transactions on Networking,* vol. 6, no. 2, pp. 186-196, Apr. 1998.

[KaKL00] K. Kar, M. Kodialam, and T. V. Lakshman, "Minimum interference routing of bandwidth guaranteed tunnels with MPLS traffic engineering applications," *IEEE Journal on Selected Areas in Communications,* vol. 18, no. 12, pp. 2566-2579, Dec. 2000.

[KaKL03] K. Kar, M. Kodialam, and T. V. Lakshman, "Routing restorable bandwidth guaranteed connections using maximum 2-route flows," *IEEE/ACM Transactions on Networking*, vol. 11, no. 5, pp. 772 – 781, Oct. 2003.

[KaKY03] D. Katz, K. Kompella, and D. Yeung, "Traffic Engineering (TE) Extensions to OSPF Version 2," Internet Engineering Task Force, Request for Comments (RFC) 3630, Sep. 2003.

[KaSG04] G. S. Kanter, A. K. Samal, and A. Gandhi, "Electronic dispersion compensation for extended reach," *Proceedings, Optical Fiber Communication (OFC'04)*, Los Angeles, CA, Feb. 22-27, 2004, Paper TuG1.

[KBBE04] D. C. Kilper, R. Bach, D. J. Blumenthal, D. Einstein, T. Landolsi, L. Ostar, M. Preiss, and A. E. Willner, "Optical performance monitoring," *Journal of Lightwave Technology*, vol. 22, no. 1, pp. 294-304, Jan. 2004.

[KiLu03] S. Kim and S. Lumetta, "Evaluation of protection reconfiguration for multiple failures in WDM mesh networks," *Proceedings, Optical Fiber Communication (OFC'03)*, Atlanta, GA, Mar. 23-28, 2003, Paper TuI7.

[KKKV04] F. Kuipers, T. Korkmaz, M. Krunz, and P. Van Mieghem, "Performance evaluation of constraint-based path selection algorithms," *IEEE Network*, vol. 18, no. 5, pp. 16-23, Sep./Oct. 2004.

[KNSZ04] P. M. Krummrich, R. E. Neuhauser, H.-J. Schmidtke, H. Zech, and M. Birk, "Compensation of Raman transients in optical networks," *Proceedings, Optical Fiber Communication (OFC'04)*, Los Angeles, CA, Feb. 22-27, 2004, Paper MF82.

[KoGr05] A. Kodian and W. D. Grover, "Failure-independent path-protecting p-cycles: Efficient and simple fully preconnected optical-path protection," *Journal of Lightwave Technology*, vol. 23, no. 10, pp. 3241-3259, Oct. 2005.

[KoKr01] T. Korkmaz and M. Krunz, "Multi-constrained optimal path selection," *Proceedings, IEEE INFOCOM 2001*, Anchorage, AK, Apr. 22-26, 2001, vol. 2, pp. 834–843.

[KoLa00] M. Kodialam and T. V. Lakshman, "Dynamic routing of bandwidth guaranteed tunnels with restoration," *Proceedings, IEEE INFOCOM 2000*, Tel-Aviv, Israel, Mar. 26-30, 2000, vol. 2, pp. 902-911.

[KoMB81] L. Kou, G. Markowsky, and L. Berman, "A fast algorithm for Steiner trees," *Acta Informatica*, vol. 15, no. 2, pp. 141-145, Jun. 1981.

[KTMT05] P. Kulkarni, A. Tzanakaki, C. M. Machuka, and I. Tomkos, "Benefits of Q-factor based routing in WDM metro networks," *Proceedings, European Conference on Optical Communication (ECOC'05)*, Glasgow, Scotland, Sep. 25-29, 2005, vol. 4, pp. 981-982.

[Kurt93] C. Kurtzke, "Suppression of fiber nonlinearities by appropriate dispersion management," *IEEE Photonics Technology Letters*, vol. 5, no. 10, pp. 1250-1253, Oct. 1993.

[KYJH06] V. Kaman, S. Yuan, O. Jerphagnon, R. J. Helkey, and J. E. Bowers, "Comparison of wavelength-selective cross-connect architectures for reconfigurable all-optical networks," *Proceedings, International Conference on Photonics in Switching*, Crete, Greece, Oct. 16-18, 2006, pp. 1-3.

[KZJP04] V. Kaman, X. Zheng, O. Jerphagnon, C. Pusarla, R. J. Helkey, and J. E. Bowers, "A cyclic MUX–DMUX photonic cross-connect architecture for transparent waveband optical networks," *IEEE Photonics Technology Letters*, vol. 16, no. 2, pp. 638-640, Feb. 2004.

[LaAa87] P. J. M. van Laarhoven and E. H. L. Aarts, *Simulated Annealing: Theory and Applications*, Boston, MA: D. Reidel Publishing Co., 1987.

[LaKe91] A. M. Law and W. D. Kelton, *Simulation Modeling and Analysis*, Second Edition, New York, NY: McGraw-Hill, Inc., 1991.

[LeJC04] J. Leuthold, J. Jaques, and S. Cabot, "All-optical wavelength conversion and regeneration," *Proceedings, Optical Fiber Communication (OFC'04)*, Los Angeles, CA, Feb. 22-27, 2004, Paper WN1.

[LiCS05] G. Li, A. L. Chiu, and J. Strand, "Failure recovery in all-optical ULH networks," *5th International Workshop on Design of Reliable Communication Networks (DRCN'05)*, Island of Ischia, Italy, Oct. 16-19, 2005.

[Lin06] C. Lin, Editor, *Broadband Optical Access Networks and Fiber-to-the-Home: Systems Technologies and Deployment Strategies*, West Sussex, England, John Wiley & Sons, 2006.

[LiRa97] C.-S. Li and R. Ramaswami, "Automatic fault detection, isolation, and recovery in transparent all-optical networks," *Journal of Lightwave Technology*, vol. 15, no. 10, pp. 1784-1793, Oct. 1997.

[LiRa01] G. Liu and K. G. Ramakrishnan, "A*Prune: An algorithm for finding K shortest paths subject to multiple constraints," *Proceedings, IEEE INFOCOM 2001*, Anchorage, AK, Apr. 22-26, 2001, vol. 2, pp. 743-749.

[LiVe02] R. Lingampalli and P. Vengalam, "Effect of wavelength and waveband grooming on all-optical networks with single layer photonic switching," *Proceedings, Optical Fiber Communication (OFC'02)*, Anaheim, CA, Mar. 17-22, 2002, Paper ThP4.

[LiWM07] W. Lin, R. S. Wolff, and B. Mumey, "A markov-based reservation algorithm for wavelength assignment in all-optical networks," *Journal of Lightwave Technology*, vol. 25, no. 7, pp. 1676-1683, Jul. 2007.

[LLBB03] O. Leclerc, B. Lavigne, E. Balmefrezol, P. Brindel, L. Pierre, D. Rouvillain, and F. Seguineau, "Optical regeneration at 40 Gb/s and beyond," *Journal of Lightwave Technology*, vol. 21, no. 11, pp. 2779-2790, Nov. 2003.

[LSTN05] M.-J. Li, M. J. Soulliere, D. J. Tebben, L. Nederlof, M. D. Vaughn, and R. E. Wagner, "Transparent optical protection ring architectures and applications," *Journal of Lightwave Technology*, vol. 23, no. 10, pp. 3388-3403, Oct. 2005.

[LWYD07] G. Li, D. Wang, J. Yates, R. Doverspike, and C. Kalmanek, "IP over optical cross-connect architectures," *IEEE Communications Magazine*, vol. 45, no. 2, pp. 34-39, Feb. 2007.

[MaLe03] B. Manseur and J. Leung, "Comparative analysis of network reliability and optical reach," *National Fiber Optic Engineers Conference (NFOEC'03)*, Orlando, FL, Sep. 7–11, 2003.

[Mann04] E. Mannie, Editor, "Generalized Multi-Protocol Label Switching (GMPLS) Architecture," Internet Engineering Task Force, Request for Comments (RFC) 3945, Oct. 04.

[Maro05] D. M. Marom, et al., "Wavelength-selective 1xK switches using free-space optics and MEMS micromirrors: Theory, design, and implementation," *Journal of Lightwave Technology*, vol. 23, no. 4, pp. 1620-1630, Apr. 2005.

[MaTo03] C. M. Machuca and I. Tomkos, "Failure detection for secure optical networks," *Proceedings, International Conference on Transparent Optical Networks (ICTON'03)*, Warsaw, Poland, Jun. 29-Jul. 3, 2003, pp. 70-75.

[Mats05] M. Matsumoto, "Regeneration of RZ-DPSK signals by fiber-based all-optical regenerators," *IEEE Photonics Technology Letters*, vol. 17, no. 5, pp. 1055-1057, May 2005.

[MaTT05] C. Mas, I. Tomkos, and O. K. Tonguz, "Failure location algorithm for transparent optical networks," *IEEE Journal on Selected Areas in Communications*, vol. 23, no. 8, pp. 1508-1519, Aug. 2005.

[MDLP05] S. Melle, R. Dodd, C. Liou, D. Perkins, M. Sosa, and M. Yin, "Network planning and economic analysis of an innovative new optical transport architecture: The digital optical network," *National Fiber Optic Engineers Conference (NFOEC'05)*, Anaheim, CA, Mar. 6-11, 2005, Paper NTuA1.

[MeCS98] M. Médard, S. R. Chinn, and P. Saengudomlert, "Attack detection in all-optical networks," *Proceedings, Optical Fiber Communication (OFC'98)*, San Jose, CA, Feb. 22-27, 1998, Paper ThD4.

[MFBG99] M. Médard, S. G. Finn, R. A. Barry, and R. G. Gallager, "Redundant trees for pre-planned recovery in arbitrary vertex-redundant or edge-redundant graphs," *IEEE/ACM Transactions on Networking,* vol. 7, no. 5, pp. 641-652, Oct. 1999.

[MMBF97] M. Médard, D. Marquis, R. A. Barry, and S. G. Finn, "Security issues in all-optical networks," *IEEE Network*, vol. 11, no. 3, pp. 42-48, May-Jun. 1997.

[MMMT03] S. Mechels, L. Muller, G. D. Morley, and D. Tillett, "1D MEMS-based wavelength switching subsystem," *IEEE Communications Magazine*, vol. 41, no. 3, pp. 88-94, Mar. 2003.

[MoAz98] A. Mokhtar and M. Azizoglu, "Adaptive wavelength routing in all-optical networks," *IEEE/ACM Transactions on Networking,* vol. 6, no. 2, pp. 197-206, Apr. 1998.

[MoBB04] A. Mokhtar, L. Benmohamed, and M. Bortz, "OXC port dimensioning strategies in optical networks – a nodal perspective," *IEEE Communications Letters*, vol. 8, no. 5, pp. 283-285, May 2004.

[MoLS02] R. Monnard, H. K. Lee, and A. Srivastava, "Suppressing amplifier transients in lightwave systems," *Proceedings, IEEE/LEOS Summer Topicals*, Mont Tremblant, Quebec, Jul. 15-17, 2002, Paper WE3.

[MORC05] J. McNicol, M. O'Sullivan, K. Roberts, A. Comeau, D. McGhan, and L. Strawczynski, "Electrical domain compensation of optical dispersion", *Proceedings, Optical Fiber Communication (OFC'05)*, Anaheim, CA, Mar. 6-11, 2005, Paper OThJ3.

[MSSD03] X. Masip-Bruin, S. Sànchez-López, J. Solé-Pareta, J. Domingo-Pascual, and D. Colle, "Routing and wavelength assignment under inaccurate routing information in networks with sparse and limited wavelength conversion," *Proceedings, IEEE Global Telecommunications Conference (Globecom'03)*, San Francisco, CA, Dec. 1-5, 2003, vol. 5, pp. 2575-2579.

[Mukh06] B. Mukherjee, *Optical WDM Networks*, New York, NY: Springer, 2006.

[NoVD01] L. Noirie, M. Vigoureux, and E. Dotaro "Impact of intermediate traffic grouping on the dimensioning of multi-granularity optical networks," *Proceedings, Optical Fiber Communication (OFC'01)*, Anaheim, CA, Mar. 17-22, 2001, Paper TuG3.

[OIF04] Optical Internetworking Forum, "User Network Interface (UNI) 1.0 Signaling Specification, Release 2," Feb. 27, 2004.

[Okam98] K. Okamoto, "Tutorial: Fundamentals, technology and applications of AWG's," *Proceedings, European Conference on Optical Communication (ECOC'98)*, Madrid, Spain, Sep. 20-24, 1998, pp. 35–37.

[OuMu05] C. Ou and B. Mukherjee, *Survivable Optical WDM Networks*, New York, NY: Springer, 2005.

[OzPJ03] T. Ozugur, M.-A. Park, and J. P. Jue, "Label prioritization in GMPLS-centric all-optical networks," *Proceedings, IEEE International Conference on Communications (ICC'03)*, Anchorage, AK, May 11-15, 2003, vol. 2 , pp. 1283-1287.

[OZSZ04] C. Ou, H. Zang, N. K. Singhal, K. Zhu, L. H. Sahasrabuddhe, R. A. MacDonald, and B. Mukherjee, "Subpath protection for scalability and fast recovery in optical WDM mesh networks," *IEEE Journal on Selected Areas in Communications,* vol. 22, no. 9, pp.1859-1875, Nov. 2004.

[OZZS03] C. Ou, K. Zhu, H. Zang, L. H. Sahasrabuddhe, and B. Mukherjee, "Traffic grooming for survivable WDM networks - shared protection," *IEEE Journal on Selected Areas in Communications*, vol. 21, no. 9, pp. 1367-1383, Nov. 2003.

[PaPP03] G. I. Papadimitriou, C. Papazoglou, and A. S. Pomportsis, "Optical switching: Switch fabrics, techniques, and architectures," *Journal of Lightwave Technology*, vol. 21, no. 2, pp. 384-405, Feb. 2003.

[PCHN03] A. R. Pratt, B. Charbonnier, P. Harper, D. Nesset, B. K. Nayar, and N. J. Doran, "40 x 10.7 Gbit/s DWDM transmission over a meshed ULH network with dynamically reconfigurable optical crossconnects," *Proceedings, Optical Fiber Communication (OFC'03)*, Atlanta, GA, Mar. 23-28, 2003, Paper PD9.

[PDCS06] M. Pickavet, P. Demeester, D. Colle, D. Staessens, B. Puype, L. Depré, and I. Lievens, "Recovery in multilayer optical networks," *Journal of Lightwave Technology*, vol. 24, no. 1, pp. 122-134, Jan. 2006.

[PrAS05] R. G. Prinz, A. Autenrieth, and D. A. Schupke, "Dual failure protection in multilayer networks based on overlay or augmented model," *5th International Workshop on Design of Reliable Communication Networks (DRCN'05)*, Island of Ischia, Italy, Oct. 16-19, 2005.

[QiXu02] C. Qiao and D. Xu, "Distributed partial information management (DPIM) schemes for survivable networks - Part I," *Proceedings, IEEE INFOCOM 2002*, New York, NY, Jun. 23-27, 2002, vol. 1, pp. 302–311.

[QiYo99] C. Qiao and M. Yoo, "Optical Burst Switching (OBS) – A new paradigm for an optical internet," *Journal of High Speed Networks*, vol. 8, no. 1, pp. 69-84, Jan. 1999.

[RaEl05] T. Rahman and G. Ellinas, "Protection of Multicast Sessions in WDM Mesh Optical Networks," *Proceedings, Optical Fiber Communication (OFC'05)*, Anaheim, CA, Mar. 6-11, 2005, Paper OTuK5.

[RaSi01] R. Ramaswami and K. N. Sivarajan, *Optical Networks: A Practical Perspective*, 2nd Edition, San Francisco, CA: Morgan Kaufmann Publishers, 2001.

[ReLG06] R. Rejeb, M. S. Leeson, and R. J. Green, "Fault and attack management in all-optical networks," *IEEE Communications Magazine,* vol. 44, no. 11, pp. 79-86, Nov. 2006.

[RLAC03] R. Ramamurthy, J.-F. Labourdette, A. Akyamac, and S. Chaudhuri,, "Limited sharing on protection channels in mesh optical networks," *Proceedings, Optical Fiber Communication (OFC'03)*, Atlanta, GA, Mar. 23-28, 2003, Paper TuI3.

[RoSt02] K. Rottwitt and A. Stentz, "Raman amplification in lightwave communication systems," in *Optical Fiber Telecommunications IV A*, I. Kaminow and T. Li, Editors, San Diego: Academic Press, 2002, pp. 213-258.

[Sala02] D. Y. Al-Salameh, "Optical switching in transport networks: Applications, requirements, architectures, technologies and solutions," in *Optical Fiber Telecommunications IV A*, I. Kaminow and T. Li, Editors, San Diego: Academic Press, 2002, pp. 295-373.

[Sale98a] A. A. M. Saleh, "Islands of transparency – an emerging reality in multiwavelength optical networking," *Proceedings, IEEE/LEOS Summer Topical Meeting on Broadband Optical Networks and Technologies*, Monterey, CA, Jul. 20-24, 1998, p. 36.

[Sale98b] A. A. M. Saleh, "Short- and long-term options for broadband access to homes and businesses," *Conference on the Internet: Next Generation and Beyond*, Cambridge, MA, Nov. 1-2, 1998.

[Sale00] A. A. M. Saleh, "Transparent optical networking in backbone networks," *Proceedings, Optical Fiber Communication (OFC'00)*, Baltimore, MD, Mar. 7-10, 2000, Paper ThD7.

[Sale03] A. A. M. Saleh, "Defining all-optical networking and assessing its benefits in metro, regional and backbone networks," *Proceedings, Optical Fiber Communication (OFC'03)*, Atlanta, GA, Mar. 23-28, 2003, Paper WQ1.

[SaMu00] L. H. Sahasrabuddhe and B. Mukherjee, "Multicast routing algorithms and protocols: A tutorial," *IEEE Network*, vol. 14, no. 1, pp. 90-102, Jan./Feb. 2000.

[SaSi99] A. A. M. Saleh and J. M. Simmons, "Architectural principles of optical regional and metropolitan access networks," *Journal of Lightwave Technology*, vol. 17, no. 12, pp. 2431-2448, Dec. 1999.

[SaSi06] A. A. M. Saleh and J. M. Simmons, "Evolution toward the next-generation core optical network," *Journal of Lightwave Technology*, vol. 24, no. 9, pp. 3303-3321, Sep. 2006.

[SaSR05] H. P. Sardesai, Y. Shen, and R. Ranganathan, "Optimal WDM layer partitioning and transmission reach in optical networks," *Proceedings, Optical Fiber Communication (OFC'05)*, Anaheim, CA, Mar. 6-11, 2005, Paper OTuP4.

[ScAF01] D. A. Schupke, A. Autenrieth, and T. Fischer, "Survivability of multiple fiber duct failures," *Proceedings, Third International Workshop on the Design of Reliable Communication Networks (DRCN)*, Budapest, Hungary, Oct. 7-10, 2001, pp. 213-219.

[SDIR04] G. Swallow, J. Drake, H. Ishimatsu, and Y. Rekhter "Generalize Multiprotocol Label Switching (GMPLS) User-Network Interface (UNI): Resource ReserVation Protocol-Traffic Engineering (RSVP-TE) Support for the Overlay Model," draft-ietf-ccamp-gmpls-overlay-05, Internet Engineering Task Force, Work In Progress, Oct. 2004.

[SeKS03] S. Sengupta, V. Kumar, and D. Saha, "Switched optical backbone for cost-effective scalable core IP networks," *IEEE Communications*, vol. 41, no. 6, pp. 60-70, Jun. 2003.

[ShGr04] G. Shen and W. D. Grover, "Segment-based approaches to survivable translucent network design under various ultra-long-haul system reach capabilities," *Journal of Optical Networking*, vol. 3, no. 1, pp. 1-24, Jan. 2004.

[ShWi06] F. B. Shepherd and P. J. Winzer, "Selective randomized load balancing and mesh networks with changing demands," *Journal of Optical Networking*, vol. 5, no. 5, pp. 320-339, May 2006.

[SiGS98] J. M. Simmons, E. L. Goldstein, and A. A. M. Saleh, "On the value of wavelength-add/drop in WDM rings with uniform traffic," *Proceedings, Optical Fiber Communication (OFC'98)*, San Jose, CA, Feb. 22-27, 1998, Paper ThU3.

[Simm99] J. M. Simmons, "Hierarchical restoration in a backbone network," *Proceedings, Optical Fiber Communication (OFC'99)*, San Diego, CA, Feb. 21-26, 1999, Paper TuL2.

[Simm02] J. M. Simmons, "Analysis of wavelength conversion in all-optical express backbone networks," *Proceedings, Optical Fiber Communication (OFC'02)*, Anaheim, CA, Mar. 17-22, 2002, Paper TuG2.

[Simm04] J. M. Simmons, "An introduction to optical network design and planning," *Optical Fiber Communication (OFC'04)*, Los Angeles, CA, Feb. 22-27, 2004, Short Course 216.

[Simm05] J. M. Simmons, "On determining the optimal optical reach for a long-haul network," *Journal of Lightwave Technology*, vol. 23, no. 3, pp. 1039-1048, Mar. 2005.

[Simm06] J. M. Simmons, "Network design in realistic 'all-optical' backbone networks," *IEEE Communications Magazine*, vol. 44, no. 11, pp. 88-94, Nov. 2006.

[Simm07] J. M. Simmons, "Cost vs. capacity tradeoff with shared mesh protection in optical-bypass-enabled backbone networks," *Proceedings, Optical Fiber Communication/National Fiber Optic Engineers Conference (OFC/NFOEC'07)*, Anaheim, CA, Mar. 25-29, 2007, Paper NThC2.

[SiSa99] J. M. Simmons and A. A. M. Saleh, "The value of optical bypass in reducing router size in gigabit networks," *Proceedings, IEEE International Conference on Communications (ICC'99)*, Vancouver, British Columbia, Jun. 6-10, 1999, vol. 1, pp. 591-596.

[SiSa07] J. M. Simmons and A. A. M Saleh, "Network agility through flexible transponders," *IEEE Photonics Technology Letters*, vol. 19, no. 5, pp. 309-311, Mar. 1, 2007.

[SiSB01] J. M. Simmons, A. A. M. Saleh, and L. Benmohamed, "Extending Generalized Multi-Protocol Label Switching to configurable all-optical networks," *Proceedings, National Fiber Optic Engineers Conference (NFOEC'01)*, Baltimore, MD, Jul. 8-12, 2001, pp. 14-23.

[SiSM03] N. K. Singhal, L. H. Sahasrabuddhe, and B. Mukherjee, "Provisioning of survivable multicast sessions against single link failures in optical WDM mesh networks," *Journal of Lightwave Technology,* vol. 21, no. 11, pp. 2587-2594, Nov. 2003.

[SMCS05] A. G. Striegler, M. Meissner, K. Cveček, K. Sponsel, G. Leuchs, and B. Schmauss, "NOLM-based RZ-DPSK signal regeneration," *IEEE Photonics Technology Letters*, vol. 17, no. 3, pp. 639-641, Mar. 2005.

[SoPe02] H. Soliman and C. Peyton, "An efficient routing algorithm for all-optical networks with turn constraints," *IEEE International Symposium on Modeling, Analysis, and Simulation of Computer and Telecommunications Systems (MASCOTS'02)*, Fort Worth, TX, Oct. 12-16, 2002, pp. 161-166.

[SrSS02] M. Sridharan, R. Srinivasan, and A. K. Somani, "Dynamic routing with partial information in mesh-restorable optical networks," *Proceedings, Sixth Working Conference on Optical Networks Design and Modelling (ONDM)*, Torino, Italy, Feb. 4-6, 2002.

[StBa99] T. E. Stern and K. Bala, *Multiwavelength Optical Networks*, Reading, MA: Addison-Wesley Longman Inc,1999.

[SuAS96] S. Subramaniam, M. Azizoglu, and A. K. Somani, "All-optical networks with sparse wavelength conversion," *IEEE/ACM Transactions on Networking*, vol. 4, no. 4, pp. 544-557, Aug. 1996.

[SuTa84] J. W. Suurballe and R. E. Tarjan, "A quick method for finding shortest pairs of disjoint paths," *Networks*, vol. 14, pp. 325-336, 1984.

[Suur74] J. W. Suurballe, "Disjoint paths in a network," *Networks,* vol. 4, pp. 125–145, 1974.

[TCFG95] R. W. Tkach, A. R. Chraplyvy, F. Forghieri, A. H. Gnauck, and R. M. Derosier, "Four-photon mixing and high-speed WDM systems," *Journal of Lightwave Technology*, vol. 13, no. 5, pp. 841-849, May 1995.

[Tekt01]   Tektronix,   *SONET   Telecommunications   Standard   Primer*,   2001, www.tek.com/Measurement/App_Notes/SONET.

[Telc05] Telcordia, *Synchronous Optical Network (SONET) Transport Systems: Common Generic Criteria*, GR–253–CORE, Issue 4, Dec. 2005.

[TeRo03] J. Teng and G. N. Rouskas, "A comparison of the JIT, JET, and Horizon wavelength reservation schemes on a single OBS node," *Proceedings, The First International Workshop on Optical Burst Switching (WOBS)*, Dallas, TX, Oct. 16, 2003.

[ThSo02] S. Thiagarajan and A. K. Somani, "Traffic grooming for survivable WDM mesh networks," *Optical Networks Magazine*, vol. 3, no. 3, pp. 88-98, May/Jun. 2002.

[TkCh94] R. W. Tkach and A. R. Chraplyvy, "Dispersion and nonlinear effects in lightwave systems," *Proceedings, 7th Annual Meeting of the IEEE LEOS*, Boston, MA, Oct. 31-Nov. 3, 1994, vol. 1, pp. 192-193.

[Tomk02] I. Tomkos, et al., "Ultra-long-haul DWDM network with 320x320 wavelength-port 'Broadcast & Select' OXCs," *Proceedings, European Conference on Optical Communication (ECOC'02)*, Copenhagen, Denmark, Sep. 8-12, 2002, vol. 5, pp. 1-2.

[TsMT05] I. Tsirilakis, C. Mas, and I. Tomkos, "Cost comparison of IP/WDM vs. IP/OTN for European backbone networks," *Proceedings, International Conference on Transparent Optical Networks (ICTON'05)*, Barcelona, Spain, Jul. 3-7, 2005, pp. 46-49.

[TzZT03] A. Tzanakaki, I. Zacharopoulos and I. Tomkos, "Optical add/drop multiplexers and optical cross-connects for wavelength routed networks," *Proceedings, International Conference on Transparent Optical Networks (ICTON'03)*, Warsaw, Poland, Jun. 29-Jul. 3, 2003, pp. 41-46.

[VAAD01] W. Van Parys, P. Arijs, O. Antonis, and P. Demeester, "Quantifying the benefits of selective wavelength regeneration in ultra long-haul WDM networks," *Proceedings, Optical Fiber Communication (OFC'01)*, Anaheim, CA, Mar. 19-22, 2001, Paper TuT4.

[VaPD04] J. Vasseur, M. Pickavet, and P. Demeester, *Network Recovery: Protection and Restoration of Optical, SONET-SDH, IP, and MPLS*, San Francisco, CA: Morgan Kaufmann, 2004.

[Verb05] S. Verbrugge, et al., "Modeling operational expenditures for telecom operators," *Proceedings, Conference on Optical Network Design and Modeling (ONDM'05)*, Milan, Italy, Feb. 7-9, 2005, pp. 455-466.

[Voss92] S. Voss, "Steiner's problem in graphs: Heuristic methods," *Discrete Applied Mathematics*, vol. 40, no. 1, 1992, pp. 45-72.

[WASG96] R. E. Wagner, R. C. Alferness, A. A. M. Saleh, and M. S. Goodman, "MONET: Multiwavelength optical networking," *Journal of Lightwave Technology*, vol. 14, no. 6, pp. 1349-1355, Jun. 1996.

[Waxm88] B. M. Waxman, "Routing of multipoint connections," *IEEE Journal on Selected Areas in Communications*, vol. 6, no. 9, pp. 1617-1622, Dec. 1988.

[WBSK03] S. T. Wilkinson, E. B. Basch, V. Shukla, P. Kubat, S. Raguram, and P. Limaye, "SONET mesh network architecture," *Proceedings, National Fiber Optic Engineers Conference (NFOEC'03)*, Orlando, FL, Sep. 7–11, 2003, pp. 293-302.

[WeCZ05] Y. Wen, V. W. S. Chan, and L. Zheng, "Efficient fault-diagnosis algorithms for all-optical WDM networks with probabilistic link failures," *Journal of Lightwave Technology*, vol. 23, no. 10, pp. 3358-3371, Oct. 2005.

[Welc06] D. F. Welch, et al., "The realization of large-scale photonic integrated circuits and the associated impact on fiber-optic communication systems," *Journal of Lightwave Technology*, vol. 24, no. 12, pp. 4674-4683, Dec. 2006.

[Will06] A. E. Willner, "The optical network of the future: Can optical performance monitoring enable automated, intelligent and robust systems?" *Optics and Photonics News*, pp. 30-35, Mar. 2006.

[WLYK04] D. Wang, G. Li, J. Yates, and C. Kalmanek, "Efficient segment-by-segment restoration," *Proceedings, Optical Fiber Communication (OFC'04)*, Los Angeles, CA, Feb. 22-27, 2004, Paper TuP2.

[WSGM03] I. Widjaja, I. Saniee, R. Giles, and D. Mitra, "Light core and intelligent edge for a flexible, thin-layered, and cost-effective optical transport network," *IEEE Communications Magazine*, vol. 41, no. 5, pp. S30-S36, May 2003.

[WuSF06] M. C. Wu, O. Solgaard, and J. E. Ford, "Optical MEMS for lightwave communication," *Journal of Lightwave Technology*, vol. 24, no. 12, pp. 4433-4454, Dec. 2006.

[XuQX07] D. Xu, C. Qiao, and Y. Xiong, "Ultrafast potential-backup-cost (PBC)-based shared path protection schemes," *Journal of Lightwave Technology,* vol. 25, no. 8, pp. 2251-2259, Aug. 2007.

[XuXQ03] D. Xu, Y. Xiong, and C. Qiao, "Novel algorithms for shared segment protection," *IEEE Journal on Selected Areas in Communications,* vol. 21, no. 8, pp.1320-1331, Oct. 2003.

[XXQL03] D. Xu, Y. Xiong, C. Qiao, and G. Li, "Trap avoidance and protection schemes in networks with shared risk link groups," *Journal of Lightwave Technology,* vol. 21, no. 11, pp. 2683-2693, Nov. 2003.

[YaRa05a] X. Yang and B. Ramamurthy, "Dynamic routing in translucent WDM optical net-works: The intradomain case," *Journal of Lightwave Technology,* vol. 23, no. 3, pp. 955-971, Mar. 2005.

[YaRa05b] W. Yao and B. Ramamurthy, "Survivable traffic grooming with path protection at the connection level in WDM mesh networks," *Journal of Lightwave Technology,* vol. 23, no. 10, pp. 2846-2853, Oct. 2005.

[YaYo05] H. Yang and S. J. B. Yoo, "All-optical variable buffering strategies and switch fabric architectures for future all-optical data routers," *Journal of Lightwave Technology,* vol. 23, no. 10, pp. 3321-3330, Oct. 2005.

[Yen71] J. Y. Yen, "Finding the K shortest loopless paths in a network," *Management Science,* vol. 17, no. 11, pp. 712–716, Jul. 1971.

[Yue91] M. Yue, "A simple proof of the inequality $FFD(L) \leq (11/9)OPT(L) + 1$, for all L, for the FFD bin-packing algorithm," *Acta Mathematicae Applicatae Sinica,* vol. 7, no. 4, pp. 321–331, Oct. 1991.

[ZaJM00] H. Zang, J. P. Jue, and B. Mukherjee, "A review of routing and wavelength assign-ment approaches for wavelength-routed optical WDM networks," *Optical Networks Maga-zine,* vol. 1, no. 1, pp. 47-60, Jan. 2000.

[ZhMu03] K. Zhu and B. Mukherjee, "A review of traffic grooming in WDM optical networks: Architectures and challenges," *Optical Networks Magazine,* vol. 4, no. 3, pp. 55-64, Mar./Apr. 2003.

[ZhQi98] X. Zhang and C. Qiao, "Wavelength assignment for dynamic traffic in multi-fiber WDM networks," *Proceedings, International Conference on Computer Communications and Networks (ICCCN'98),* Lafayette, LA, Oct. 12-15, 1998, pp. 479 – 485.

[ZhZM05] K. Zhu, H. Zhu, and B. Mukherjee, *Traffic Grooming in Optical WDM Mesh Net-works,* New York, NY: Springer, 2005.

[ZhZM06] J. Zhang, K. Zhu, and B. Mukherjee, "Backup reprovisioning to remedy the effect of multiple link failures in WDM mesh networks," *IEEE Journal on Selected Areas in Commu-nications,* vol. 24, no. 8, pp. 57-67, Aug. 2006.

[ZTTD02] Y. Zhang, K. Taira, H. Takagi, and S. K Das, "An efficient heuristic for routing and wavelength assignment in optical WDM networks," *Proceedings, IEEE International Con-ference on Communications (ICC'02),* New York, NY, Apr. 28-May 2, 2002, pp. 2734-2739.

[ZZZM03] H. Zhu, H. Zang, K. Zhu, and B. Mukherjee, "A novel generic graph model for traffic grooming in heterogeneous WDM mesh networks," *IEEE/ACM Transactions on Networking,* vol. 11, no. 2, pp. 285-299, Apr. 2003.

Index

1310-nm wavelength, 12, 13, 22, 25, 44, 46, 51, 55

access network, 3, 4, 5
  passive optical network, 5
add/drop port
  colorless, 43, 44
  multi-wavelength, 34, 44, 48
  single-wavelength, 35, 43, 49
alarms, 231
algorithm
  greedy, 64
  heuristic, 62
all-optical network, 15, 28
all-optical switch. *See* switch, all-optical
American National Standards Institute (ANSI), 7
amplifier hut, 67
  spacing, 105, 107, 108
arrayed waveguide grating (AWG), 18, 19, 21
Asynchronous Transfer Mode (ATM), 5
attack detection, 232
Automatically Switched Optical Network (ASON), 9
availability, 63, 80, 82, 168, 183, 220

backbone network, 4, 5, 11, 36, 55, 61, 116, 147–49, 234, 235, 241, 259
backhauling. *See* grooming, backhauling
bin packing, 156, 169
broadcast-and-select architecture, 34, 38, 47

candidate paths
  bottleneck avoidance, 69, 70, 91
  generating, 68–70

K-shortest paths, 68, 69
  least loaded, 72
  selecting one, 72, 74, 130
  shared protection, 220
capital expenditure, 15, 236–37
carrier office, 12, 26
client layer, 5, 12, 17, 20, 23, 51, 55, 164, 190–91
client-server model, 9
configurability, 5, 8–10, 17, 24, 29, 240, 253
  edge, 32–33, 42, 44, 47, 50–53, 56, 121, 123, 161, 240
connected dominating sets, 116, 117
control plane, 9, 10
cross-connect,
  optical (OXC), 44, 236, 242, 245, 252
  wavelength-selective (WSXC), 19
  See also switch
cross-layer bandwidth optimization, 10
cross-phase modulation, 103
crosstalk, 104, 138

demand, 11
  aggregate, 176, 244
  bi-directional, 11
  multicast. *See* multicast
demultiplexer, 18
dispersion, 107, 113
  affect on optical reach, 144, 145
  chromatic, 54, 103
  polarization-mode, 54, 103
  relation to optical impairments, 103, 104
  slope, 104
dispersion compensation, 233
  electronic, 104
  fiber-based, 103, 113, 114
  MLSE, 104

PMD, 104
distributed computing, 10
domain, 10, 84, 98
drop-and-continue, 33, 42, 97, 182
dual homing, 164–66
dynamic channel equalizer, 41
dynamic spectral equalizer, 41

edge network, 265–68
equipment costs, 236–37
Erbium doped fiber amplifier (EDFA), 2,
    54, 105, 108, 113, 207, 233
Ethernet, 5, 8, 153, 260
    switch, 159
External Network-Network Interface (E-
    NNI), 10

fault isolation, 25, 200, 231–32
    optical supervisory channel, 232
fault localization. *See* fault isolation
fiber
    attenuation, 12, 108
    bypass, 49, 259
    dispersion compensating, 103
    dispersion level, 107
    non dispersion-shifted fiber (NDSF),
        107
    nonlinearities, 103
    non-zero dispersion-shifted fiber (NZ-
        DSF), 107
    refractive index, 103
    splicing loss, 107
    type, 107, 145
filter narrowing, 104
First Fit Decreasing bin packing, 156
forward error correction (FEC), 8, 54, 106,
    246
four-wave mixing, 54, 103

GMPLS, 9
graph transformation
    routing in O-E-O network, 75–77
    routing in optical-bypass-enabled
        network, 77–78
    routing with limited regenerator sites,
        116
    routing with SRLGs, 87–90
    single-step RWA, 133–35
greenfield network, 62, 256
grid computing, 10
grooming, 130, 157–59
    algorithm, 169–75
    backhauling, 84, 163, 164–66, 170, 178

core switch, 159–60
dual homing, 164–66
edge switch, 160–62
efficiency, 176–79
node architecture, 159–62
node failure, 225, 227
optical domain, 180, 181–82, 265–68
parent node, 164
protection, 164–66, 175, 224–31
relation to line-rate, 260
relation to regeneration, 150, 154, 163,
    173, 174, 247, 255, 257
run-time, 175
site selection, 163
subset of nodes, 163, 178–79, 246
switch, 154, 158, 159–62, 163, 171
tradeoffs, 166–69
vs. multiplexing, 158
with optical bypass, 154, 163, 173, 177,
    178
grooming connection, 170
    fill-rate, 175, 176, 178, 226
    operations on, 170–75
    protected, 224, 228
    regeneration, 171, 173, 174

heat dissipation, 2, 26, 28, 45, 57, 154

integrated transceiver, 55, 56, 162
interface
    intermediate-reach, 13
    short-reach, 13, 24, 26, 44, 122, 159
interference length, 74
Internal Network-Network Interface (I-
    NNI), 10
International Telecommunication Union, 7,
    8, 9
Internet Engineering Task Force (IETF), 9,
    98
Internet Protocol (IP), 5, 153, 156, 167
    flow, 159
    over optical, 229–31, 243, 244
    over OXC, 242, 243
    protection, 229–31
    router, 159, 160, 179, 182, 236, 244–
        45, 266
    transport technology, 242–46
islands of transparency, 114–16

jitter, 167, 168, 169, 180

K-shortest paths. *See* shortest path
    algorithm, K-shortest paths

lambda, 12
lasing, 196, 214
latency, 167, 168, 169
launch power, 106
lighttrail, 182
line-rate
    affect on switch size, 260, 264
    cost increase factor, 260, 261, 264
    effect on grooming, 260
    effect on optical reach, 260, 261
    optimal, 259–64
link, 10
    bi-directional, 10, 64
    removal, 256–59
link engineering, 113–14
local access port, 47, 48

MAC protocol, 181, 182
make-before-break, 171
management plane, 9
mesh protection.
    See protection, mesh-based
    See shared protection, mesh
metro-core network, 3, 4, 5, 37, 55, 61,
            116, 130, 149–51, 256–59, 265–
            68
micro-electro-mechanical-system
            (MEMS). *See* switch, MEMS
modulation format, 106
multicast, 11
    minimum spanning tree, 94
    network element support, 33, 42, 47
    protection, 97
    regeneration point, 96
    routing, 93–97
    Steiner tree, 94
multi-fiber-pair system, 57–59, 136, 137
multilayer protection, 229–31
    backoff timer, 230
    bottom-up escalation, 230
    uncoordinated, 229, 230
multiplexer, 18
multiplexing
    bin packing, 156
    end-to-end, 14, 153, 154, 155–57, 177
    inverse, 14, 155
    quad-card, 155, 157
    statistical, 179
    vs. grooming, 158
multi-vendor environment, 25, 28, 46, 56,
            114, 115, 162

network churn, 9, 107, 137, 157, 172
network cost
    capital cost, 15, 236–37
    operating cost, 15, 237–38
network planning, 14
    long-term, 14, 62, 93, 131, 133, 141
    real-time, 14, 61, 73, 75–78, 133, 135
    traffic engineering, 14, 64, 65
Network-Network Interface (NNI), 10
node, 10
    amplifiers, 41, 236, 237, 262
    degree, 11, 36, 37, 59, 163, 257
    parent, 164
noise
    ASE, 102
    optical-signal-to-noise-ratio (OSNR),
            102, 105, 107, 108, 113, 232
noise figure, 108–10
    cumulative, 109
    network element, 110, 111
    routing metric, 109
    units, 109
non-return-to-zero (NRZ), 106

OADM, 17, 27–36
    add/drop limit, 29, 249–51
    architecture, 34–36
    drop-and-continue, 33
    East/West separability, 34
    edge configurability, 32–33
    granularity, 30
    reconfigurable, 29
    upgrade path, 32, 40, 48
    with wavelength reuse, 30, 36
    without wavelength reuse, 31, 32, 111–
            12, 193
OADM-MD, 17, 38–44, 121, 122, 123,
            161, 194, 199, 236
    add/drop limit, 44, 249–51
    adding edge configurability, 50–53
    architecture, 41–44
    economics, 251–53
    upgrade path, 40
OADM-only architecture, 37, 251–53
    non-optimal orientation, 253
O-E-O architecture, 22–26, 146, 159–60
    configurable, 24
    degree-two node, 22, 23
    higher-degree nodes, 23, 24
    non-configurable, 22, 23, 238, 240
    with extended reach, 240–42
O-E-O switch. *See* switch, electrical

O-E-O-at-the-hubs, 37, 251–53
on-off keying, 106, 125
operational expenditure, 15, 236, 237–38, 240, 249
optical amplifier transients, 184, 207–9, 216–17, 266
optical burst switching (OBS), 181, 266
    Just Enough Time (JET), 181
    Just in Time (JIT), 181
optical bypass, 2, 3, 5, 15, 17, 27, 38, 46
    economics, 238–42
    network element limits, 105, 107, 114
optical channel shared protection ring (OChSPRing), 197
optical cross-connect (OXC). See cross-connect, optical
optical flow switching, 180–81
optical frequency, 2, 12
optical grooming. See grooming, optical domain
optical impairment, 102–3
    mitigation, 103–4
Optical Internetworking Forum (OIF), 10
optical multiplex section shared protection ring (OMS-SPRing), 197, 198
optical packet switching (OPS), 182
optical performance monitor (OPM), 232
optical reach, 17, 53, 54, 101, 105, 106, 108, 234
    cost increase factor, 237, 238, 239, 245, 246, 247, 254
    optimal, 246–49
    reduced, 144–46
    relation to line-rate, 260, 261
optical supervisory channel, 232
optical switch. See switch, optical
optical terminal, 17, 19–21
    colorless, 21
    fixed, 21
    shelf density, 20
Optical Transport Network (OTN), 8
    digital wrapper, 8
    optical channel transport unit (OTU), 8
OSNR. See noise, OSNR

packet services, 154
passive combiner, 18
passive coupler, 18
passive splitter, 18
patch-panel, 22
path computation element (PCE), 98
p-cycle, 214–15
performance monitoring, 25, 28, 200, 231

optical, 232
photonic integrated circuit (PIC), 56, 57
PMD. See dispersion, polarization-mode
polarization dependent loss, 104
power consumption, 2, 26, 28, 45, 57, 154, 182, 240
power equalization, 32
pre-cross-connected trail (PXT), 215–16
predeployed equipment, 33, 47, 51, 53, 78, 79, 118, 122
predeployed subconnection, 209–13
primary path, 185
protect path, 185, 219
protection
    1:1, 186, 208
    1+1, 186, 205, 208
    algorithms, 217–23
    capacity requirements, 186, 188, 198
    client-side, 190–91
    dedicated, 186, 187–90, 217–18
    fault-dependent, 200–201, 202, 231
    fault-independent, 201–2, 205, 231
    groups, 226
    hierarchical, 210, 211
    hub, 212, 252
    link, 200–201
    M:N, 187
    mesh-based, 198–99
    mixing working and protect paths, 228
    multilayer, 229–31
    multiple concurrent failures, 184, 205–7
    network-side, 191, 192–94
    nodal, 185
    non-revertive, 186
    OChSPring, 197
    OMS-SPRing, 197, 198
    optical amplifier transients, 208, 210, 211, 216–17
    path, 201–2
    revertive, 186, 187
    ring-based, 194–98
    routing. See routing, disjoint paths
    segment, 202–4
    shared. See shared protection
    sub-path, 204
    subrate-level, 226–29
    transponder, 191–92
    wavelength assignment, 128, 192–93, 196, 198, 199, 217, 218
    wavelength-level, 224–26, 228–29
provisioning, 11, 26, 28
pump power, 246, 262

Q-factor, 110, 232

Raman amplification, 54, 105, 106, 107,
        108, 113, 137, 207, 233
receiver sensitivity, 106
reconfigurability. *See* configurability
regeneration, 25, 53, 67, 112, 130, 145
    2R, 125
    3R, 25, 101, 120
    adding to alleviate wavelength
                contention, 131–32, 148,
                149, 150
    affect on wavelength assignment, 102,
                128–30, 133, 192
    all-optical, 125, 126
    architecture, 119–26
    back-to-back transponders, 120–22
    designated site, 116–18
    function of optical reach, 246, 247
    islands of transparency, 114–16
    multi-wavelength, 126
    selective, 118–19
    system rules, 107
regenerator card, 122–24
    all-optical, 125, 126
    flexible, 124
    tunable, 122, 129
regional network, 4, 55, 116, 259, 265–68
reliability, 26, 28, 164, 169
request for information (RFI), 269–70
request for proposal (RFP), 269–70
restoration. *See* protection
return-to-zero (RZ), 106
ring protection. *See* protection, ring-based
ROADM. *See* OADM
routing
    alternative-path, 71–73, 92, 130, 170
    contention avoidance, 99
    contention resolution, 99
    demand order, 92–93
    disjoint paths, 79–92
    distributed, 98, 99
    dynamic, 73, 74
    fixed-alternate, 73
    fixed-path, 71
    load balancing, 69–74, 130, 178
    multicast, 93–97
    probabilistic, 98
    round-robin, 93
    trap topology, 81
    with inaccurate information, 97–99

routing and wavelength assignment
        (RWA), 127
    multi-step, 127, 130–32
    single-step, 127, 132–36, 151
    See also routing
    See also wavelength assignment

secondary path, 185
selective randomized load balancing, 180
self-phase modulation, 103
service level agreement (SLA), 183
shared protection, 186–90, 206, 218–23
    candidate paths, 220
    distributed algorithms, 222–23
    hierarchical, 210, 211, 212
    mesh, 198–99, 209–17
    p-cycle, 214–15
    potential backup cost, 221–22
    pre-cross-connected bandwidth, 184,
                213–17
    pre-cross-connected trail (PXT), 215–
                16
    predeployed subconnection, 184, 209–
                13, 247, 252, 255
    regeneration, 130, 211, 247, 252, 255
    ring, 194–98
    shareability metric, 220–21
    speed, 189, 210, 214, 215
    using partial information, 222–23
shared risk group (SRG), 87, 217
shared risk link group (SRLG), 87–90,
        217, 218, 259
    bridge configuration, 90
    fork configuration, 87, 88
    general routing heuristics, 90
shortest path algorithm, 63–65
    Breadth-First-Search, 64, 271
    constrained, 64
    Dijkstra, 63, 64
    disjoint pair of paths, 82–87, 217
    disjoint paths (Bhandari), 82, 271
    disjoint paths (Suurballe), 82
    dual sources/dual destinations, 84–87,
                164
    K-shortest paths, 64, 271
    link-and-node-disjoint paths, 82
    link-disjoint paths, 82
    maximally disjoint paths, 82, 271
    metric, 65–68
    minimum regeneration, 67, 69, 91, 109
    N disjoint paths, 82, 205, 271
    noise figure metric, 109
    restricted, 65

undirected network, 64
silent failure, 186
SONET/SDH, 5, 7, 8, 153, 259, 260
   add/drop multiplexer (ADM), 27
   bi-directional line-switched ring
       (BLSR), 198
   grooming switch, 158, 236, 246
   multiplexing, 155, 156
   performance monitoring, 25, 200
span, 10
   distance, 105, 107, 108
   engineering. *See* link engineering
subconnection, 119, 129, 136, 139, 141–
       43, 145, 151, 204
   bi-directional, 143
   group, 218
   predeployed, 209–13
switch
   add/drop limit, 249, 251
   all-optical, 17, 44, 46–49, 97, 121, 122,
       162, 194, 199, 236
   all-optical upgrade path, 48
   core, 24, 159–60, 161
   dual fabric, 162
   edge, 51, 121, 123, 160–62, 191, 192,
       199, 209, 210, 211, 236
   electrical, 24, 44, 45
   fabric, 19, 264
   grooming, 46, 154, 158, 159–62, 163,
       236
   hierarchical, 49
   local access port, 47, 48
   make-before-break, 171
   MEMS, 19, 24, 36, 46, 48, 49, 124
   optical, 19, 44–49
   photonic, 45, 46, 51
   waveband, 49
   wavelength-selective (WSS), 19, 21,
       42, 43, 44, 46
system margin, 106, 107, 113, 119, 138

time-domain wavelength interleaved
       networking (TWIN), 181–82,
       266
topology, 11
   backbone, 36, 147, 234, 235, 241, 259
   discovery, 9
   interconnected ring, 36, 37, 39, 150
   mesh, 11, 36, 39, 147
   metro-core, 37, 150, 256–59
   optimal cost, 256–59
   regional, 259
   ring, 11, 39

trap, 81
   virtual, 6
traffic, 11
   add/drop, 23, 38, 55, 160, 161
   best-effort, 153, 183
   bursty, 153, 179, 244
   churn. *See* network churn
   circuit, 154, 179
   hose model, 180
   line-rate, 13, 61, 153
   packet, 154
   preemptible, 230
   statistics, 236
   subrate, 13, 61, 153
   through, 23
transients. *See* optical amplifier transients
transmission band, 12, 137
   C-band, 12
   L-band, 12
   S-band, 12
transmission cost
   function of fiber capacity, 262
   function of line-rate, 261
   function of optical reach, 237
transparency, 6
transponder. *See* WDM transponder
turn constraint, 77, 135

User-Network Interface (UNI), 10

Virtual Concatenation (VCAT), 155

waveband, 30, 74, 119
   grooming, 49
wavelength assignment
   algorithm, 127, 136–41
   bi-directional, 143–44
   First-Fit, 137–38, 141, 144, 145, 152
   Least-Loaded, 137
   Most-Used, 139, 141
   protected paths, 192–93, 196, 198, 199,
       217, 218
   relation to optical reach, 144–46
   relation to regeneration, 128–30, 131–
       32, 133, 192, 193
   Relative Capacity Loss, 139–41, 142
   subconnection ordering, 141–43
wavelength blocker, 41
wavelength contention
   affect on network efficiency, 146–52
   alleviating, 131–32, 148, 149, 150
wavelength conversion, 26, 120, 122, 129,
       130, 146

all-optical, 29, 126, 128
wavelength division multiplexing, 2, 4, 11
    channel spacing, 12, 106, 116
    spectrum, 12
wavelength grating router, 18
wavelength reuse. *See* OADM, with
        wavelength reuse
wavelength service, 153, 160, 161, 162
wavelength-continuity constraint, 29, 66,
        127
wavelength-selective architecture, 35, 48
wavelength-selective switch. *See* switch,
        wavelength-selective

WDM transponder, 12, 13, 20, 22, 233,
        246
    client-side, 12, 13, 155
    flexible, 52, 191
    integrated. *See* integrated transceiver
    network side, 13
    N-way, 52
    optical filter, 13, 21
    protection, 191–92
    quad-card, 155, 157
    reduced-reach, 253–55
    tunable, 13, 129, 148, 193, 199
    two-way (flexible), 191
working path, 185, 218, 219

Printed in the United States of America